W9-CZA-485

Guidelines for Laboratory Quality Auditing

QUALITY AND RELIABILITY

A Series Edited by

EDWARD G. SCHILLING

Coordinating Editor
Center for Quality and Applied Statistics
Rochester Institute of Technology
Rochester, New York

W. GROVER BARNARD
Associate Editor for
Human Factors
Vita Mix Corporation
Cleveland, Ohio

RICHARD S. BINGHAM, JR.
Associate Editor for
Quality Management
Consultant
Brooksville, Florida

LARRY RABINOWITZ
Associate Editor for
Statistical Methods
College of William and Mary
Williamsburg, Virginia

THOMAS WITT
Associate Editor for
Statistical Quality Control
Rochester Institute of Technology
Rochester, New York

ADDITIONAL VOLUMES IN PREPARATION

Guidelines for Laboratory Quality Auditing

Donald C. Singer
SmithKline Beecham Pharmaceuticals
King of Prussia, Pennsylvania

Ronald P. Upton
Consultant
Newton, North Carolina

ASQC Quality Press　　　　　　　　　　　　　　**Milwaukee**

Marcel Dekker, Inc.　　　　　**New York • Basel • Hong Kong**

Library of Congress Cataloging-in-Publication Data

Singer, Donald C.,
  Guidelines for laboratory quality auditing / Donald C. Singer,
Ronald P. Upton.
    p.   cm.—(Quality and reliability ; 39)
  Includes bibliographical references and index.
  ISBN 0-8247-8784-6
  1. Medical laboratories—Quality control. 2. Biological
laboratories—Quality control. 3. Laboratories—Quality control.
I. Upton, Ronald P., II. Title. III. Series.
R850.S37 1993
362.1'77—dc20   92-29093
                                                      CIP

This book is printed on acid-free paper.

ASQC
Quality Press
611 East Wisconsin Avenue
Milwaukee, Wisconsin 53202

Marcel Dekker, Inc.
270 Madison Avenue
New York, New York 10016

Current printing (last digit):
10 9 8 7 6 5 4 3 2 1

PRINTED IN THE UNITED STATES OF AMERICA

ABOUT THE SERIES

The genesis of modern methods of quality and reliability will be found in a simple memo dated May 16, 1924, in which Walter A. Shewhart proposed the control chart for the analysis of inspection data. This led to a broadening of the concept of inspection from emphasis on detection and correction of defective material to control of quality through analysis and prevention of quality problems. Subsequent concern for product performance in the hands of the user stimulated development of the systems and techniques of reliability. Emphasis on the consumer as the ultimate judge of quality serves as the catalyst to bring about the integration of the methodology of quality with that of reliability. Thus, the innovations that came out of the control chart spawned a philosophy of control of quality and reliability that has come to include not only the methodology of the statistical sciences and engineering, but also the use of appropriate management methods together with various motivational procedures in a concerted effort dedicated to quality improvement.

This series is intended to provide a vehicle to foster interaction of the elements of the modern approach to quality, including statistical applications, quality and reliability engineering, management, and motivational aspects. It is a forum in which the subject matter of these various areas can be brought together to allow for effective integration of appropriate techniques. This will promote the true benefit of each, which can be achieved only through their interaction. In this sense, the whole of quality and reliability is greater than the sum of its parts, as each element augments the others.

The contributors to this series have been encouraged to discuss fundamental concepts as well as methodology, technology, and procedures at the leading edge of the discipline. Thus, new concepts are placed in proper perspective in these evolving disciplines. The series is intended for those in manufacturing, engineering, and marketing and management, as well as the consuming public, all of whom have an interest and stake in the improvement and maintenance of quality and reliability in the products and services that are the lifeblood of the economic system.

The modern approach to quality and reliability concerns excellence: excellence when the product is designed, excellence when the product is made, excellence as the product is used, and excellence throughout its lifetime. But excellence does not result without effort, and products and services of superior quality and reliability require an appropriate combination of statistical, engineering, management, and motivational effort. This effort can be directed for maximum benefit only in light of timely knowledge of approaches and methods that have been developed and are available in these areas of expertise. Within the volumes of this series, the reader will find the means to create, control, correct, and improve quality and reliability in ways that are cost effective, that enhance productivity, and that create a motivational atmosphere that is harmonious and constructive. It is dedicated to that end and to the readers whose study of quality and reliability will lead to greater understanding of their products, their processes, their workplaces, and themselves.

Edward G. Schilling

PREFACE

The purpose of this book is twofold: to provide information that can help you develop confidence in the way your laboratory works, and to give insight into better ways of assuring that your laboratory performs at its best and provides your customers with consistent and trustworthy results.

A laboratory can exist for the purpose of analyzing samples and nothing more. Or, a laboratory can be a site of investigation, creativity and development, achievement, training and education, as well as a trusted analytical testing site. A combined effort by management, employees, and customers is necessary to ensure that the product is of the highest quality. Indeed, a laboratory must meet the expectations of each customer in order to assure continued existence. When the product is numerical data, new or revised techniques, identification, or education, a laboratory must provide its customers with nothing less than a truly consistent and trustworthy process to obtain that product.

The trust and consistency of a laboratory's results are achieved by developing attitudes and an environment that are based on proper management of human resources, ongoing control (validation and/or qualification) of materials, instruments and techniques, and successful communication between all internal and external customers. These three factors—management, control, and communication—must work together like a well-tuned race car engine. Each integral part depends on the action and result of the other. When all work together successfully, the customer receives a high quality product.

The authors have shared both employee and management responsibilities in a variety of consumer products laboratories during the past fourteen years. Communication and human relations are called the energy source of a trustworthy laboratory. The energy source drives the engine, the laboratory. Costs of state-of-the-art equipment, salaries, and the desire for additional working space are limiting factors to the capabilities of a laboratory. Nothing can be better justified than a thorough, ongoing training process, and a performance-management process that stresses communication, achievement, and reward. These processes are the basis of employee empowerment, which can provide a foundation for ongoing quality improvement.

This book contains information, resources, and discussion about evaluating the control aspects of a laboratory. These relate to adequate procurement, quality control, use of standardized and validated materials and techniques, current regulations, methods for evaluating a laboratory, and other areas of concern that may need to be evaluated and improved to meet the rigors of today's customer demand for high-quality testing.

In addition, the basic requirements for a laboratory to start on the road towards quality improvement are discussed. Further enhancement of this process will depend on the ability of all employees to develop an understanding that successful performance and communication are mandatory.

The next time you perform an audit or have an audit performed on your laboratory, be sure you can answer a resounding YES! to this question: Would you trust the results from the laboratory if your life depended on them?

Donald C. Singer
Ronald P. Upton

ACKNOWLEDGMENTS

Many people have contributed to the writing of this book, including those who have toiled in our laboratories to produce the solid results we as managers have relied upon for years.

I also would like to acknowledge the support, patience and constant encouragement of my wife, Sue, and our children, Chelsea and Trevor. A special expression of appreciation must go to Jeanine Lau at ASQC for her encouragement and professional guidance in the areas of publishing.

Donald C. Singer

I wish to acknowledge the constant support and encouragement of my wife, Mary Ann, and our children.

Ronald P. Upton

CONTENTS

INTRODUCTION

One service industry area that generated much publicity during the 1980s and where quality improvement can have a profound effect is laboratory testing in the health-related fields. Quality improvement programs that increase the reliability of test results from a laboratory provide business security and enhance clients' confidence. Quality control and quality assurance programs have become accepted parts of the operation of commercial and private laboratories, especially where accreditation or licensing is involved. Governmental agencies and professional groups have begun defining program requirements (Alvarez, 1982).

This book will combine the proven strategies of procurement quality control and quality auditing into an informative and useful manual for evaluating laboratories that serve the clinical, biomedical, pharmaceutical, food, cosmetic, and biotechnological industries. It is of utmost importance that a person, hospital, or company audit the laboratory to which they contract testing, and assure themselves that the laboratory is meeting their quality requirements. A business contract should depend on the acceptable results of this audit. In addition, it is strongly recommended that the laboratory conduct its own internal quality audit with sufficient frequency to ensure that test analyses provide continuously reliable results. An internal audit will also provide the laboratory with knowledge of how well it follows its own quality program and prepare it for audits by clients.

Many laboratories are audited by more than one organization. The standards expected are similar in some areas and different in others (See Appendices). There are many manuals which state quality assurance guidelines, but they are industry or field specific. The guidelines in this book are a standardization of the guidelines used by various laboratory auditing groups in the industries and fields of study previously mentioned.

Laboratory testing for the medical field and the health industries is provided by thousands of private and commercial laboratories in the United States and throughout the world. There are an estimated additional 100,000 physician's office laboratories. Clients require this service to be competent, efficient, reliable and professional. In other words, high quality service must be demanded from all laboratories. The end result could mean the difference between life and death.

1

THE NEED FOR QUALITY CONTROL AND QUALITY ASSURANCE IN A LABORATORY

A quality assurance (QA) program must be planned and implemented to provide confidence in a laboratory's execution of its business. The most effective means of evaluating a QA program is auditing.

The quality audit has been defined as a "management tool used to evaluate, confirm, or verify activities related to quality." It is a constructive process (Mills, 1989).

Performing a quality audit has become a routine activity of any business that seeks quality improvement. The results of a quality audit are carefully evaluated and used for developing objectives which will assist a business in improving the quality process that is already in place. Alternatively, the results can also be used to develop a quality improvement plan that will improve the strengths and reduce or remove the weaknesses that could create future problems.

Every product that is manufactured by food, cosmetic, pharmaceutical, medical device, or biotechnology firms has characteristics that need to be quantified or qualified by laboratory testing. Every sample of body tissue or fluid submitted by or to a physician, veterinarian, clinic or hospital for diagnosis requires laboratory testing. Diagnostic test kits for use at home by the public are batch sampled and tested by the manufacturers' laboratories prior to their release for sale to the public. Just about anything that is used to improve, prevent deterioration of, maintain, or diagnose health in humans or animals is subjected to testing in a laboratory.

Quality control and quality assurance are the necessary processes that play the role of a check and balance system in a laboratory.

Analytical testing for known characteristics should have corresponding known standards for comparison. Analytical testing for unknown characteristics must have both known standards and controls that can ensure that whatever result is obtained is reliable. The reliability of analytical testing is the means for building trust in the customer. Customers demand trustworthy, consistent analytical practices, which result from solid quality control and quality assurance processes.

The laboratory environment consists of people, facility, instrumentation, chemicals, supplies, and samples submitted for analysis. There has to be a logical and scientific manner of organization and management that can drive the laboratory system. Each laboratory works in an individual manner, so it requires a customized approach and attitude. Every laboratory depends on consistency, and the development of a system which will ensure consistency is dependent on the following factors: well-managed and adequately trained people, well-maintained facilities, calibrated instrumentation, high quality chemicals, adequate supplies, and proper handling of samples. These factors will help form a laboratory environment that can meet the requirements of any kind of laboratory quality audit.

The laboratory quality audit can be used as a tool to help increase or substantiate trust and confidence of the customers for the laboratory's capabilities.

The laboratory quality audit will evaluate the strengths and weaknesses of the quality control/quality assurance processes. Then, the audit can be used to improve the processes and build a better system for the benefit of the laboratory owners, employees, and customers.

What is quality control in a laboratory? Quality control is defined as the operational techniques and the activities that sustain quality of a product or service that will satisfy given needs (ASQC, 1983).

Each instrument requires periodic calibration, by physical or chemical means, where appropriate. Chemicals used in analytical testing should be of the purity required by the procedures. Known standards, where available, are routinely used to check instrument and method variation. When testing unknown materials, known standards can be used for benchmarks. Consistency in the preparation of testing materials, the use of instrumentation, following methodology, and documenting results are results of a quality control process.

What is quality assurance in a laboratory? Quality assurance is defined as all those planned or systematic actions necessary to provide adequate confidence that a product or service will satisfy given needs (ASQC, 1983).

Written procedures and adequate documentation of all quality control practices, training, and analytical results make up one part of a laboratory quality assurance process. The other part of the process is an experienced quality assurance staff which manages and performs internal audits of the quality control and quality assurance processes.

The quality control and quality assurance processes in a laboratory are usually defined by internal and external regulatory requirements. The minimum criteria for quality control and quality assurance processes should not differ significantly from regulatory requirements, and thus can provide a more defined basis for an audit. Employees of the laboratory directly affect the quality control and quality assurance processes. Hiring and training criteria have become an important part of quality control and quality assurance and should be a significant measurement in an audit.

Employees, quality control/quality assuance, and customer relations are the three most important areas in an audit of a laboratory. If these three areas are well developed and documented, and if the laboratory can satisfactorily perform the testing of which it claims to have the capability, then having a larger facility and higher levels of instrumentation technology are advantages and not requirements.

2

THE REGULATED INDUSTRIES

Many consumer product or service industries are regulated. Where laboratory testing is utilized in these industries, predetermined guidelines are followed for evaluating the safety, efficacy, and overall quality of the products or services. These guidelines originate either internally or externally. There are many external organizations which develop guidelines that have become industry standards, such as a government agency, accreditation group, or industry forum.

The Food and Drug Administration (FDA) enforces the 1938 Food, Drug and Cosmetic Act, which was revised in 1976. The FDA inspectors follow guidelines set forth in an Inspection Operations Manual (1983). The FDA monitors how industry follows guidelines set forth in documents called Current Good Manufacturing Practice in Manufacturing, Processing, Packing, or Holding of Drugs (1978), and Current Good Manufacturing Practice in Manufacturing, Processing, Packing, or Holding of Foods for Human Consumption (1979), and Good Laboratory Practices for Nonclinical Laboratory Studies (1978).

FDA inspections occur either on a periodic basis or with higher frequency based on the following:

1. Customer complaints or reported adverse reactions
2. Voluntary recalls by a firm
3. FDA product sampling program finds a deviation from product quality or product claim

4. Raw material or packaging component manufacturer problems
5. Pre-approval inspection for a new drug application (NDA)
6. Current GMP inspection for manufacturing of new clinical supplies, medical devices, or diagnostics products
7. Approval of a sterile product manufacturing facility
8. GLP inspection for support of nonclinical testing

The United States Department of Agriculture (USDA) oversees the agricultural industry, which includes meat, poultry, egg, and dairy products. Testing for microbiological quality attributes, nutritional labeling, and pesticide residues are part of product evaluation programs. USDA regulated laboratories follow standard methodology such as the Official Methods of Analysis (AOAC, 1990).

The U.S. Environmental Protection Agency (EPA) has the authority to monitor for air, soil, and water pollution and to enforce standards to protect natural habitats. Sampling techniques and methodology for testing samples have been approved by the EPA for public drinking water and wastewater (American Public Health Association, 1985). Maximum allowable levels of contaminants in public drinking water have been set (EPA, 1988).

The Health Care Financing Administration (HCFA) is part of the U.S. Department of Health and Human Services. The HCFA regulates laboratories under the Medicare, Medicaid and Clinical Laboratories Improvement Act of 1967. Evaluation criteria for clinical laboratories have been developed under this act. These criteria, when met, provide supporting evidence for continued financing of insurance by the government agency.

The United States Pharmacopeial Convention is a forum of professionals who represent the pharmaceutical industry. The "bible" of this forum is the United States Pharmacopeia and National Formulary (USP-NF XXII, 1990). The USP-NF is a reference for the biological, chemical and physical attributes that are used to determine the purity of pharmaceutical raw ingredients or compounds. The USP-NF also recommends test protocols for evaluating the attributes of each pharmaceutical material. Other countries have similar references of pharmaceutical standards, such as the British Pharmacopeia, the European Pharmacopeia, and the Japanese Pharmacopeia.

The World Health Organization is a forum of professionals who represent various nations' interests in health. Expert committees have written documents which include pharmaceutical good manufacturing practices (to be published) and general guidelines for national control laboratories (WHO, 1966).

According to Bell (1989), "there has been an increasing trend towards the development of broad spectrum accreditation programs that apply the same principles of good laboratory practices to laboratories working in any field of science or technology." Thus, the International Standards Organization (ISO) published a set of documents which were developed by the International Laboratory Accreditation Conference. The ISO Guide 25 is considered to be a generic accreditation document of criteria (Bell, 1989) for technical competence of a testing laboratory. The ISO 9000 series, published in 1987, is becoming the single source of standardized requirements for the design and implementation of a quality system. These requirements were initially interpreted and adapted in an industry-specific manner (Marquardt, 1991). To prevent movement away from standardization, the ISO Technical Committee TC176 met in 1990. They agreed on a strategy for developing worldwide acceptance of global standardization. One of their goals was to seek harmonization between ISO guides and European Community standards (EN 45000 series) dealing with, in part, operation, assessment, and accreditation of laboratories.

3

GOALS OF A LABORATORY QUALITY AUDIT

The objective of auditing a laboratory's quality program is to evaluate the activities and existing documentation and to determine if they meet predetermined standards. A total quality program in a laboratory is usually developed to assure that all activities are performed with the objective of meeting certain standards, both internal and external. Some standards are generated internally, e.g. corporate quality improvement process, routine quality assurance/quality control protocols, or annual accreditation. A corporate quality improvement program may, in part, require that certain laboratory operations become routine and that customers are given information when they need it (for example, by a designated customer service contact person, facsimile capabilities, and a computerized database that can track samples). Quality assurance programs often provide standards for receiving, handling, coding, and testing samples, and for recording and reporting results. Quality control programs usually provide the standards for validating instrumentation, reagents, and test procedures, and provide requirements for training of analysts.

A laboratory quality program should have an established set of goals which every effort is made to reach. Objectives are set to provide the means for achieving the established goals. These objectives must be measurable. It is these measurable objectives on which an audit is based.

Competency is the key objective of any laboratory quality program. Laboratory accreditation is a "formal recognition that a testing laboratory is competent to carry out specific tests or specific types of tests" (Schock, 1989). The accreditation of a laboratory can be either an internal or external standard, or both. The audit for accreditation is an on-site examination of the laboratory to determine if it meets the accreditation criteria (Schock, 1989).

Some sources of standards or criteria that are generated externally are government agencies, customer requirements for contracted laboratory services, and procurement quality control requirements. The standards generating agencies of the U.S. government that are concerned with foods, drugs, cosmetics, and medical devices are at all levels of government, i.e. federal, state and local. Some of the federal agencies are the Food and Drug Administration, Environmental Protection Agency, U.S. Department of Agriculture, and the Health Care Finance Administration. Internationally, as mentioned in Chapter 2, a standards generating group is developing a basis for laboratory competency which will become a part of the ISO (International Standards Organization) series of standards. Laboratories in the healthcare services industry and in the food, drug, cosmetic, and medical device industries must meet specific criteria developed and enforced by the U.S. government's Food and Drug Administration. The Environmental Protection Agency has set standards for air, water and ground contamination levels, as well as standards for testing for contamination.

The Health Care Financing Administration has provided healthcare laboratories with standards for quality control and proficiency testing.

When a contract laboratory provides testing support for a manufacturing facility, the manufacturer usually provides the laboratory with the criteria it must meet to ensure a continued business relationship. Routinely, some overlap exists in the criteria that a laboratory must meet originating from a manufacturer of food, drug, cosmetic or medical devices and the criteria generated by government agencies related to the same products.

The users of laboratory services in the healthcare services industry, i.e. hospitals, clinics, and physicians' offices, may also develop standards for the laboratories to meet. Many of the standards are already integral parts of the Health Care Financing Administration standards, which are called Good Laboratory Practices, or GLP. These standards have been incorporated into national standards in numerous other countries.

Procurement quality programs usually require sampling and testing of purchased materials, ingredients and finished products. Testing and documentation criteria are commonly set by the clients, including requirements for calibration of instruments, use of reference standards, etc.

4

WORKING KNOWLEDGE AND AUDIT DOCUMENTS

A properly conducted audit of a laboratory should be documented thoroughly in a format that will make information simple to extract. Documenting data and recording comments are the most effective permanent record of pertinent information. An auditor should rely on memory only for a very short time (minutes or hours) and only if carrying pen and notepaper or a microcassette recorder is not possible; for example, in a sterile testing area.

Prior to conducting an audit, an agenda must be prepared, listing the areas to be evaluated and specific testing to be observed. Results of previous audits should be reviewed to determine if follow-up to previous concerns is necessary. The laboratory is then contacted and the agenda is reviewed by the auditor and client. This initial preparation will ensure that time and personnel are available when the actual audit takes place.

There are three areas in a laboratory from which audit information should be derived: personnel, documentation, and observation of testing procedures. Even under the best circumstances and relationships, there is usually a limit to an auditor's contact with the personnel who actually perform the testing. Almost every analyst, when correctly performing a test procedure, does it in a unique way. These unique differences provide a challenge for the auditor. An auditor must observe the performance of a procedure.

First, the auditor reads and interprets the procedure. Then the auditor observes the actual conditions and manipulation of the test procedure by the individual who is routinely assigned to perform that procedure.

The preparation of supplies, materials, test site, and instruments that will be used are all significant parts of a properly performed test procedure. As the analyst performs the test, reads the results and records the results, the auditor should note any aspects of the test that could lead to error. There are a few test procedures which are repeated on consecutive days or require more than one day to complete. Since most laboratories carry out routine testing daily, enough overlap of tests occurs that an auditor can observe the different steps of a single procedure in one or two days, even if they are performed on different samples (e.g., microbiological testing for Salmonella species may require three transfer steps over a four or five day period; many microbiology laboratories are performing this test daily so that on some days all three transfers can be observed).

If a reference standard calibration is a preparatory step in a procedure, the auditor observes where the standard was stored and how it was handled before and during the calibration. It is also important to observe the type of documentation kept to assure the stability and confidence in the standard, as well as consistency in the calibration.

The laboratory director, or a supervisor of the specialty area (toxicology, microbiology, chemistry, etc.) usually responds to questions from the auditor. It is not only logical, but imperative that audit questions be directed to the most knowledgeable person in a laboratory area. Those key persons are the only sources of complete, up-to-date information about the test procedures that are being followed. Those individuals know the background of the changes that have been made to procedures in their fields of expertise. They are also familiar with the precision and sensitivity of each procedure performed in their areas. Nowhere else could an auditor expect to find a resource more familiar with the specialized synergy of the laboratory, procedures, and personnel in their area of scientific testing.

Observation permits the auditor to see the actions required to perform a test. The analyst should carry out the test as it is written. If an analyst deviates from the written procedure, the deviation(s) can be a source of error and invalidate the test. A deviation must be documented and reviewed before any further testing is performed. A procedure should be revised, reviewed, and revalidated before changes are implemented.

If an auditor is to observe and accurately evaluate an experienced analyst performing a test, the auditor should understand the technical intricacies of the test. In other words, the most accurate evaluation is made by a qualified auditor who has performed the test enough times to become intimately familiar with it. The auditor who is experienced in the test methodology will be the best person to make a fair, accurate evaluation of the client laboratory's test conditions. Thus, it is strongly suggested that a chemist audit chemistry testing, a microbiologist audit microbiology testing, a toxicologist audit toxicology testing, and so on.

Since there are many laboratory audits that involve a variety of testing areas, it is not uncommon to have more than one specialist perform an audit. In fact, a team audit is common and very productive when a variety of specialty areas are involved. A team composed of qualified individuals with different fields of expertise (e.g. chemistry, microbiology, toxicology, metrology), can audit many areas in a laboratory at the same time. The team concept is effective, efficient, and professional. The preparation for a team audit is very important. A plan must be well thought out and agreed upon in order to accomplish daily objectives.

5

PARAMETERS FOR EVALUATION

QUALITY PROGRAM

There should be an established program that monitors and evaluates all laboratory procedures and the competence of the laboratory staff. The program should be well-documented and provide the auditor with evidence that the laboratory is effective in the following areas:

1. Accuracy of test data
2. Timely reporting of test data
3. Action plans to correct problems
4. Proficiency testing
5. Adequate training of analysts

Accuracy of Data

There must be an approved written procedure for each step in the handling, storage, preparation, and testing of a sample (or specimen). The review of all procedures at least annually is part of the quality program.

An instrument calibration and maintenance program must be

well documented. There should be evidence that reference standards have been used where appropriate and necessary.

All calculations of raw data are written into a bound laboratory notebook along with a signature and are checked and signed by a second person. Calculations should be based on a validated procedure.

One procedure of particular importance in government regulated laboratories is for the chain of custody of specimens. The protocol should include guidelines for the documentation of names and dates for each person-to-person transfer of a specimen, and the tracking of use quantities for each sample removed from the original specimen. This type of protocol is a requirement for the tracking of controlled (drug) substances, but can be applied to any specimen.

Timely Reporting

Determining if routine time allotments for testing and analysis are appropriate requires good judgement. Depending on the use of a manual, semi-automated, or fully automated procedure, testing time can be rapid (results in seconds) or lengthy (results in hours or days). The chosen method must be appropriate for precise results, and in many cases for the speed with which results are available to make crucial decisions, e.g. medical decisions.

There should be a record of the starting date and time and the completion date and time of tests and analyses. A review is performed on completed reports and procedures which define the test period and the amount of time required for calculations and reporting results. This can provide adequate information to evaluate the timeliness of reporting test data.

Action Plans to Correct Problems

When a problem is identified in the quality program, there must be documentation of the assessment of the problem, corrective action taken, and a follow-up review assessing the effect of the corrective action.

Proficiency Testing

Proficiency testing will be described in more detail later in the book. The primary consideration is that some type of periodic interlaboratory or intralaboratory "check sample" testing program exists and is followed. Without this type of program, a laboratory cannot measure itself relative to the status quo. Confidence in the accuracy and precision of test procedures in a laboratory should be partly based on the comparison of results from testing reference specimens in two or more laboratories.

Training

Education and experience of analysts are important for providing confidence in their capabilities in the laboratory. Performance, though, is significantly affected by training and supervision. A training program must be established and documented. Training must occur before new or revised techniques and procedures are implemented. Periodic proficiency testing should be used to validate each analyst's performance. Although it is difficult to determine the quality of supervision in a one to three day audit, observation of an analyst performing a procedure can provide an accurate assessment of the success of a training program and the type of supervision that exists.

DOCUMENTATION

How critical is documentation? It is extremely critical in a laboratory. From the recording of test results to the explanations written in a preventative maintenance log for an instrument, documenting information carries a lot of weight.

Documentation is the accepted method for recording information for future reference. Some types of information that are routinely recorded are:

- Procedures for:
    Quality assurance/quality control program
    Receipt and storage of samples
    Sampling

       Analytical testing
       Validation
       Calibration
       Data recording
       Data reporting
       Operation of instruments
       Reagent preparation

- Training records
- Organizational charts
- Sampling schedules
- Analytical testing assignments
- Supplies inventory
- Expiration dating of reagents
- Instrument calibration data
- Instrument maintenance
- Methods validation data
- Analytical testing results
- Test conditions: time, temperature, etc.
- Facility maintenance and cleaning

The above list is only a sample. If data can be recorded for future analysis, tracking, trending or evidence of any laboratory activity, it is appropriate and necessary that it be documented.

An auditor should check to see that records are written in indelible ink. It is preferable that most records be kept in permanently bound notebooks. In general, original data (raw data) is recorded in permanently bound laboratory notebooks which have been permanently coded in some numerical system. When in use, notebooks should be protected from damage. When in permanent storage, conditions must exist which prevent loss of records by damage or tampering.

The auditor should observe the entries in a laboratory notebook. Each entry must be accompanied by the signature or initials of the author and the date the data was entered. It is of utmost importance that the data be traceable to the time and person who made the entry. In many regulated industries it is mandatory that a second signature also accompany data entries as proof that a review of the data has taken place.

A change control system is necessary to acknowledge procedural revisions and allow for corrections to be made where problems or errors are found. Revisions to procedures should be reviewed and approved by the appropriate authorities in the laboratory. Whenever a procedure is changed, a document describing the change must accompany the procedure and become part of the permanent file. The permanent file contains the history of all procedure revisions.

Computers are being used as alternative means of keeping permanent records. Original data can be entered automatically (from an instrument) or manually into a database. The software program can provide data storage, retrieval, calculations, and much more. Users of a computerized database must be cautioned about the regulatory concerns (EPA draft, 1990). It is important that if computer systems are used to store original laboratory data, the auditor should ask if there is a built-in security system to prevent tampering with the data. Also, investigate how corrections to original data are controlled. For example, some programs have been written to either prevent changes to original data or to allow changes to be made accompanied by an automatically entered permanent marking adjacent to the data (e.g. # or *). Another technological advance that is being considered is a legal alternative to the handwritten signature, the electronic identification system. Compatible with computers, retinal optical scanning and fingerprinting are recent systems used for data entry identification. Part of an effective audit of a laboratory computer system is to observe original data entry and ask the operator to attempt to make a correction to the data after it has been entered.

It is recommended that a computer system be validated (PMA, 1986; 1990) before it is used to handle critical data. An auditor should review the validation protocol and results.

Documentation is mentioned again in various other sections of this book. A successful quality assurance program in a laboratory is proven, in part, by the thoroughness of the laboratory's documentation. The requirement for documentation is absolute.

PERSONNEL

The single most influential factor for a successful laboratory is the hiring and management of qualified people. Laboratory accreditation programs and federal regulations for Good Laboratory Practices (Appendix F) specify the education required for a laboratory director, laboratory supervisor and laboratory analyst.

Written position descriptions should be available for evaluation. The requirements for education and specialist training, where necessary, should at least meet laboratory accreditation requirements. Individual personnel files should contain a curriculum vitae which can be reviewed by an auditor and compared to the description of the position currently held by the individual.

An auditor should be aware of individuals without adequate scientific background performing complex tasks that require more technical experience than they possess. A list of assigned duties should be requested to determine which tests certain individuals are performing, and which individuals are assigned quality control responsibilities. This list should be used with personnel files to compare with the experience, education and training of the individuals who perform the tests. Complex tests should not be performed under any circumstances by individuals without adequate scientific background.

Training records should also be a part of each individual's personnel file. Documentation should exist that can show the ability of an individual to perform specified tests. This documentation should provide evidence of the training, experience or skills of an individual. Training records include internal checklists of procedures or instrumentation reviewed by the supervisor and approved by the lab director. Copies of school course records with grades and workshop completion certificates are other acceptable records of an individual's training.

An up-to-date organization chart should be requested for review of the reporting relationships that exist. Adequate reporting relationships are those that ensure immediate and adequate review and action are taken to resolve a problem, and that results are reported in a timely manner.

The size of a staff is a product of laboratory space and budget. A laboratory may have a small number of people performing a small number of routine tests; or a laboratory may have a small number of people performing a large variety of tests. The latter situation could be cost effective and highly competitive if the personnel have appropriate skills and education. If not, the situation is stressful and can take a toll on accuracy of results. An auditor should ascertain whether the staffing and experience of the personnel are congruent with the intended output of the laboratory.

ORGANIZATIONAL STRUCTURE

The importance of an effective organizational structure should not be overlooked. It has been stated that a primary objective of an analytical laboratory is to produce "high quality analytical data through the use of analytical measurements that are accurate, reliable, and adequate for the intended purpose" (Garfield, 1984). The type of working environment and the reliability of data is a direct result of the effectiveness of the organizational structure.

The auditor should request an up-to-date organizational chart of the laboratory staff. The chart should be accompanied by descriptions of the communication that occurs for all laboratory procedures. Effective lines of communication are the source of timely actions and timely reporting. Any procedures that specify information reporting responsibilities should coincide with the present organizational structure. Laboratory management should review procedures after any organizational change that may affect reporting responsibilities. At least two procedures that state action response guidelines should be compared to the organizational chart. If discrepancies exist, a determination must be made as to why they exist and what is necessary to correct them.

There are an unlimited number of possibilities for a laboratory organizational structure. Regardless of the titles chosen (group leader, supervisor, section head, manager, director) the key elements for strong communication in the laboratory are the reporting relationships between the analysts and their immediate supervisor, and between the supervisor and his or her manager. All raw data and calculations from analysts should be reviewed by the laboratory supervisor. When problems occur, e.g. instrumentation error or loss of integrity of a client's specimen, all actions taken by an analyst to resolve problems must first be reviewed by the laboratory supervisor. The supervisor's boss should be notified before or concurrent with the action plan, and it is his or her responsibility to notify the client at the appropriate time. Written procedures must exist that list the steps to be taken, the internal communication that must occur, and the documentation required to describe problems that develop, and also the action taken to resolve the problems. Notification of the client should be part of the procedure.

Timely and reliable problem resolution depends on all the appropriate personnel following the written procedures as described above. Problems that appear in the areas of a laboratory that are audited (calculations, instrument calibration, training, procedures, etc.) can usually be related to poor communications.

FACILITY

Maintenance and housekeeping of the building enclosing the laboratory and upkeep of the grounds surrounding the building are a reflection of laboratory management.

There must be adequate space and facilities (water, electric, lighting) to assure that laboratory workers can perform all procedures as safely and efficiently as possible. The quality and quantity of available water and electricity directly affect the efficiency and quality of work performed.

As a general rule, at least 15 linear feet of bench space should be available for each laboratory analyst (EPA, 1978). Storage space for specimens, reagents, and documents should be sufficient and accessible. An adequate number of electrical outlets, designed with safety in mind, should be available to ensure continuous use of instrumentation. Backup battery power packs are commonly used to prevent loss of instrument analysis time during power surges or power outages. Sufficient lighting is very important in areas where visual measurements are performed. A source of laboratory water, the purity of which is determined by its intended use (e.g. USP purified water, NCCLS type I reagent grade water, distilled water), must be easily accessible and monitored routinely for quality and for any required specifications.

The building should have an appropriate solid and liquid waste management system, including proper drainage of nonhazardous liquid waste and a routine trash pickup program.

The design of the laboratory includes two critical features that affect sanitation and housekeeping: 1) floors, walls and ceilings materials, and 2) layout. The results of a routine schedule for floor cleaning (and waxing where appropriate) are usually obvious to an auditor and any visitor. Walls and ceilings should be made of inert building materials and kept in good condition, i.e. without signs of chipping or peeling paint, water leaks, or any type of debris.

A well-designed laboratory can provide impetus for efficiency and success. There should be a physical separation between areas where critical measurements are performed and areas where airborne debris is expected (e.g. handling of dehydrated microbiological medium). Also, there should be a physical separation of sample receipt, testing and storage. Any testing that requires very strict environmental conditions should be performed in a separate, environmentally controlled enclosure (e.g. sterility testing).

Just as an HVAC (heating, ventilation and air cooling) system is specified and maintained by a building's maintenance staff (Belsky, 1991), the heating and other environmental control needs of many tests should be monitored and maintained by the laboratory staff. The laboratory should have adequate heating and refrigeration space for all of its testing requirements. The size and number of ovens, environmental chambers, and refrigeration units will be a function of the testing requirements and available laboratory space. During an audit, it is important to assess the oven, environmental, or refrigeration space available and determine if the specified conditions are being met. Adequate space helps assure that environmental conditions will be met. All environmentally controlled units (ovens, incubators, water baths, chambers, etc.) must be monitored on a routine basis. Procedures and documentation should be available demonstrating the monitoring process, i.e., comparing routine measurements to the appropriate predetermined specifications (e.g. temperature, humidity, air flow). Also, it is advisable to keep permanent records of monitored measurements by using recorders and charts, or computers, on a daily basis. Many laboratories manually record measurements once or twice each day, Monday through Friday, but there is no assurance that the specifications are met over 24 hours or on Saturday and Sunday.

SAMPLE CONTROL

An audit should cover the procedures that laboratory personnel perform that detail the receipt, documentation, storage and handling of samples of analysis.

Representative sampling procedures will not be discussed here, although if laboratory personnel perform any initial samples collection, written procedures should exist, and the procedures should be reviewed in an audit. Procedures should include size of samples collected, how they should be collected, frequency of collection, and how the sample is subsequently preserved and shipped to the laboratory. Conditions of preservation for each sample type should be specified, i.e. refrigeration, use of sterile bags or amber bottles.

The sample receiving procedures must be documented by a recordkeeping system. Whether a computerized system or a manually written log is used, the records must be current. Information necessary for identifying each sample should be required before a sample is accepted for testing. Every sample must be given a distinct identification number and labeled with that number and any other information for proper storage and handling. Chain-of-custody type documentation is a legal requirement in many government regulated laboratories, but is also advantageous in private laboratories as a preventative measure to assist in criminal court actions.

Procedures that state the method of handling and storage of different types of samples are based on the initial identification provided by the originators of the sample and logical preservation techniques. It is obvious that the facility must have adequate space and environmental condition flexibility to store samples under specified conditions such as refrigeration or absence of light.

A detailed description of subsampling of samples for analysis should exist for each type of sample. Reserve samples for additional testing should be a consideration if an adequate original quantity is received.

In certain situations a tracking system might be used to detail quantities of each subsample used in a test by recording in a book or by recording quantities used or remaining on the label of the sample storage container. Presently, it is common practice to record quantities of controlled drug substances as they are used in testing. It would also be prudent to track rare materials.

SUPPLIES ORGANIZATION

An efficient laboratory requires a defined organization of its supplies. A laboratory should have a documented procurement process to ensure that all supplies purchased meet requirements for their intended use. Microbiological media, for example, should not be accepted without completed certificates of analysis from the vendor. Also, all sterile supplies must be inspected upon receipt and must have documentation of the sterilization cycle used. Vendor audits should be performed, where appropriate, to certify the manufacture of supplies.

Written procedures should describe how reagents, pure chemicals, and any combinations therefrom are inspected or tested to confirm their identities. A permanent record should be kept listing the name of each material received, the manufacturer's expiration date (if applicable), the laboratory's assigned expiration date, and any testing or inspection performed to confirm its identity. In assigning expiration dates, the laboratory should determine the criteria and have a written procedure describing the criteria, e.g. first use date, receipt date or manufacturer's recommendations.

In addition to assigning expiration dates to materials, it is a strongly recommended practice that the laboratory follow a first in, first out (FIFO) inventory system to assure that any supply having a limited shelf stability is used before its expiration date. A lack of control in this area can cause delays in testing and possibly the mistaken use of expired materials.

An observation of the organization of supplies in the laboratory can be a significant indicator of the concern laboratory personnel have for efficiency, inventory control, and laboratory cleanliness. When conducting an audit or when being given a preliminary tour of a facility, it is very useful to observe and ask about the organization of supplies. Organization of supplies is another area that reflects the concerns and attitudes of the laboratory management.

INSTRUMENTATION AND CALIBRATION

Instrumentation and calibration will be discussed in further detail in Chapter 6 of this book. For the instrumentation and calibration program to function there are three essential elements: qualified personnel, detailed procedures, and raw documentation. Personnel must be qualified and experienced with the instrumentation. There must be detailed procedures describing the frequency and steps of each calibration. Finally, raw data demonstrating the results of past calibration efforts must be available. If the audit is prospective, review existing data on the appropriate instrumentation. Do not rely on future activities to demonstrate capabilities in instrument calibration. Instrument calibration is an important aspect of overall laboratory accuracy and precision. The competency with which this program is carried out is a general indication of the overall capability in a laboratory.

PROFICIENCY TESTING

Proficiency testing was developed as a tool to evaluate clinical laboratory accuracy (Belk and Sunderman, 1947). Concerns have been raised relating to the use of proficiency testing programs as criteria for accreditation (Daniel, 1990). Many laboratories, e.g. clinical laboratories and federally regulated environmental laboratories, are required to pass proficiency testing in order to become accredited and maintain accreditation.

Where proficiency testing is a requirement for laboratory accreditation it is wise to inspect the records of the laboratory's proficiency testing program. The frequency of testing should at least meet the minimal requirements of the accreditation program. The laboratory should have a procedure which details how "check" samples are handled, stored, tested, and reported, and where results are filed.

Proficiency test records must show that "check" samples were handled according to routine sample control procedures, testing and document review. This is to assure that the objective of the proficiency testing program was met (Daniel, 1990).

Where proficiency testing is not a requirement, it is strongly recommended as an evaluation tool to improve the accuracy of testing.

Some of the criteria that can be used to measure the usefulness of a "check" sample are: use of a national standard analyte, stability of an analyte in test sample during shipping, number of internal repeat tests performed, and number of external tests performed.

The recommended testing program should be corroborated with an external, unbiased laboratory. But a successful program can also be developed among multiple laboratories of an individual business, which provides discretion and confidentiality in this decade of corporate competitiveness. In this situation, one central laboratory acts as the "check" sample library and also utilizes an outside contract laboratory to confirm sample identification.

ACCREDITATION

Laboratory accreditation is defined as the "determination of the competence of a testing laboratory to conduct testing services according to specific test methods, by a qualified evaluation agency, and the issuance of a public statement attesting to that fact by the agency" (Gladhill, 1989).

Actually, accreditation is a form of the laboratory quality audit. If a laboratory satisfactorily meets all the criteria for accreditation set by a reputable technical agency it can become accredited by that agency. "The agency must be technically qualified to perform the evaluation and must provide an unbiased third party evaluation of the laboratory" (Gladhill, 1989).

Inquire about accreditation of the laboratory being audited. Accreditation by a reputable agency is a sign of competence. Determine if the accreditation was given for specific tests or for general laboratory services. If accreditation was given for specific tests, it is advantageous if those tests are required as part of your audit. If accreditation was given for general laboratory services and covered most of the routine testing performed, the laboratory has thus demonstrated competency to another evaluator.

Discussions of accreditation practices on a national and international basis may be found in Accreditation Practices for Inspections, Tests, and Laboratories (Schock, 1989).

CUSTOMER SATISFACTION

One of the laboratory's most important responsibilities is to satisfy the needs of their customers. Competitiveness, technical competence and growth are influenced by the concern given to satisfying customers.

Part of an audit should be to determine if a customer satisfaction program exists. If a program exists, determine if the employees are aware of the needs of the customers and whether the program is successful.

A laboratory should periodically solicit comments and requirements from all its internal and external customers. The responses should be evaluated and action taken where appropriate. The latter is a simplistic form of quality function deployment (Bossert, 1990). In simple terms, the laboratory should ask its clients how service can be improved. Internally, employees should periodically be asked for feedback that can be used to help improve customer relations, conditions of the testing facility, employee development and training, and more.

It is sufficient to say there definitely is a place for total quality processes (customer satisfaction, total quality improvement, quality function deployment, statistical quality control, etc.) in the laboratory.

LABORATORY HEALTH AND SAFETY

Health and safety of the laboratory personnel are usually areas of concern to the laboratory management and not to the client who contracts testing from the laboratory. But a smart auditor should investigate the design of the facility and any modifications to protect employees from chemical and biological contaminants. The laboratory should have a safety program for handling chemicals, biologicals, needles, etc. It should be a well-documented program with information easily accessible to all employees at all times (not only normal working hours, but after normal working hours and on weekends as well). Material safety data sheets should be readily accessible and familiar to all laboratory personnel who handle hazardous materials.

Why is the area of health and safety important to an auditor? A well-run laboratory shows observable concern for its employees' health and safety. When health and safety concerns are given low priority, be sure that other areas will show signs of weak management.

6

LABORATORY INSTRUMENTATION: CALIBRATION AND PREVENTATIVE MAINTENANCE

The laboratory program for performing and monitoring equipment calibration and preventative maintenance must assure that instruments and equipment meet allowable tolerances and are operating properly. Units failing to meet requirements are identified and taken out of service. There are three absolute requirements for such a program: qualified personnel, detailed procedures and raw documentation. These three requirements form a troika. One missing requirement will result in a program that fails totally.

The first element of the calibration and preventative maintenance program is the qualifications of the personnel carrying out the program. Personnel training and education must be carefully checked, including continuing education and training, such as off-site seminars and on-the-job training.

The second element of the calibration and preventative maintenance program starts with the general description of the overall program, or standard operating procedure (SOP). This SOP should describe the overall objectives of the program, the instruments, calibration standards and limits, calibration intervals for each instrument, means of instrument identification, recordkeeping requirements and logs, instrument calibration stickering, and removal from service of equipment that fails to meet requirements. At this time, also review the master inventory of all instruments (assets) to be calibrated. Note that each has an asset number and a reference to a calibration and maintenance procedure.

The general description should contain several major parts: a description of the source, storage, and use of the standard materials, a rationale for the calibration intervals, a numerical and descriptive means of identifying the unit to be calibrated, tolerance limits specified by the manufacturer, a description of the recordkeeping requirements, a requirement for and description of instrument calibration stickering, and a description of actions to be taken where problems are found. There should be an explicit description of the calibrating material and the material's relationship to compendial/NITS standards. An objective of the overall program is that calibration standards be directly traceable to compendial/NBS standards. The control of calibration standards and supplies is an important aspect of the overall program. A procedure should exist for the responsibility, inventory, purchase, storage and usage (first in, first out) of these standard materials.

The third element of the calibration and preventative maintenance program is the records of calibrations as they are carried out. Log books are the form these records usually take. These log books must adhere to all the requirements of laboratory notebooks.

There exists an important linkage between instrument calibration and preventative maintenance. The interval between calibration periods cannot practically encompass all possible insults to the measuring system. Preventative maintenance is critical in assuring that the instrument is in good operating condition and capable of holding the calibration within the normal expectations for that instrument.

The first step in an audit is to obtain an inventory of instruments along with all SOP's controlling the calibration and preventative maintenance program. Next ascertain the validity of this inventory by checking it with the instruments in the laboratory. Note any new instruments not on the company's inventory and find out the time frame for adding new instrumentation to the program and the official inventory.

The second step is to check out the laboratory's handling of glassware and reagents. It is not apparent that glassware ever could be considered instrumentation, however there are strong reasons in favor of the argument that a buret is simply yesterday's instrument. The care and treatment of laboratory glassware will be a harbinger of instrument calibration and maintenance. Also, an instrument calibration program cannot very well be successful when the necessary glassware is in questionable or abused condition. The four important concerns are class, washing, storage and responsibility.

Quantitative glassware must be Class A. A procedure should exist for washing glassware. The storage area must be appropriate to minimize wear and tear. Responsibility for ordering, keeping inventory, and evaluation of glassware must be clearly communicated. The laboratory's handling of reagents (quantitative and qualitative) must be assessed, as many of these are used in the calibration and standardization of instruments. All reagents should be labeled with contents, date made, expiration date, and procedural reference (where appropriate). There are four important concerns: source, use conditions, storage, and responsibility. The source of material must be appropriate to its intended use; the conditions of use must be clear; the storage conditions must be spelled out; and responsibility for ordering, storage, and inventory must be clear. Finally, a procedure must be available describing all these activities.

The calibration of specific instruments can now be considered. The general procedure for laboratory calibration, as described above, outlines the systematic approach: qualified personnel, relevant, detailed procedures, and recordkeeping. The next step is the detailed procedure that gives step by step instructions for calibration. This combines with the general procedure to yield a record of calibration that is readily auditable. The following elements can be checked: identity of instrument, identity of personnel (for qualifications check), frequency of calibration, source (traceability) and appropriateness of calibration standard, and action taken as a result of calibration. The instrument calibration sticker has the following information: asset number, date calibration expires, name of person calibrating the unit, and a reference to the notebook (logbook) containing the calibration data. Check the logbook and sticker for agreement.

There are two broad types of laboratory instruments to be calibrated: those whose calibration is absolute (measurement of time), and those whose calibration is relative to a specific standard (pH). The consequences of miscalibration of the first class is easy to perceive. If an analyst is using a stopwatch that is off by several seconds, measuring in minute intervals is worthless. A gas chromatograph whose oven temperature indicator is off $5°C$ may or may not present an immediate problem to an analyst, but future attempts to reproduce the work could be futile. The operation of the gas chromatograph requires the daily preparation of a standard curve. An inappropriate temperature will not affect quantitation, though it may effect separation or tailing.

Contract and outside instrument calibration present requirements very similar to those for an internal program. In reality, though, contract and outside instrument calibrations are much more difficult to audit. Calibration services done by others yield many different types of final reports and often it is not possible to determine the exact procedure used. The written calibration procedure used by the outside firm is usually quite general to allow them to handle a wide variety of equipment. This problem is not easily overcome. It is best to carefully review the outside service's final report for the following: exact identity of the unit being calibrated, reference to previous and future calibrations, standards used for acceptance, availability of raw data and procedures, and qualifications of the person performing the calibration. Finally, check the firm's audit of its outside calibration services. At the very least, verify that the firm has audited the outside calibration service.

Having said so much about calibration it is time to move on to preventative maintenance. As stated above, there is a linkage between instrument calibration and preventative maintenance. Periodic calibration cannot take into account all possible failures of the measuring system. Preventative maintenance is needed to assure that the instrument is in good operating condition.

The elements of a good preventative maintenance program are very similar to those of a good calibration program: an inventory of instruments, a general procedure, a schedule of preventative maintenance, specific preventative maintenance procedures for each of the instruments, and logs for recording results.

In practical terms the preventative maintenance program is often a weak sister of the calibration program. A key aspect is the person responsible for the preventative maintenance. The laboratory manager is usually responsible for the calibration program, while preventative maintenance will usually be carried out by someone outside the laboratory who has responsibility for preventative maintenance for the entire facility.

Since the preventative maintenance records must be detailed, the key preventative maintenance activities to check include: the general procedure, the schedule, the specific procedure, and the records of preventative maintenance performed. One key source material for the auditor will be the instrument's manuals supplied by the manufacturer. These manuals usually give a real indication of the expectations for the unit, and often give a good idea of what the preventative maintenance schedule should be.

The instrument's preventative maintenance log should also include unscheduled maintenance. Every time an instrument is repaired, cleaned, or adjusted, a record should be made in the preventative maintenance log. Using the maintenance log, compare scheduled and unscheduled maintenance. The comparison should reinforce the preventative maintenance schedule's validity. A maintenance log of this type gives a real indication of the functioning of the instrument.

7

CLOSING CONFERENCE, FINAL REPORT, AND FOLLOW-UP

At the conclusion of an audit all findings should be summarized and discussed. It is very important that this is a summary. There should be no surprises at this closing conference or meeting. Each individual who accompanied the auditor during the audit should have been made aware of concerns when they were made apparent. The auditor and the client's management who are intimately responsible for implementing any changes in the audited process should attend the meeting. In other words, the discussion should involve top management representation who have the authority to initiate and support any recommended changes. Personnel who are held accountable by management for the supervision of the audited areas should also attend this meeting, because they must understand the reasons for and actually implement any recommended changes.

At the closing meeting the auditor reiterates each observation made, assigns levels of concern to each observation, and listens to the client respond to each observation. This is the key to the success of the closing conference. The auditor must exhibit a balance in leadership by showing good listening skills and providing clear and concise details of the audit.

The closing conference is the first opportunity for the client to formally respond to the audit findings by prioritizing any problems observed and stating their plans for resolving problems. If contradictions develop in the understanding or interpretation of observations, they should be discussed at this time. It will take an auditor's strong skills to properly and professionally deal with any misunderstandings at that time. It is the auditor's responsibility, for his profession and for the organization he represents, to assure that observations are weighted in a reasonable and appropriate manner based on the reasons or requirements originally set forth in planning to perform the audit. The auditor must then assure that the client provides a schedule of and plans to resolve each problem discussed. The most important point agreed upon at the closing conference should be that resolving the discussed problems will ensure continued accuracy and confidence in the laboratory's business.

The auditor should give a list of observations to the client's top management at the closing conference. A final report is written by the auditor after the closing conference. The report should be written to the laboratory's top official and should contain the list of observations that were stated at the closing conference. The observations should be classified by importance. In addition to the summary of the observations and their classifications, the final report should contain a statement of the time required for a response. Usually a period of 30 days is allowed for a client to respond. An additional 60 to 90 days is expected for the implementation of changes.

Follow-up to an audit can be performed in two ways. A physical reinspection should be performed within 30 days, if possible, of any change made to facility design, housekeeping and maintenance practices, or changes in instrumentation. Most other areas where changes are made can be documented, and a written statement of such changes should be required from the client's top official within the specified period of time. In many cases, procedural or documentation changes can be approved without an immediate physical reinspection, and in the next scheduled audit, particular attention should first be directed toward those changes.

8

FREQUENCY OF AUDIT

In general, a standard frequency for quality auditing of a laboratory does not exist, except where it is specified in the accreditation criteria by which a laboratory is evaluated. Planning an audit schedule is very individualistic, and it is customized to the requirements of the quality assurance program (if internal auditors are utilized) or of the firm contracted to perform the audit (if external auditors are contracted).

Table 8.1 is an example of a list of criteria which can be used to determine how often to audit.

It is prudent to perform a quality audit of each laboratory at least annually. Laboratory auditing can be prioritized by using the higher concern factors for increasing the frequency of audits and the lower concern factors for decreasing the frequency of audits. The above table is only an example of prioritizing criteria. It is important that each firm develop its own selection criteria. They should use those criteria consistently to determine the frequency of quality audits of client laboratories.

Table 8.1　Criteria for Determining Audit Frequency

Factor	Higher concern	Lower concern
1. Quantity of tests performed:		
High or moderate number	X	
Low number		X
2. Nature of testing:		
Health related	X	
Legal related	X	
Routine		X
Research	X	
3. Laboratory performance history:		
Excellent		X
Average	X	
4. Test methodology:		
Custom or new	X	
Standard		X
5. Laboratory audit history:		
Accreditation		X
No accreditation	X	
Regulatory audits (e.g., FDA, EPA):		
Met requirements		X
Did not meet requirements	X	

EPILOG

A laboratory quality audit should always be performed with the primary intention of improving the laboratory. Since many areas of quality improvement exist, the audit becomes an important management tool for evaluating technical competence and competitiveness. Good communications can drive organizational improvement. When a continuous quality improvement process is in place, a laboratory can become more productive, more competitive, and a better place to be employed.

Audits performed by government agencies, accreditation groups, and international standards certification groups are usually very thorough. These audits should be incorporated into a total quality laboratory audit (TQLA) plan, which combines internal and external audits to evaluate the laboratory as often as possible. Add these audits to an effective customer satisfaction process and the laboratory will be well prepared for the journey toward continuous quality improvement.

Appendix A

COMPARISON OF DOMESTIC LABORATORY STANDARDS (FDA, EPA, AND HCFA)

Appendix A Comparison of Domestic Laboratory Standards

Organization/ Standard	Facilities		Environment, Storage	Materials Control: Labeling, Performance Testing
	Space, Utilities	Equipment		
Food and Drug Administration, HHS (4-1-88) 21 CFR Chapter 1 Part 58 Good Laboratory Practices for Non-clinical Laboratory Studies (FDA-GLP)	58.41 ... degree of separation ... prevent any activity from having adverse effect on the study. 58.43 ... sufficient number of animal rooms ... to assure ... separation ... isolation ... and ... housing of animals.	58.61 ... of appropriate design and capacity to function according to protocol ... 58.63 ... shall be adequately ... cleaned and maintained. ... shall be adequately tested, calibrated and/or standardized.	58.45 There shall be storage areas ... separated from areas housing test systems ... 58.49 Space shall be provided for ... storage and retrieval of ... raw data and specimens from completed studies.	58.105 The identity, strength, purity, and composition ... shall be determined ... The stability of each test or control article shall be determined ... Each container shall be labeled by name, ... code number, ... expiration date ... storage conditions ... 58.83 All reagents and solutions ... shall be labeled ... identity ... liter or concentration ... expiration date.
Environmental Protection Agency (State of Illinois) Title 35 Part 183 Certification and Operation of Environmental Laboratories (12-1-83) (EPA-EL)	183.125, 183.315 ... Minimum of 150 sq. ft. of floor space for each analyst. Walls and ceilings should be covered with ... which is easily cleanable and disinfected.	183.315 ... shall include a vacuum source ... 183.220, 183.320 ... balance ... refrigerator ... stirrer ... oven ... pH meter ... conductivity meter	183.215, 183.315 Shall be well ventilated and free from dusts, drafts ... temperature changes.	183.240 Chemicals shall be dated upon receipt ....

	Part 493.223	Part 493.225	Part 493.227	Part 493.229
Health Care Finance Administration DHHS (8-5-88) Clinical Laboratory Improvement Act Proposed Rules Part 493 (Reprinted in Appendix) (HCFA)	Laboratory lighting . . . minimum of 100 footcandles . . . . . Adequate electrical supply . . . . . Source of laboratory pure water . . . distilled water or deionized water or both.	Spectrophotometer, atomic absorption spectrophotometer, gas chromatograph, mass spectrometer . . . . . . Autoclave, incubation unit, hot plate, water bath, membrane filter equipment . . .		
Food and Drug Administration Current Good Manufacturing Practices CFR Title 21 (FDA-GMP)				
Environmental Protection Agency Manual for Interim Certification of Laboratories Involved in Analyzing Public Drinking Water Supplies PB 287-118 (May 1978) (EPA-PDW)	Laboratory space should be . . . 150 to 200 ft^2 per person . . . and include . . . hot and cold running water, electricity . . . source of distilled and/or deionized water . . . . exhaust hood . . .	Minimum requirements are listed in text		Chemicals are dated upon receipt . . . .

Appendix A Continued

Org./Std.	Maintenance	Equipment		Test Procedure Manual	Methods Validation
		Calibration Frequency	Calibration Standards		
(FDA-GLP)	58.63 . . . standard operating procedures shall detail . . . methods, materials . . . schedules to be used . . . maintenance . . .	58.63 . . . standard operating procedures shall detail . . . methods, materials . . . schedules to be used . . . testing, calibration and/or standardization . . .		58.81 Each laboratory area shall have . . . laboratory manuals and SOP's relative to the laboratory procedures being performed. A historical file of standard operating procedures and all revisions . . . dates . . . shall be maintained.	
(EPA-EL)	183.240 . . . service contract . . . on all analytical balances. 183.355 . . . current service contracts or in-house protocols . . . on balances, autoclaves, . . . ovens, incubators . . . service records . . .	183.240 . . . periodic checks on balances. (Inorganic): . . . standard reagent curve . . . daily checks . . . (Organic): . . . minimum of three calibration standards . . . each day . . . trihalomethane control standard each day. 183.355 . . . pH meter . . . calibrated each use period . . . spore strips, or ampoules should be used weekly . . . to verify steriliza-tion temperatures . . .	183.240 . . . class S weights shall be available . . . thermometer certified by NBS . . . . . . shall conduct analyses on known reference samples once per quarter . . .	183.240, 183.350 A laboratory manual . . . instructions for each parameter . . .	183.350 When . . . performance evaluation samples . . . at least one per year . . . If . . . more than one analyst in the laboratory . . . once per month . . . each shall perform parallel analysis . . . 183.440 . . . shall participate at least twice per year in . . . U.S. EPA intercomparison studies . . .

	Part 493.233 (HCFA)	Part 493.237	Part 493.231	Part 493.235
(FDA-GMP)		211.160 . . . calibration . . . at suitable intervals . . . with an established program . . . Instruments . . . not meeting established specifications shall not be used. 211.194 Complete records shall be maintained of . . . calibration		
(EPA-PDW)	Current service contract is in effect on all balances.	Class S weights NBS thermometer Color standards		Minimum requirements: Laboratory must analyze an unknown performance sample once per year for parameters measured. . . . standard curves must be verified . . . daily . . . . . . a known reference sample once per quarter . . . At least one duplicate sample should be run every 10 samples to verify precision of the method.

Appendix A Continued

| Org./Std. | Equipment | | | Test Procedure Manual | Methods Validation |
	Maintenance	Calibration Frequency	Calibration Standards		
(HCFA)	Part 493.239	Part 493.240	Parts 493.241 to 493.315	Subpart G Appendix A	Subpart G Appendix B Appendix C	Subpart H
(FDA-GMP)			211.173 *Animals used in testing* . . . shall be maintained and controlled in a manner that assures their suitability for use. 211.194 Laboratory records shall include complete data . . . description of sample . . . method used . . . weight for measure of sample . . . calculations. . . . results . . . and how results			211.160 . . . establishment of standards, specifications, sampling plans, test procedures . . shall be drafted . . . reviewed and approved by the quality control unit.

(EPA-PDW)	compare with established standards . . . initials or signature of person who performs each test . . . date the test(s) were performed . . . initials or signature of a second person . . .	Director: Minimum of bachelor's degree and 5 years experience. Supervisor: Minimum of bachelor's degree and 1-2 years experience (2 years for chemistry, 1 year for microbiology)	Analyst: Minimum of high school diploma or equivalent and 2 years experience in specific testing. . . . 30 days on the job training for a microbiological analyst.	EPA will provide two types of samples to Regional and Principal Laboratory. Known-value quality control samples . . . appropriate performance evaluation sample will be furnished . . .
	Enforcement data should be kept for 3 years			

Appendix A Continued

Org./Std.	Action Response Program	QC Document Retention	Areas of Special Requirements	Personnel		Quality Assurance Program
				Lab Management Requirements	Technical Personnel Requirements	
(FDA-GLP)		58.190 All raw data . . . reports . . . specimens . . . shall be retained. There shall be archives . . . 58.195 . . . records, raw data . . . shall be retained for at least: 2 years following the date . . . permit is approved . . . 5 years following the date . . . results are submitted . . . In other situations . . . 2 years following the date . . . the study is completed, terminated or discontinued.	58.120 Each study shall have an approved written protocol . . . All changes and revisions of . . . protocol . . . shall be documented, signed . . . 58.130 The . . . study shall be conducted in accordance with the protocol.	58.31 . . . testing facility management shall designate a study director before the study is initiated. 58.33 For each . . . study . . . a scientist or other professional of appropriate education, training, and experience . . . shall be identified as the study director. . . . has overall responsibility for . . . the study . . .	58.29 Each individual engaged . . . supervision of . . . study shall have education, training and experience . . . to perform assigned functions . . . facility shall maintain a current summary of training . . . for each individual . . . There shall be a sufficient number of personnel for . . . conduct of the study.	58.35 . . . a quality assurance unit . . . shall be responsible for monitoring each study . . . . . . quality assurance unit shall be entirely separate from . . . personnel engaged in the direction and conduct of that study . . . shall maintain a copy of master schedule . . . . shall inspect each . . . study . . .

| (EPA-EL) | 83.255, 183.450 When results indicate . . . parameter has been exceeded . . . facility shall be notified within 48 hours . . . . 83.370 . . . results are to be reported . . . without waiting for membrane filtration verification . . . | 183.245 . . . chemical analyses shall be kept for not less than one year . . . . public water supplies records . . . not less than 10 years . . . . 183.365 . . . sample report shall be maintained for at least 5 years | 183.210 . . . laboratory director . . . minimum of a bachelor's degree . . . and shall have had a minimum of 3 years experience in an environmental lab. . . . Laboratory supervisor . . . minimum of a bachelor's degree . . . and . . . minimum of 2 years experience . . . . 183.310 . . . Laboratory supervisor . . . minimum of a bachelor's degree . . . and minimum of 1 year bench experience in an environmental lab . . . | 183.210 An analyst . . . high school diploma or its equivalent and has completed a basic chemistry course . . . and at least 1 year experience in an analytical laboratory . . . . 183.310 . . . or laboratory oriented vocational courses . . . and a minimum of 3 months experience in a microbiological analytical laboratory. |

Appendix B

SAMPLE LABORATORY QUALITY AUDIT DOCUMENT

A. Business Data

1. Name and address of firm _____
   _____
   _____

2. Lab Director, name _____
   Phone and Fax number _____

3. Contact person (if other than Lab Director) _____

4. Name of tests performed for (hospital or company name)
   _____

5. Materials submitted for testing, annual requirements _____
   _____

6. Accreditations, organizational memberships _____
   _____

7. Date of last FDA inspection _____
   (State whether GMP or GLP inspection) _____
   Are results available? _____

8. Results of this audit, approval status _____

   Date of Audit _____ Auditor _____

The following audit questions are categorically separated to simplify information gathering. The grading scheme is based on three levels of competency, and the auditor should mark the appropriate number for the corresponding level of competency: 3 = excellent, 2 = average, 1 = unsatisfactory.

B. Personnel

All contract laboratory testing should be performed or supervised by a scientist(s) whose education, experience and knowledge lend credence to the performance of testing and the analysis of results. The scientific understanding of the test methodology is an important qualification, so that discussion of analysis results is effective and efficient. Since microbiological testing can be affected by poor hygiene, it is necessary that an adequate health program be administered in the laboratory.

Questions:

Data

Who is responsible for supervision of the laboratory? _____
_____

What is their educational and training background? _____
_____

How many analysts/microbiologists are available in the laboratory? _____
_____

Information

Are the educational and training backgrounds of all analysts sufficient to perform testing on submissions?	3	2	1
Is there a training program for GMP's, or GLP's?	3	2	1
Is there a training program to maintain job competency?	3	2	1
Is there an instrumentation/test procedure training program with annual recertification?	3	2	1

Is the program offered internally?	3	2	1

Is there documentation of qualified training and certification of each analyst who performs tests on submissions?   3   2   1

Are analysts certified (e.g. Microbiology)?   3   2   1

Comments: _____
_____
_____

C. Facilities

Physical facilities should be adequate with respect to space, equipment, environmental control, and maintenance. Good housekeeping should be evident. Special requirements, such as clean rooms, should be adequately maintained for the respective use.

Questions:

Data

Who is responsible for administering the general housekeeping
program? _____
_____

Information

Is adequate space available for the type of testing
performed?   3   2   1

Does the laboratory appear organized?   3   2   1

Is there a routine housekeeping schedule?   3   2   1

Comments: _____
_____
_____

D. Instrumentation and Calibration

The Laboratory should follow a routine preventive maintenance and calibration program. Instrumentation used in the preparation and analysis of samples should be maintained and calibrated at appropriate time intervals and with sufficient frequency to assure high quality performance. Program procedures should be written, and there should be documentation of each maintenance and calibration performed on an instrument.

Each instrument that is calibrated must be labeled with the date when the most recent calibration was performed. Calibrations should be performed by a person experienced with the instrument using appropriate certified standards.

Questions:

Data

List the main instrument(s) used to perform testing on samples submitted. _____

_____

Who are the manufacturers of the listed instruments? _____

_____

List the names of any outside firms that perform calibration of any of the listed instruments. _____

_____

What standards are used for calibration? _____

_____

_____

Information

Does a preventive maintenance (PM) program exist for
the instrument(s)?   3   2   1

Is the person who performs PM qualified?   3   2   1

Are there written procedures for the operation and cali-
bration of the instrument(s)?   3   2   1

Are there written procedures for the preventive maintenance program?	3	2	1
Is there a written schedule for calibration?	3	2	1
Are all calibrations documented?	3	2	1
Is each instrument visibly labeled with its most recent calibration date?	3	2	1
Are standards used to calibrate an instrument?	3	2	1
If reference standards are used, is there a written procedure which specifies the vendor of the standard and its required handling?	3	2	1
If nonreference standards are used, is there a written procedure detailing the required preparation of the standard?	3	2	1
Are thermometers, if used, calibrated against standard NITS thermometers?	3	2	1
Are all instruments calibrated by qualified personnel?	3	2	1
Are all instruments operated by experienced personnel?	3	2	1
Are instruments calibrated frequently enough to ensure reproducibility and reliability?	3	2	1

Comments: _____

_____

_____

E. Quality Assurance

1. Program

A formal quality program should exist with appropriate documents that state the responsibilities and objectives of the program. If a formal program does not exist, the requirements of this audit in the specific areas that follow must be met, in addition to the acquisition of a statement by the lab director to initiate a formal quality program within six months of the audit.

Quality control must be performed in the laboratory. Appropriate protocols should be written and available for review. The laboratory must provide a signed statement which says that they comply with GMP and, where applicable, with GLP regulations.

Questions:

Data

Who administers the quality assurance program? _____

Who is responsible for laboratory document control? _____

Information

Does the laboratory follow a quality assurance program?	3	2	1
Are there written protocols for the quality assurance program?	3	2	1
Is there a written procedure for document control?	3	2	1

Comments: _____

_____

_____

2. Test Procedures

Procedures used for testing of materials should be specified by (company name) Quality Assurance. Any modifications to the procedures must be reviewed and approved by (company name) Quality Assurance. All procedures must be in written format, reviewed and signed by personnel with appropriate levels of authority, on periodic basis. Procedures should state or reference the required materials and standards necessary to perform the test(s). Each modified procedure must be validated, reviewed and approved by (company name) Quality Assurance.

Where appropriate, a routine qualification of the testing capability of a laboratory can be performed. (Company name) can submit to the contract laboratory a combination of standard, spiked samples along with placebo samples. Routine qualification should be performed at least on an annual basis.

Questions:

Data

What compendial methods are followed? _____

From where are reference standards purchased for procedures? _____

Information

Do written procedures exist for testing all materials submitted?	3	2	1
Are written procedures reviewed, and are they up-to-date?	3	2	1
Do procedures represent compendial methods?	3	2	1
Does a protocol exist which requires that client be notified and asked to review any changes made to test procedures?	3	2	1
Have procedures been validated for precision and reproducibility?	3	2	1
Does a protocol exist that states what changes would cause a revalidation to occur?	3	2	1
Are procedures qualified periodically after the initial validation?	3	2	1
Are reference standards specified in the procedure(s)?	3	2	1
When testing a material submitted, if a procedure is repeated for any reason, is (company name) formally notified?	3	2	1

Does the laboratory take part in a proficiency testing
program on a routine basis?   3   2   1

3. Chemicals, Biologicals, References Standards

a. Chemicals and Reagents, Purchased

Chemicals and reagents should be used only within labeled or posted expiration dates. They should be purchased from reputable manufacturers,
and inspected periodically to confirm their identification. They should be
stored appropriately.

A satisfactory program should employ a system that alerts users to obtain
new supply before the expiration date of the chemical or reagent is reached.

Questions:

Data

From where are chemicals and reagents purchased?

_____

_____

_____

_____

_____

What manufacturers' brands are purchased?

_____

_____

_____

_____

_____

Information

Are vendors approved through a vendor certification
program?   3   2   1

Is a first-in first-out (FIFO) system followed?   3   2   1

If yes, for what materials?

_____

_____

Are chemicals inspected for proper labeling and expiration dating when they are received?	3	2	1
Are expiration dates, where applicable, logged-in?	3	2	1
Are chemicals tested to confirm their identification when they are received?	3	2	1
Is there a written procedure for each identification test?	3	2	1
Are chemicals and reagents stored appropriately?	3	2	1
If refrigeration is required, is the temperature monitored?	3	2	1
Is there a DEA approved program for narcotic storage?	3	2	1

b. Chemical or Biological Reference Standards, Purchased

All chemical or biological standards should be purchased from a qualified source. Each standard should be accompanied by a certificate of analysis stating the source of testing, date of testing, and the standard to which the product was compared. All standards must be stored under conditions stated by the supplier. Expiration dating must be present on the standard, as well as on any further products prepared by these standards.

Questions:

Data

From whom are Reference Standards purchased? _____
_____
_____

Information

Is a certificate of analysis obtained with each Standard?	3	2	1
Is each Standard tested internally to confirm its quality?	3	2	1
Are Standards stored appropriately?	3	2	1

Does each Standard have a legible expiration date?   3   2   1

c. Chemical or Biological Reference Standards, Internally Prepared

All chemical or biological standards that are prepared internally should be tested to confirm their quality with known, purchased Standards or with standards supported by highly reliable data and testing. All standards must be stored under the best conditions to support their stability. All standards must be dated for their expiration. Water used in test procedures and for cleaning equipment/supplies must be of acceptable purity for its intended use.

Questions:

Data

What Standards used in testing materials are prepared internally? _____
_____

What is the purity (or grade) of water used for analyses and preparation of standards? _____
_____

What is the purity (or grade) of water used for cleaning glassware and other equipment? _____
_____

What tests are performed on the water to determine its purity (or grade)? _____
_____

Information

Are standards prepared frequently?   3   2   1

Are internally prepared standards tested routinely to confirm their quality?   3   2   1

Are they tested frequently?   3   2   1

Are the standards properly stored?   3   2   1

Are the standards dated for expiration?   3   2   1

Have the expiration dates been validated?	3 2 1
Are water purity tests performed frequently?	3 2 1
Are tests on water adequate to determine its purity?	3 2 1
If a treatment system is used to prepare water for laboratory use, does a written procedure exist for monitoring that system?	3 2 1

4. Sample Receipt, Handling, Storage, Documentation

A written standard procedure should exist for documentation and labeling of samples and for storage conditions. The procedure should assure that all samples can be tracked from receipt to testing. Samples should be stored under conditions which are specified by (company name), where appropriate. Storage facilities should be controlled and monitored for temperature and humidity where necessary. The disposition of all samples after testing should be agreed upon by client and the contract laboratory.

Questions:

Data

Who is responsible for receiving samples for testing? _____

What documentation and coding is used to track a sample from receipt to test? _____

Where are samples stored before and after testing? _____

How long are samples kept after test results are reported? _____

What happens to a sample after testing and final report are complete?

Information

Is there a written procedure for documentation, label-
ing and storage of samples?   3   2   1

Is the storage facility monitored and controlled for temperature? For humidity?   3   2   1

Comments:_____
_____
_____

5. Data Recording and Reporting

All raw data should preferably be recorded in bound notebooks. Data
should have proper identification to track original data to final report.
Data and calculations should be reviewed by a second analyst or by a
supervisor before the final report is written. A formal written explanation
should be provided whenever a retest is performed. A single person should
be named by the contract laboratory as the contact if any questions arise
regarding testing and data. Confidentiality must be assured for all client-
related data.

Questions:

Data

Who should be contacted for any questions regarding results? _____
_____

Who transfers raw data to the final report forms? _____
_____

Information

Is raw data recorded in bound books?   3   2   1

Is the instrument used identified with the raw data?   3   2   1

Is there a way to track a final report to the original raw
data?   3   2   1

Is the format of a final report complete and thorough?   3   2   1

Is there a review of the data, calculations and the final report?	3	2	1
Do all reviewers sign the data or final report?	3	2	1
Are all test data reported to client?	3	2	1
If a retest is performed, is a formal written explanation of the reason for the retest provided automatically?	3	2	1

F. Audit Considerations for Special Testing

This section will be used by an auditor experienced in the scientific discipline appropriate to the methodology performed by the laboratory. Specific review questions should be developed by a person directly responsible for and experienced in the type of testing being audited. (The following sections contain a list of sample questions for specific areas of testing. The list is not all inclusive)

1. Microbiological Bioburden Testing

Have sterilizer cycles been validated for media preparation?
Have test methodologies been validated to assure that no inhibitory or cidal activity is present?
Can microbiological media prepared internally be tracked to the respective lot number and sterilizer cycle used?
How are specimens collected, transported and preserved?
Where is microbial testing performed?
Is there an environmental monitoring program for the area where testing is performed?
What controls are used to validate media, methods, and test environment?
When was the last proficiency test performed? Results?

2. Sterility Testing

Where is testing performed? (hood, cleanroom, both, neither)
Is testing performed manually, semi-automatically, or is the system totally automated?
What is the frequency of environmental testing in the test area?
What are the environmental action levels?
What action is taken responding to a result which is above the action level?

Are there written procedures for environmental testing and action level response?

Are personnel tested routinely for hand sanitation?

What gowning are used by testing personnel?

What procedure is followed when a positive test occurs in test sample?

What procedure is followed when a positive test occurs in negative control or if positive controls don't grow adequately?

If a test is repeated at any time, is client notified?

3. Bacterial Endotoxins Testing

Where is the testing performed?

Is testing performed manually, semi-automatically, or is the system totally automated?

What method is used to depyrogenate testing supplies?

Are any supplies purchased with depyrogenation certification?

If water is required to reconstitute sample, what is the source and quality of the water?

Is water for injection tested routinely? How often?

If a negative control results in a positive endotoxin test, what procedure is followed?

If inhibition or enhancement occurs in testing, what procedure is followed?

4. Pyrogen Testing, and Other Safety Tests

Is the animal facility managed to comply to GLP's?

Are animals cared for by experienced personnel?

Does a veterinarian routinely check the health of the animals?

How often are new animals purchased?

What time intervals are allowed between testing on the same animals?

How often are the thermometers calibrated?

Is the animal facility inspected frequently for sanitation?

What standard of reference is used for absorption from site of injection?

REFERENCES

Garfield, F.M. (1984) Quality Assurance Principles for Analytical Laboratories. AOAC.

Taylor, J.K. (1980) Quality Assurance of Chemical Measurements. Lewis Publishers, Inc.

Appendix C

THE OECD PRINCIPLES OF GOOD LABORATORY PRACTICE

Chemicals control laws passed in OECD Member countries in recent decades call for testing and assessing of chemicals to determine their potential hazards. A basic principle of this legislation is that assessments of hazards associated with chemicals should be based on test data of assured quality.

Good Laboratory Practice (GLP) is intended to promote the quality and validity of test data. It is a managerial concept covering the organisational process and the conditions under which laboratory studies are planned, performed, monitored, recorded and reported.

The application of GLP is of crucial importance to national authorities entrusted with the responsibility of assessing test data and evaluating chemical hazards. The issue of data quality also has an international dimension. If countries can rely on test data developed in other countries, duplicative testing can be avoided and costs to government and industry saved. Moreover, common principles and procedures for GLP facilitate the exchange of information and prevent the emergence of non-tariff barriers to trade while contributing to environmental and health protection.

The OECD Principles of Good Laboratory Practice were developed by an Expert Group on GLP established in 1978 under the Special Programme

Environment Directorate, Organisation for Economic Co-operation and Development, Copyright OECD, 1992 on the Control of Chemicals. The GLP regulations for non-clinical laboratory studies, published by the US Food and Drug Administration in 1976, provided the basis for the work of the Expert Group. The Group was chaired by Dr. Carl Morris, United States Environmental Protection Agency. The following countries and organisations participated in the Expert Group: Australia, Austria, Belgium, Canada, Denmark, France, the Federal Republic of Germany, Greece, Italy, Japan, the Netherlands, New Zealand, Norway, Sweden, Switzerland, the United Kingdom, the United States, the Commission of the European Communities, the World Health Organization and the International Organization for Standardization (ISO/ CERTICO).

The OECD Principles of GLP as set out in Part One of this publication were reviewed in the relevant policy bodies of the Organisation and were formally recommended for use in Member countries by the OECD Council in 1981. They are an integral part of the Council Decision on Mutual Acceptance of Data, which states "that data generated in the testing of chemicals in an OECD Member country in accordance with OECD Test Guidelines* and OECD Principles of Good Laboratory Practice shall be accepted in other Member countries for purposes of assessment and other uses relating to the protection of man and the environment" [C(81)30(Final)].

The OECD Principles of GLP were first published in 1982 in *Good Laboratory Practice in the Testing of Chemicals.*** This publication also contained guidance provided in the final report of the Expert Group. Since the early 1980's OECD has continued to elaborate and refine this guidance, and has undertaken further work on national and international aspects of compliance with the GLP Principles and monitoring of such compliance. The results of that work are being published (or reprinted) in this OECD Series on Good Laboratory Practice and Compliance Monitoring, beginning in 1991. It is therefore appropriate that the OECD Principles of GLP and the Council Acts concerning GLP be the subject of this first number of the series.

---

* *OECD Guidelines for the Testing of Chemicals* (1981 and continuing series).
** OECD, 1982, out of print.

PART ONE: THE OECD PRINCIPLES OF GLP*

Section I: Introduction

Preface

A number of OECD Member countries have recently passed legislation to control chemical substances and others are about to do so. This legislation usually requires the manufacturer to perform laboratory studies and to submit the results of these studies to a governmental authority for assessment of the potential hazard to human health and the environment.

Government and industry are increasingly concerned with the quality of studies upon which hazard assessments are based. As a consequence, several OECD Member countries have, or plan to establish, criteria for the performance of these studies.

To avoid different schemes of implementation that could impede international trade in chemicals, OECD Member countries have recognised the unique opportunity for international harmonization of test methods and good laboratory practices.

During 1979–80, an international group of experts established under the Special Programme on the Control of Chemicals developed this document concerning the "Principles of Good Laboratory Practice (GLP)" utilising common managerial and scientific practices and experience from various national and international sources.

The purpose of these Principles of Good Laboratory Practice is to promote the development of quality test data. Comparable quality of test data forms the basis for the mutual acceptance of test data among countries.

If individual countries can confidently rely on test data developed in other countries, duplicative testing can be avoided, thereby introducing economies in test costs and time. The application of these Principles should help avoid the creation of technical barriers to trade, and further improve the protection of human health and the environment.

1. Scope

These Principles of Good Laboratory Practice should be applied to testing of chemicals to obtain data on their properties and/or their safety with respect to human health or the environment.

---

* The OECD Principles of Good Laboratory Practice are contained in Annex II of the Decision of the Council concerning the Mutual Acceptance of Data in the Assessment of Chemicals [C(81)30(Final)].

Studies covered by Good Laboratory Practice also include work conducted in field studies.

These data would be developed for the purpose of meeting regulatory requirements.

2. Definitions of Terms

2.1 *Good Laboratory Practice*

1. Good Laboratory Practice (GLP) is concerned with the organisational process and the conditions under which laboratory studies are planned, performed, monitored, recorded, and reported.

2.2 *Terms Concerning the Organisation of a Test Facility*

1. *Test facility* means the persons, premises and operational unit(s) that are necessary for conducting the study.
2. *Study Director* means the individual responsible for the overall conduct of the study.
3. *Quality Assurance Programme* means an internal control system designed to ascertain that the study is in compliance with these Principles of Good Laboratory Practice.
4. *Standard Operating Procedures (SOPs)* means written procedures which describe how to perform certain routine laboratory tests or activities normally not specified in detail in study plans or test guidelines.
5. *Sponsor* means a person(s) or entity who commissions and/or supports a study.

2.3 *Terms Concerning the Study*

1. *Study* means an experiment or set of experiments in which a test substance is examined to obtain data on its properties and/or its safety with respect to human health and environment.
2. *Study plan* means a document which defines the entire scope of the study.
3. *OECD Test Guideline* means a test guideline which the OECD has recommended for use in its Member countries.
4. *Test system* means any animal, plant, microbial, as well as

other cellular, sub-cellular, chemical, or physical system or a combination thereof used in a study.

5. *Raw data* means all original laboratory records and documentation, or verified copies thereof, which are the result of the original observations and activities in a study.

6. *Specimen* means any material derived from a test system for examination, analysis, or storage.

2.4 *Terms Concerning the Test Substance*

1. *Test substance* means a chemical substance or a mixture which is under investigation.

2. *Reference substance (control substance)* means any well defined chemical substance or any mixture other than the test substance used to provide a basis for comparison with the test substance.

3. *Batch* means a specific quantity or lot of a test or reference substance produced during a defined cycle of manufacture in such a way that it could be expected to be of a uniform character and should be designed as such.

4. *Vehicle (carrier)* means any agent which serves as a carrier used to mix, disperse, or solubilise the test or reference substance to facilitate the administration to the test system.

5. *Sample* means any quantity of the test or reference substance.

Section II: Good Laboratory Practice Principles

1. Test Facility Organisation and Personnel

1.1 *Management's Responsibilities*

1. Test facility management should ensure that the Principles of Good Laboratory Practice are complied with in the test facility.

2. At a minimum it should:
   a) ensure that qualified personnel, appropriate facilities, equipment, and materials are available;
   b) maintain a record of the qualifications, training, experience and job description for each professional and technical individual;

c) ensure that personnel clearly understand the functions they are to perform and, where necessary, provide training for these functions;

d) ensure that health and safety precautions are applied according to national and/or international regulations;

e) ensure that appropriate Standard Operating Procedures are established and followed;

f) ensure that there is a Quality Assurance Programme with designated personnel;

g) where appropriate, agree to the study plan in conjunction with the sponsor;

h) ensure that amendments to the study plan are agreed upon and documented;

i) maintain copies of all study plans;

j) maintain a historical file of all Standard Operating Procedures;

k) for each study ensure that a sufficient number of personnel is available for its timely and proper conduct;

l) for each study designate an individual with the appropriate qualifications, training, and experience as the Study Director before the study is initiated. If it is necessary to replace a Study Director during a study, this should be documented;

m) ensure that an individual is identified as responsible for the management of the archives.

1.2 *Study Director's Responsibilities*

1. The Study Director has the responsibility for the overall conduct of the study and for its report.

2. These responsibilities should include, but not be limited to, the following functions:

   a) should agree to the study plan;

   b) ensure that the procedures specified in the study plan are followed, and that authorisation for any modification is obtained and documented together with the reasons for them;

   c) ensure that all data generated are fully documented and recorded;

   d) sign and date the final report to indicate acceptance of responsibility for the validity of the data and to confirm compliance with these Principles of Good Laboratory Practice;

e) ensure that after termination of the study, the study plan, the final report, raw data and supporting material are transferred to the archives.

1.3 *Personnel Responsibilities*

1. Personnel should exercise safe working practice. Chemicals should be handled with suitable caution until their hazard(s) has been established.
2. Personnel should exercise health precautions to minimise risk to themselves and to ensure the integrity of the study.
3. Personnel known to have a health or medical condition that is likely to have an adverse effect on the study should be excluded from operations that may affect the study.

2. Quality Assurance Programme

2.1 *General*

1. The test facility should have a documented quality assurance programme to ensure that studies performed are in compliance with these Principles of Good Laboratory Practice.
2. The quality assurance programme should be carried out by an individual or by individuals designated by and directly responsible to management and who are familiar with the test procedures.
3. This individual(s) should not be involved in the conduct of study being assured.
4. This individual(s) should report any findings in writing directly to management and to the Study Director.

2.2 *Responsibilities of the Quality Assurance Personnel*

1. The responsibilities of the quality assurance personnel should include, but not be limited to, the following functions:
   a) ascertain that the study plan and Standard Operating Procedures are available to personnel conducting the study;
   b) ensure that the study plan and Standard Operating Procedures are followed by periodic inspections of the test

facility and/or by auditing the study in progress. Records of such procedures should be retained;

c) promptly report to management and the Study Director unauthorised deviations from the study plan and from Standard Operating Procedures;

d) review the final reports to confirm that the methods, procedures, and observations are accurately described, and that the reported results accurately reflect the raw data of the study;

e) prepare and sign a statement, to be included with the final report, which specifies the dates inspections were made and the dates any findings were reported to management and to the Study Director.

3. Facilities

3.1 *General*

1. The test facility should be of suitable size, construction and location to meet the requirements of the study and minimise disturbances that would interfere with the validity of the study.

2. The design of the test facility should provide an adequate degree of separation of the different activities to assure the proper conduct of each study.

3.2 *Test System Facilities*

1. The test facility should have a sufficient number of rooms or areas to assure the isolation of test systems and the isolation of individual projects, involving substances known or suspected of being biohazardous.

2. Suitable facilities should be available for the diagnosis, treatment and control of diseases, in order to ensure that there is no unacceptable degree of deterioration of test systems.

3. There should be storage areas as needed for supplies and equipment. Storage areas should be separated from areas housing the test systems and should be adequately protected against infestation and contamination. Refrigeration should be provided for perishable commodities.

3.3 *Facilities for Handling Test and Reference Substances*

1. To prevent contamination or mix-ups, there should be separate areas for receipt and storage of the test and reference substances, and mixing of the test substances with a vehicle.
2. Storage areas for the test substances should be separate from areas housing the test systems and should be adequate to preserve identity, concentration, purity, and stability, and ensure safe storage for hazardous substances.

3.4 *Archive Facilities*

1. Space should be provided for archives for the storage and retrieval of raw data, reports, samples and specimens.

3.5 *Waste Disposal*

1. Handling and disposal of wastes should be carried out in such a way as not to jeopardise the integrity of studies in progress.
2. The handling and disposal of wastes generated during the performance of a study should be carried out in a manner which is consistent with pertinent regulatory requirements. This would include provision for appropriate collection, storage and disposal facilities, decontamination and transportation procedures, and the maintenance of records related to the preceding activities.

4. Apparatus, Material, and Reagents

4.1 *Apparatus*

1. Apparatus used for the generation of data, and for controlling environmental factors relevant to the study should be suitably located and of appropriate design and adequate capacity.
2. Apparatus used in a study should be periodically inspected, cleaned, maintained, and calibrated according to Standard Operating Procedures. Records of procedures should be maintained.

4.2 *Material*

1. Apparatus and materials used in studies should not interfere with the test systems.

4.3 *Reagents*

1. Reagents should be labelled, as appropriate, to indicate source, identity, concentration, and stability information and should include the preparation date, earliest expiration date, specific storage instructions.

5. Test Systems

5.1 *Physical/Chemical*

1. Apparatus used for the generation of physical/chemical data should be suitably located and of appropriate design and adequate capacity.
2. Reference substances should be used to assist in ensuring the integrity of the physical/chemical test systems.

5.2 *Biological*

1. Proper conditions should be established and maintained for the housing, handling and care of animals, plants, microbial as well as other cellular and sub-cellular systems, in order to ensure the quality of the data.
2. In addition, conditions should comply with appropriate national regulatory requirements for the import, collection, care and use of animals, plants, microbial as well as other cellular and sub-cellular systems.
3. Newly received animal and plant test systems should be isolated until their health status has been evaluated. If any unusual mortality or morbidity occurs, this lot should not be used in studies and, when appropriate, humanely destroyed.
4. Records of source, date of arrival, and arrival condition should be maintained.
5. Animal, plant, microbial, and cellular test systems should be acclimatised to the test environment for an adequate period before a study is initiated.

6. All information needed to properly identify the test systems should appear on their housing or containers.
7. The diagnosis and treatment of any disease before or during a study should be recorded.

6. Test and Reference Substances

6.1 *Receipt, Handling, Sampling, and Storage*

1. Records including substance characterisation, date of receipt, quantities received and used in studies should be maintained.
2. Handling, sampling, and storage procedures should be identified in order that the homogeneity and stability is assured to the degree possible and contamination or mix-up are precluded.
3. Storage container(s) should carry identification information, earliest expiration date, and specific storage instructions.

6.2 *Characterization*

1. Each test and reference substance should be appropriately identified (e.g. code, chemical abstract number (CAS), name).
2. For each study, the identity, including batch number, purity, composition, concentrations, or other characterisations to appropriately define each batch of the test or reference substances should be known.
3. The stability of test and reference substances under conditions of storage should be known for all studies.
4. The stability of test and reference substances under the test conditions should be known for all studies.
5. If the test substance is administered in a vehicle, Standard Operating Procedures should be established for testing the homogeneity and stability of the test substance in that vehicle.
6. A sample for analytical purposes from each batch of test substance should be retained for studies in which the test substance is tested longer than four weeks.

7. Standard Operating Procedures

7.1 *General*

1. A test facility should have written Standard Operating Procedures approved by management that are intended to ensure the quality and integrity of the data generated in the course of the study.
2. Each separate laboratory unit should have immediately available Standard Operating Procedures relevant to the activities being performed therein. Published text books, articles and manuals may be used as supplements to these Standard Operating Procedures.

7.2 *Application*

1. Standard Operating Procedures should be available for, but not limited to, the following catagories of laboratory activities. The details given under each heading are to be considered as illustrative examples.
    a) *Test and Reference Substance*: Receipt, identification, labelling, handling, sampling, and storage.
    b) *Apparatus and Reagents*: Use, maintenance, cleaning, calibration of measuring apparatus and environmental control equipment; preparation of reagents.
    c) *Record Keeping, Reporting, Storage, and Retrieval*: Coding of studies, data collection, preparation of reports, indexing systems, handling of data, including the use of computerised data systems.
    d) *Test system (where appropriate)*:
        i) Room preparation and environmental room conditions for the test system.
        ii) Procedures for receipt, transfer, proper placement, characterisation, identification and care of test system.
        iii) Test system preparation, observations examinations, before, during and at termination of the study.
        iv) Handling of test system individuals found moribund or dead during the study.
        v) Collection, identification and handling of specimens including necropsy and histopathology.
    e) *Quality Assurance Procedures*: Operation of quality as-

surance personnel in performing and reporting study audits, inspections, and final study report reviews.

f) *Health and Safety Precautions*: As required by national
and/or international legislation or guidelines.

8. Performance of the Study

8.1 *Study Plan*

1. For each study, a plan should exist in a written form prior
   to initiation of the study.
2. The study plan should be retained as raw data.
3. All changes, modifications, or revisions of the study plan,
   as agreed to by the Study Director, including justification(s),
   should be documented, signed and dated by the Study Director, and maintained with the study plan.

8.2 *Content of the Study Plan*
The study plan should contain, but not be limited to the following
information:

1. *Identification of the Study, the Test and Reference Substance*
   a) A descriptive title;
   b) A statement which reveals the nature and purpose of
   the study;
   c) Identification of the test substance by code or name
   (IUPAC; CAS number, etc.);
   d) The reference substance to be used.
2. *Information Concerning the Sponsor and the Test Facility*
   a) Name and address of the Sponsor;
   b) Name and address of the Test Facility;
   c) Name and address of the Study Director.
3. *Dates*
   a) The date of agreement to the study plan by signature of
   the Study Director, and when appropriate, of the sponsor and/or the test facility management;
   b) The proposed starting and completion dates.
4. *Test Methods*
   a) Reference to OECD Test Guideline or other test guideline to be used.
5. *Issues (where applicable)*
   a) The justification for selection of the test system;

      b) Characterisation of the test system, such as the species, strain, substrain, source of supply, number, body weight range, sex, age, and other pertinent information;

      c) The method of administration and the reasons for its choice;

      d) The dose levels and/or concentrations(s), frequency, duration of administration;

      e) Detailed information on the experimental design, including a description of the chronological procedure of the study, all methods, materials and conditions, type and frequency of analysis, measurements, observations and examinations to be performed.

6. *Records*

      a) A list of records to be retained.

8.3 *Conduct of the Study*

1. A unique identification should be given to each study. All items concerning this study should carry this identification.
2. The study should be conducted in accordance with the study plan.
3. All data generated during the conduct of the study should be recorded directly, promptly, accurately, and legibly by the individual entering the data. These entries should be signed or initialled and dated.
4. Any change in the raw data should be made so as not to obscure the previous entry, and should indicate the reason, if necessary, for change and should be identified by date and signed by the individual making the change.
5. Data generated as a direct computer input should be identified at the time of data input by the individual(s) responsible for direct data entries. Corrections should be entered separately by the reason for change, with the date and the identity of the individual making the change.

9. Reporting of Study Results

9.1 *General*

1. A final report should be prepared for the study.
2. The use of the International System of Units (SI) is recommended.

3. The final report should be signed and dated by the Study Director.
4. If reports of principal scientists from co-operating disciplines are included in the final report, they should sign and date them.
5. Corrections and additions to a final report should be in the form of an amendment. The amendment should clearly specify the reason for the corrections or additions and should be signed and dated by the Study Director and by the principal scientist from each discipline involved.

9.2 *Content of the Final Report*
The final report should include, but not be limited to, the following information:

1. *Identification of the Study, the Test and Reference Substance*
    a) A descriptive title;
    b) Identification of the test substance by code or name (IUPAC; CAS number, etc.);
    c) Identification of the reference substance by chemical name;
    d) Characterisation of the test substance including purity, stabiliy and homogeneity.
2. *Information Concerning the Test Facility*
    a) Name and address;
    b) Name of the Study Director;
    c) Name of other principal personnel having contributed reports to the final report.
3. *Dates*
    a) Dates on which the study was initiated and completed.
4. *Statement*
    a) A Quality Assurance statement certifying the dates inspections were made and the dates any findings were reported to management and to the Study Director.
5. *Description of Materials and Test Methods*
    a) Description of methods and materials used;
    b) Reference to OECD Test Guidelines or other test guidelines.
6. *Results*
    a) A summary of results;
    b) All information and data required in the study plan;

    c) A presentation of the results, including calculations and statistical methods;

    d) An evaluation and discussion of the results and, where appropriate, conclusions.

7. *Storage*

    a) The location where all samples, specimens, raw data and the final report are to be stored.

10. Storage and Retention of Records and Material

10.1 *Storage and Retrieval*

1. Archives should be designed and equipped for the accommodation and the secure storage of:
   a) The study plans;
   b) The raw data;
   c) The final reports;
   d) The reports of laboratory inspections and study audits performed according to the Quality Assurance Programme;
   e) Samples and specimens.
2. Material retained in the archives should be indexed so as to facilitate orderly storage and rapid retrieval.
3. Only personnel authorised by management should have access to the archives. Movement of material in and out of the archives should be properly recorded.

10.2 *Retention*

1. The following should be retained for the period specified by the appropriate authorities:
   a) The study plan, raw data, samples, specimens, and the final report of each study;
   b) Records of all inspections and audits performed by the Quality Assurance Programme;
   c) Summary of qualifications, training, experience and job descriptions of personnel;
   d) Records and reports of the maintenance and calibration of equipment;
   e) The historical file of Standard Operating Procedures.
2. Samples and specimens should be retained only as long as the quality of the preparation permits evaluation.

3. If a test facility or an archive contracting facility goes out of business and has no legal successor, the archive should be transferred to the archives of the sponsor(s) of the study(s).

PART TWO: OECD COUNCIL ACTS ON GLP PRINCIPLES AND COMPLIANCE MONITORING

DECISION OF THE COUNCIL
concerning the Mutual Acceptance of Data
in the Assessment of Chemicals
[C(81)30(Final)]

(Adopted by the Council at its 535th Meeting on 12th May, 1981)

The Council,
Having regard to Articles 2(a), 2(d), 5(a) and 5(b) of the Convention on the Organisation for Economic Co-operation and Development of 14th December, 1960;

Having regard to the Recommendation of the Council of 26th May, 1972, on Guiding Principles concerning International Economic Aspects of Environmental Policies [C(72)128];

Having regard to the Recommendation of the Council of 14th November, 1974, on the Assessment of the Potential Environmental Effects of Chemicals [C(74)215];

Having regard to the Recommendation of the Council of 26th August, 1976, concerning Safety Controls over Cosmetics and Household Products [C(76)144(Final)];

Having regard to the Recommendation of the Council of 7th July, 1977, establishing Guidelines in respect of Procedure and Requirements for Anticipating the Effects of Chemicals on Man and in the Environment [C(77)97(Final)];

Having regard to the Decision of the Council of 21st September, 1978, concerning a Special Programme on the Control of Chemicals and the Programme of Work established therein [C(78)127(Final)];

Having regard to the Conclusions of the First High Level Meeting of the Chemicals Group of 19th May, 1980, dealing with the control of health and environmental effects of chemicals [ENV/CHEM/HLM/80.M/1];

Considering the need for concerted action amongst OECD Member countries to protect man and his environment from exposure to hazardous chemicals;

Considering the importance of international production and trade in chemicals and the mutual economic and trade advantages which accrue to OECD Member countries from harmonization of policies for chemicals control;

Considering the need to minimise the cost burden associated with testing chemicals and the need to utilise more effectively scarce test facilities and specialist manpower in Member countries;

Considering the need to encourage the generation of valid and high quality test data and noting the significant actions taken in this regard by OECD Member countries through provisional application of OECD Test Guidelines and OECD Principles of Good Laboratory Practice;

Considering the need for and benefits of mutual acceptance in OECD countries of test data used in the assessment of chemicals and other uses relating to protection of man and the environment;

On the proposal of the High Level Meeting of the Chemicals Group, endorsed by the Environment Committee;

PART I

1. DECIDES that data generated in the testing of chemicals in an OECD Member country in accordance with OECD Test Guidelines and OECD Principles of Good Laboratory Practice shall be accepted in other Member countries for purposes of assessment and other uses relating to the protection of man and the environment.
2. DECIDES that for the purposes of this decision and other Council actions the terms OECD Test Guidelines and OECD Principles of Good Laboratory Practice shall mean guidelines and principles adopted by the Council.
3. INSTRUCTS the Environment Committee to review action taken by Member countries in pursuance of this Decision and to report periodically thereon to the Council.
4. INSTRUCTS the Environment Committee to pursue a programme of work designed to facilitate implementation of this Decision with a view to establishing further agreement on assessment and control of chemicals within Member countries.

PART II

To implement the Decision set forth in Part I:
1. RECOMMENDS that Member countries, in the testing of chemicals, apply the OECD Test Guidelines and the OECD Principles of Good

Laboratory Practice, set forth respectively in Annexes I and II* which are integral parts of this text.

2. INSTRUCTS the Management Committee of the Special Programme on the Control of Chemicals in conjunction with the Chemicals Group of the Environment Committee to establish an updating mechanism to ensure that the aforementioned test guidelines are modified from time to time as required through the revision of existing Guidelines or the development of new Guidelines.

3. INSTRUCTS the Management Committee of the Special Programme on the Control of Chemicals to pursue its programme of work in such a manner as to facilitate internationally-harmonized approaches to assuring compliance with the OECD Principles of Good Laboratory Practice and to report periodically thereon to the Council.

COUNCIL DECISION-RECOMMENDATION
on Compliance with Principles of Good Laboratory Practice
[C(89)87(Final)]

(Adopted by the Council as its 717th Meeting on 2nd October 1989)

The Council,

Having regard to Articles 5 a) and 5 b) of the Convention on the Organisation for Economic Co-operation and Development of 14th December 1960;

Having regard to the Recommendation of the Council of 7th July 1977 Establishing Guidelines in Respect of Procedure and Requirements for Anticipating the Effects of Chemicals on Man and in the Environment [C(77)97(Final)];

Having regard to the Decision of the Council of 12th May 1981 concerning the Mutual Acceptance of Data in the Assessment of Chemicals [C(81)30(Final)] and, in particular, the Recommendation that Member countries, in the testing of chemicals, apply the OECD Principles of Good Laboratory Practice, set forth in Annex 2 of that Decision;

Having regard to the Recommendation of the Council of 26th July 1983 concerning the Mutual Recognition of Compliance with Good Laboratory Practice [C(83)95(Final)];

Having regard to the conclusions of the Third High Level Meeting of the Chemicals Group (OECD, Paris, 1988);

---

* Annex I to the Council Decision (the OECD Test Guidelines) was published separately. Annex II (the OECD Principles of Good Laboratory Practice) will be found on pages 71-85 (Part One).

Considering the need to ensure that test data on chemicals provided to regulatory authorities for purposes of assessment and other uses related to the protection of human health and the environment are of high quality, valid and reliable;

Considering the need to minimise duplicative testing of chemicals, and thereby to utilise more effectively scarce test facilities and specialist manpower, and to reduce the number of animals used in testing;

Considering that recognition of procedures for monitoring compliance with good laboratory practice will facilitate mutual acceptance of data and thereby reduce duplicative testing of chemicals;

Considering that a basis for recognition of compliance monitoring procedures is an understanding of, and confidence in, the procedures in the Member country where the data are generated;

Considering that harmonized approaches to procedures for monitoring compliance with good laboratory practice would greatly facilitate the development of the necessary confidence in other countries' procedures;

On the proposal of the Joint Meeting of the Management Committee of the Special Programme on the Control of Chemicals and the Chemicals Group, endorsed by the Environment Committee;

PART I

GLP Principles and Compliance Monitoring

1. DECIDES that Member countries in which testing of chemicals for purposes of assessment related to the protection of health and the environment is being carried out pursuant to principles of good laboratory practice that are consistent with the OECD Principles of Good Laboratory Practice as set out in Annex 2 of the Council Decision C(81)30(Final) (hereinafter called "GLP Principles") shall:
    i) establish national procedures for monitoring compliance with GLP Principles, based on laboratory inspections and study audits;
    ii) designate an authority or authorities to discharge the functions required by the procedures for monitoring compliance; and
    iii) require that the management of test facilities issue a declaration, where applicable, that a study was carried out in accordance with GLP Principles and pursuant to any other provisions established by national legislation or administrative procedures dealing with good laboratory practice.
2. RECOMMENDS that, in developing and implementing national procedures for monitoring compliance with GLP Principles, Member countries apply the "Guidelines for Compliance Monitoring Proce-

dures for Good Laboratory Practice" and the "Guidance for the Conduct of Laboratory Inspections and Study Audits," set out respectively in Annexes I and II which are integral part of this Decision-Recommendation.*

PART II

Recognition of GLP Compliance among Member countries

1. DECIDES that Member countries shall recognise the assurance by another Member country that test data have been generated in accordance with GLP Principles if such other Member country complies with Part I above and Part II paragraph 2 below.
2. DECIDES that, for purposes of the recognition of the assurance in paragraph 1 above, Member countries shall:
   i) designate an authority or authorities for international liaison and for discharging other functions relevant to the recognition as set out in this Part and in the Annexes to this Decision-Recommendation;
   ii) exchange with other Member countries relevant information concerning their procedures for monitoring compliance, in accordance with the guidance set out in Annex III which is an integral part of this Decision-Recommendation;** and
   iii) implement procedures whereby, where good reason exists, information concerning GLP compliance of a test facility (including information focussing on a particular study) within their jurisdiction can be sought by another Member country.
3. DECIDES that the Council Recommendation concerning the Mutual Recognition of Compliance with Good Laboratory Practice [C(83)95(Final)] shall be repealed.

PART III

Future OECD Activities

1. INSTRUCTS the Environment Committee and the Management Committee of the Special Programme on the Control of Chemicals to ensure

---

* Annexes I and II of the Council Act will be found in Numbers 2 and 3, respectively, of this OECD series on Principles of GLP and Compliance Monitoring.
** Annex III of the Council Act will be found in Number 2 of this OECD series on Principles of GLP and Compliance Monitoring.

that the "Guidelines for Compliance Monitoring Procedures for Good Laboratory Practice" and the "Guidance for the Conduct of Laboratory Inspections and Study Audits" set out in Annexes I and II are updated and expanded, as necessary, in light of developments and experience of Member countries and relevant work in other international organisations.

2. INSTRUCTS the Environment Committee and the Management Committee of the Special Programme on the Control of Chemicals to pursue a programme of work designed to facilitate the implementation of this Decision-Recommendation, and to ensure continuing exchange of information and experience on technical and administrative matters related to the application of GLP Principles and the implementation of procedures for monitoring compliance with good laboratory practice.

3. INSTRUCTS the Environment Committee and the Management Committee of the Special Programme on the Control of Chemicals to review actions taken by Member countries in pursuance of this Decision-Recommendation.

Appendix D

STATE OF ILLINOIS RULES AND REGULATIONS: EPA RULES FOR ENVIRONMENTAL LABORATORIES

**TITLE 35: ENVIRONMENTAL PROTECTION
SUBTITLE A: GENERAL PROVISIONS
CHAPTER II: ENVIRONMENTAL PROTECTION
AGENCY**

**PART 183: JOINT RULES OF THE ILLINOIS
ENVIRONMENTAL PROTECTION AGENCY AND THE
ILLINOIS DEPARTMENT OF PUBLIC HEALTH:
CERTIFICATION AND OPERATION OF
ENVIRONMENTAL LABORATORIES**

SUBPART A: GENERAL PROVISIONS

SUBPART B: CHEMICAL ANALYSES OF PUBLIC WATER SUPPLY SAMPLES

SUBPART C: MICROBIOLOGICAL ANALYSES OF PUBLIC WATER SUPPLY SAMPLES

SUBPART D: RADIOCHEMICAL ANALYSES OF PUBLIC WATER SUPPLY SAMPLES

Section
183.405   Scope and Applicability
183.410   Personnel
183.415   Physical Facilities
183.420   Laboratory Equipment
183.425   General Laboratory Practices
183.430   Methodology and Required Equipment
183.435   Sample Collecting, Handling and Preservation
183.440   Quality Control
183.445   Record Maintenance
183.450   Action Response to Laboratory Results

Appendix A Methodology and Required Equipment for Chemical Analyses of Public Water Supply Samples

AUTHORITY: Implementing the Safe Drinking Water Act (42 U.S.C. 300f et seq.), Subpart C of the National Interim Primary Drinking Water Regulations (40 CFR 141.21 through 141.30 (1982)), the Environmental Protection Act (Ill. Rev. Stat. 1981, ch. 111 1/2, pars. 1001 et seq.) and the Civil Administrative Code of Illinois (Ill. Rev. Stat. 1981, ch. 127, pars. 1 et seq.) and authorized by Sections 4(o) and 4(p) of the Environmental Protection Act (Ill. Rev. Stat. 1981, ch. 111 1/2, pars. 1004(o) and 1004(p)) and Sections 55.10 through 55.12 of the Civil Administrative Code of Illinois (Ill. Rev. Stat. 1981, ch. 127, pars. 55.10 through 55.12).
SOURCE: Adopted at 3 Ill. Reg. 34, p. 103, effective August 19, 1979; codified at 6 Ill. Reg. 14657; amended at 7 Ill. Reg. 13523, effective September 28, 1983.

SUBPART A: GENERAL PROVISIONS

Section 183.105   Authority

Pursuant to the authority contained in Ill. Rev. Stat. 1981, ch. 127, pars. 55.10–.12 which authorizes the Illinois Department of Public Health to establish and enforce minimum standards, and establish certification procedures for laboratories making examinations in connection with the diagnosis of disease or tests for the evaluation of health hazards, and also to enter into contracts with other public agencies for the exchange of health services which may benefit the health of the people; and pursuant to the

authority contained in Section 4 (o and p) of the Environmental Protection
Act, adopted 1970, as amended (Ill. Rev. Stat. 1981, ch. 111 1/2, par. 1004
(o and p)), which authorizes the Illinois Environmental Protection Agency
to "establish and enforce minimum standards for the operation of laboratories relating to analyses and laboratory tests for air pollution, water
pollution, noise emissions, contaminant discharges onto land and sanitary,
chemical, and mineral quality of water distributed by a public water supply", and to "issue certificates of competency to persons and laboratories
meeting the minimum standards established by the Agency . . . and to
promulgate and enforce regulations relevant to the issuance and use of
such certificates", and to "enter into formal working agreements with other
departments or agencies of state government under which all or portions
of this authority may be delegated to the cooperating department or agency",
the Illinois Department of Public Health and the Illinois Environmental
Protection Agency jointly adopt the following rules and regulations.

Section 183.110   Scope and Applicability

a) This Subpart A establishes general provisions applicable to the
   certification program for environmental laboratories administered
   under this Part 183.
b) Nothing in this Part 183 shall prevent uncertified laboratories from
   performing any quality control or other tests when the state has
   not required such tests to be performed by a certified laboratory.
c) Unless the contrary is clearly indicated, all references to "Sections"
   in this Part 183 are to Ill. Adm. Code, Title 35: Environmental
   Protection. For example, "Section 183.230" is 35 Ill. Adm. Code
   183.230.

(Source: Amended at 7 Ill. Reg. 13523, effective September 28, 1983.)

Section 183.115   Definitions

For purposes of this Part 183:
   "Agency" means either the Illinois Department of Public Health or the
Illinois Environmental Protection Agency, whichever is applicable based
on the division of authority specified in Section 183.120.
   "Analyst" means any person who performs analyses for certain or all
parameters on samples submitted to the environmental laboratory and who
meets the qualifications set forth in the applicable subpart of this Part 183.

"Certification" means a status of approval granted to an environmental laboratory which meets the criteria established by this Part 183. Certification is not a guarantee of the validity of the data generated.

"Certification Officer" means any person who is designated by the Agency to inspect and evaluate environmental laboratories for compliance in meeting the criteria set forth in this Part 183. Certification officers shall meet the educational and experience qualifications for laboratory directors as set forth in the applicable subpart of this Part 183.

"Consultant" means a person who is retained by a written agreement to provide professional consultation services.

"Environmental Laboratory" means any facility which performs analyses on environmental samples in order to determine the quality of food, milk, public water supplies, surface water, ground water, recreational waters, wastewater, air, or land.

"Laboratory Director" means the person who is responsible for the operation of an environmental laboratory and who meets the qualifications set forth in the applicable subpart of this Part 183.

"Laboratory Pure Water" means water meeting the standards set forth in Section 183.345.

"Laboratory Supervisor" means a person who supervises the performance of the analytical procedures within an environmental laboratory and who meets the qualifications set forth in the applicable subpart of this Part 183.

"Major remodeling" means remodeling of the laboratory facility which requires the acquisition of a local building permit.

"Maximum Allowable Concentration" means a maximum permissible concentration of a contaminant in finished water as established by 35 Ill. Adm. Code 604.101–604.303 (prior to codification Rule 304 of the Illinois Pollution Control Board Rules and Regulations, Chapter 6: Public Water Supply).

"Provisional Certification" means certification status granted to an environmental laboratory in order to allow time for the correction of deviations. Failure to correct deviations during the provisional certification period allows the Agency to revoke certification as specified in Section 183.130(g)(1). While on provisional certification, an environmental laboratory remains approved for the analyses covered by its certification.

"Public Water Supply" means all mains, pipes and structures through which water is obtained and distributed to the public, including wells and well structures, intakes and cribs, pumping stations, treatment plants, reservoirs, storage tanks and appurtenances, collectively or severally, actually used or intended for use for the purpose of furnishing water for drinking or general domestic use and which serve at least 15 service connections or which regularly serve at least 25 persons at least 60 days per year.

(Source: Amended at 7 Ill. Reg. 13523, effective September 28, 1983.)

Section 183.120   Division of Authority

a) The Illinois Environmental Protection Agency shall administer these rules and regulations with respect to the analysis of organic and inorganic chemical parameters.

b) The Illinois Department of Public Health shall administer these rules and regulations with respect to the analysis of microbiological and radiochemical parameters.

(Source: Amended at 7 Ill. Reg. 13523, effective September 28, 1983).

Section 183.125   Certification Procedure

a) An environmental laboratory which meets or exceeds the minimum criteria for certification may receive certification from the Agency for any mircobiological, radiological, and organic or inorganic chemical parameters for which methodologies have been specified in this Part 183.

b) The operational aspects of an environmental laboratory that will be evaluated in considering certification are:
1) physical facilities,
2) personnel,
3) methodology and instrumentation,
4) data handling, and
5) quality control.

c) In seeking certification, the petitioning environmental laboratory must:
1) Submit a formal request for certification from the Agency;
2) File the applicable administrative questionnaires furnished by the Agency giving complete information on the five categories listed in Section 183.125(b);
3) Analyze performance evaluation samples to be provided by the Agency and report the results of the analyses to the Agency; and
4) Permit an on-site visit by Agency authorized certification officers. Certification officers shall provide the environmental lab-

oratory with official identification and credentials. The initial visit will be arranged at the mutual convenience of both parties. The Agency reserves the right to make subsequent visits without prior notice during regular working hours.

d) An environmental laboratory seeking certification from the Illinois Environmental Protection Agency and the Illinois Department of Public Health only needs to file a single request for certification and a single set of administrative questionnaires with either agency.

e) Approval or denial of certification will be made after the procedure described in Section 183.125(c) has been completed. Denial of certification shall be in the form of a narrative, giving complete information as to how deviations may be corrected, along with a completed survey form on which all items in deviation are clearly marked.

(Source: Amended at 7 Ill. Reg. 13523, effective September 28, 1983.)

Section 183.130 Conditions Governing the Use of Certificates

a) Certification shall be effective for a two year period from date of issue, unless modified or revoked by the Agency. Application for timely renewal of certification shall be made to the Agency no later than 90 days prior to the expiration date. Approval of a renewal application shall be contingent upon the environmental laboratory meeting all of the factors considered in granting the original approval, including acceptable results on performance evaluation samples. When an environmental laboratory has made timely and sufficient application for renewal of certification or certification for additional parameters, the existing certification shall continue in full force and effect until the final decision of the Agency on the application has been made unless a later date is fixed by order of a reviewing court.

b) Whenever deviations from the applicable requirements are found, a certified environmental laboratory may be placed on provisional certification. Provisional certification may be granted for the following periods:

1) From seven to 30 days if the deviation could compromise the quality of analytical data generated by the environmental laboratory; or

2) From 90 days to one year in the case of any other type of deviation.

c) The Illinois Environmental Protection Agency may grant written preliminary certification to an environmental laboratory which has demonstrated satisfactory capability after completion of the procedures specified in Section 183.125(c)(1–3). Preliminary certification would be available in instances where it would be impractical for the Illinois Environmental Protection Agency to schedule an on-site visit within six months from the date of a laboratory's submission of satisfactory analysis results for performance evaluation samples. Preliminary certification shall remain in effect until certification has been approved or denied in accordance with Section 183.125.

d) Certification shall not be transferrable. In the event of change of ownership, director, supervisor, analysts, or relocation or major remodeling of the physical plant of an environmental laboratory, the Agency shall be notified in writing within 15 days.

e) After receiving notification of any of the changes listed in Section 183.130(d), unless otherwise stated for a specific parameter, the Agency will request a resume (as to any new owners, directors, supervisors, or analysts), send a quality control sample for analysis by any new analyst, and make an on-site visit. However, the Agency may waive any of these actions if it appears unwarranted in a specific case. Examples of when such waivers would be appropriate include the following circumstances:

1) Waiver of submittal of a summary of education and experience when personnel transferring from one certified laboratory to another are responsible for dealing with the same analytical methods and equivalent equipment; and

2) Waiver of an on-site visit if the pertinent test procedures involve simple techniques and equipment.

f) An environmental laboratory may cancel its certification voluntarily by notifying the Agency and returning the certificate.

g) The Agency may revoke certification for cause as to all or any part of an environmental laboratory's certification. Any of the following shall be cause for partial or total revocation of certification:

1) Failure to pass any inspection, provided the laboratory has not corrected the deviations after being placed on provisional certification in accordance with the provisions of Section 183.130(b);

2) Unsatisfactory analyses of performance evaluation samples as specified in Section 183.140;

3) Failure to notify the Agency within 15 days after any of the changes listed in Section 183.130(d) have occurred; or

4) Violation of the requirements regarding advertising as specified in Section 183.130(k).

h) Certification shall be limited to those analytical procedures for which an environmental laboratory has been approved and which are listed on the certificate of approval.

i) The certificate of approval shall be posted or displayed in a prominent place in the laboratory facility.

j) Information related to the certification of an environmental laboratory shall be clearly defined in any advertising and shall prominently include the statement that, "Certification is not a guarantee of the validity of the data generated." Such information shall also include the analyses for which the environmental laboratory has been certified. The advertising shall not include any representation that the environmental laboratory is certified to perform a type of analysis for which it lacks proper certification.

k) The following factors shall be taken into account by the Agency in determining what action should be taken against a certified environmental laboratory when deviations from these rules and regulations are found:

1) The length of time during which the deviation has existed;

2) The laboratory's prior record of deviations and response in correcting deviations noted by the Agency;

3) Whether the laboratory knowingly caused or allowed the deviations; and

4) The potential effect of the deviation on the quality of analytical data generated by the laboratory.

(Source: Amended at 7 Ill. Reg. 13523, effective September 28, 1983.)

Section 183.135   Subcontracting by Certified Laboratories

a) The name of the laboratory actually performing the analyses shall be specified on all reports of analytical results.

b) For those tests that are required to be performed under certification, any laboratory with which a certified environmental laboratory subcontracts shall also be a certified environmental laboratory.

(Source: Amended at 7 Ill. Reg. 13523, effective September 28, 1983.)

Section 183.140   Performance Evaluation Samples

An environmental laboratory is required to participate in performance evaluation sample analysis relevant to the analytical parameters for which it seeks or wishes to maintain certification in accordance with the certification procedures of Section 183.125(c), the certification renewal procedures of Section 183.130(a), and the quality control requirements contained in the applicable subpart of this Part 183. Within 30 days of receipt, the environmental laboratory shall analyze such samples and report the test results to the Agency. There shall be no fee charged to the Agency for such analyses. Failure to provide results proving satisfactory precision and accuracy in two successive samples shall be cause for revocation of certification for the parameters not within satisfactory limits. Acceptance limits for trahalomethanes shall be plus or minus 20 percent of the mean value. Acceptance limits for all other performance evaluation samples shall be plus or minus two standard deviations from the mean value.

(Source: Amended at 7 Ill. Reg. 13523, effective September 28, 1983.)

Section 183.145   Authority of Certification Officers

Certification officers shall have all of the following authority with regard to environmental laboratories:

a) To inspect such laboratories in on-site visits;

b) To require information relevant to the technical operation of such laboratories;

c) To inspect quality assurance records and any other pertinent records;

d) To be permitted to observe and question analysts at work on parameters for which certification is sought; and

e) To submit oral and written reports for granting or denying certification based upon the completion of the evaluation process.

Section 183.150   Hearing, Decision and Appeal

The following procedures are established for Agency certification actions which are required by law to be preceded by notice and opportunity for hearing:

a) Prior to revocation or partial revocation, the Agency shall give written notice to the laboratory director or owner. This notice shall include the facts or conduct upon which the Agency will rely to support its proposed action and the procedures for requesting a hearing.

b) Notice given under Section 183.150(a) and any hearing requested following issuance of such notice shall be in accordance with the "Rules of Practice and Procedure in Administrative Hearings" as adopted by the Illinois Department of Public Health. A single joint hearing may be conducted when a hearing is requested concerning actions of both the Illinois Department of Public Health and the Illinois Environmental Protection Agency. With respect to the Illinois Environmental Protection Agency, the "Rules of Practice and Procedure in Administrative Hearings" (77 Ill. Adm. Code 100) are applicable only to hearings under this Section 183.150 and the included definitions of "department" and "director" are modified as follows:

1) "Department" shall mean the Illinois Environmental Protection Agency.

2) "Director" shall mean the Director of the Illinois Environmental Protection Agency.

c) If, however, the Agency finds that an emergency situation warrants immediate action, summary suspension as provided for by Section 16(c) of the Illinois Administrative Procedure Act (Ill. Rev. Stat. 1981, ch. 127, par. 1016(c) may be ordered pending revocation proceedings. An emergency situation warrants immediate action if there is substantial risk to public health, safety, or welfare resulting from laboratory deficiencies that are compromising the analytical results obtained.

d) Final decisions adopted by the Director of the Illinois Department of Public Health are appealable to the Circuit Courts under the Illinois Adminstrative Review Act (Ill. Rev. Stat. 1981, ch. 110, pars. 264 et seq.). Final decisions adopted by the Director of the Illinois Environmental Protection Agency may be contested before the Pollution Control Board under the Illinois Environmental Protection Act (Ill. Rev. Stat. 1981, ch. 111 1/2, pars. 1001 et seq.) with subsequent appeal to the Appellate Courts available.

(Source: Amended at 7 Ill. Reg. 13523, effective September 28, 1983.)

Section 183.155   Liability

Representatives of the Agency shall not waive the right to seek recovery for injuries incurred while inspecting an environmental laboratory facility.

Section 183.160   Reciprocity Agreements

The Director of the Agency may elect to enter into agreements with the governments of other states or with federal governmental units for recognition of their environmental laboratory inspections and certifications if such certification program uses equivalent controls over sample collection, data handling, quality control, analytical methods, and personnel as required of environmental laboratories within Illinois. Environmental laboratories in jurisdictions not having reciprocal agreements with Illinois which ask that their results be accepted by Illinois shall request certification from the Agency and agree to pay all of the expenses incurred by the Agency, including travel expenses, in evaluating the laboratory.

Section 183.165   Reporting (repealed)

(Source: Repealed at 7 Ill. Reg. 13523, effective September 28, 1983.)

Section 183.170   Public Inspection of Records

All files, records, and data of the Illinois Department of Public Health and the Illinois Environmental Protection Agency in relation to the administration of these rules and regulations shall be open to public inspection and may be copied upon payment of the actual cost of reproducing the original except for:
  a) Information which constitutes a trade secret;
  b) Information privileged against introduction in judicial proceedings;
  c) Internal communications of the Agency;
  d) Information concerning secret manufacturing processes or confidential data submitted by any person under these rules and regulations.

SUBPART B: CHEMICAL ANALYSES OF PUBLIC WATER SUPPLY SAMPLES

Section 183.205   Scope and Applicability

This Subpart B establishes standards applicable to environmental laboratories involved in chemical analyses of samples of water from public water supplies and their sources.

Section 183.210   Personnel

  a) The laboratory director shall be a person holding a minimum of a bachelor's degree in natural or physical sciences with at least 24 semester hours in chemistry or microbiology or both, and shall have had a minimum of three years experience in an environmental laboratory. The laboratory director shall be either a full-time employee or a consultant.

b) A laboratory supervisor shall be a person holding a minimum of a bachelor's degree in natural or physical sciences with at least 16 semester hours of course work in the analytical area of responsibility and shall have had a minimum of two years experience in the area of analytical responsibility. A laboratory supervisor shall be a full-time employee.

c) An analyst is a person who holds a high school diploma or its equivalent and has completed a basic chemistry course. In addition, an analyst shall have had at least one year of experience in an analytical laboratory and shall demonstrate ability to properly perform representative test procedures with which he or she is involved while under observation by the certification officer.

d) A person who, as of July 1, 1979, is serving in an environmental laboratory in any capacity as defined in Section 183.210(a–c) and does not meet the educational requirements or experience requirements or both for said position may be recommended to continue to serve in said position by the certification officer. In recommending that an existing laboratory director, laboratory supervisor, or analyst continue to serve in that position, the certification officer shall take into account the following factors:

1) Length of experience as an offset for not meeting educational requirements;
2) Extent of education as an offset for not meeting experience requirements; and
3) For analysts, demonstration of ability to properly perform representative test procedures with which he or she is involved while under observation by the certification officer.

Section 183.215 Physical Facilities

The laboratory's physical facilities shall meet the following specifications:

a) A minimum of 150 square feet of floor space shall be provided for each analyst.
b) A minimum of 15 linear feet of useable bench space shall be provided for each analyst.
c) The laboratory shall include a sink with hot and cold running water. All water supply outlets shall be protected by approved vacuum breakers.

d) An adequate electrical supply for operation of instruments and mechanical needs shall be provided. The certification officer may require verification from an official inspector or other qualified person that the laboratory meets local and national electrical codes.
e) All electrical outlets shall be properly grounded.
f) Instruments shall be properly grounded with an internal or external regulated power supply available to each instrument.
g) All plumbing shall meet local and state plumbing codes. The certification officer may require verification from an official inspector or other qualified person that the laboratory meets such codes.
h) The laboratory shall include a vacuum source if the analyses performed so require.
i) The laboratory shall have a readily available source of distilled water or deionized water or both.
j) The laboratory shall include at least one fume hood for analyses of organic chemicals and trace metals.

Section 183.220   Laboratory Equipment

Only those instruments that are needed to analyze for the parameters for which the laboratory is being certified are required, but those instruments shall meet the following minimum specifications. A laboratory doing all the analyses described in Section 183.230 shall have, or have access to, all of the equipment listed in this Section with the minimum specifications cited.

a) An analytical balance shall provide a sensitivity of at least 0.1 mg.
b) A spectrophotometer shall have a useable wavelength range of 400 to 700 nm, a maximum spectral band width of no more than 20 nm, and a wavelength accuracy of 0 ± 2.5 nm. The photometer shall be capable of using several sizes and shapes of absorption cells providing a sample path length varying from approximately 1 to 5 cm.
c) A filter photometer (abridged spectrophotometer) shall be capable of measuring radiant energy in range of 400 to 700 nm. Relatively broad bands (10 to 75 nm) of this radiant energy are isolated by use of filters at or near the maximum absorption of the colorimetric methods. The photometer shall be capable of using several sizes and shapes of absorption cells.
d) A magnetic stirrer shall be of variable speed and use a Teflon-coated stirring bar.
e) A pH meter shall have an accuracy of at least ± 0.5 units and a

scale readability of at least ± 0.1 units. The pH meter may be either line/bench or battery/portable operated and also should be capable of functioning with specific ion electrodes.

f) A specific ion meter shall have an accuracy and scale readability of at least ± 1 mV, and shall have expanded scale millivolt capability. The specific ion meter may be either line/bench or battery/portable operated.

g) An atomic absorption spectrophotometer shall be a single- or multichannel, single- or double-beam instrument having a grating monochromator, photomultiplier detector, adjustable slits, a wavelength range of 190 to 800 nm. Provision for interfacing with a strip chart recorder or other device for generating a permanent record shall be provided.

h) A readout system for atomic absorption shall have a response time capable of measuring the atomic absorption signal generated and shall include the capability to detect positive interference on the signal from intense non-specific absorption. In furnace analysis, a strip chart recorder shall be used for verification of adequate background correction if a CRT video readout or hard copy plotter is not available. The strip chart recorder shall have a recorder width of at least 25 cm, a full scale response time of 0.5 seconds or less, a 10- or 100-mV input to match the instrument, and variable chart speeds of at least .5 to 5 cm/min or equivalent.

i) A gas chromatograph shall be a commercial or custom designed gas chromatograph with a column oven capable of isothermal temperature control to at least 210° ± 0.2°C. Additional accessories and specifications are listed below by methodology.

1) For chlorinated hydrocarbons, the gas chromatograph shall be equipped with a glass lined injection port suitable for chlorinated hydrocarbon pesticides with a minimum of decomposition, and equipped with either an electron capture, microcoulometric titration, or electrolytic conductivity detector.

2) For chlorophenoxys, the gas chromatograph shall be equipped with a glass lined injection port and either an electron capture, microcoulometric titration, or electolytic conductivity detector.

3) For trihalomethanes by purge and trap, the gas chromatograph shall be temperature programmable from 45° to 220°C, at rates specified in the methodology and equipped with either microcoulometric titration or electrolytic conductivity detector.

4) For trihalomethanes by liquid/liquid extraction, the gas chromatograph shall be equipped with a linearized (frequency modulated) electron capture detector.

5) For trihalomethanes by gas chromatography/mass spectrometry, the gas chromatograph shall be temperature programmable from 45° to 220°C at rates specified in the methodology and interfaced to the mass spectrometer with an all glass enrichment device and an all glass transfer line.

j) A recorder for gas chromatography shall be a strip chart recorder with a recorder width of at least 25 cm, a full scale response time of 1 second or less, a 1-mV (-0.05 to 1.05) signal to match the instrument, and variable chart speeds with a range of at least .5 to 5 cm/min or equivalent. Computer generated chromatograms are accepable where a record of the data is required.

k) A mass spectrometer for trihalomethanes by gas chromatography/ mass spectrometry shall include an interfaced data system to acquire, store, reduce and output mass spectral data. The data system shall be equipped with software to acquire and manipulate data for only a few ions that were selected as characteristic of trihalomethanes and the internal standard (or surrogate compound). Mass spectral data shall be obtained with electron-impact ionization at a nominal electron energy of 70 eV. The mass spectrometer shall meet all of the following criteria when 50 ng or less of p-promo-fluorobenzene is introduced into the gas chromatograph:

p-Bromofluorobenzene Key Ions
and Ion Abundance Criteria

Mass	Ion Abundance Criteria
50	15 to 40% of mass 95
75	30 to 60% of mass 95
95	base peak, 100% relative abundance
96	5 to 9% of mass 95
173	less than 2% of mass 174
174	greater than 50% of mass 95
175	5 to 9% of mass 174
176	96 to 100% of mass 174
177	5 to 9% of mass 176

l) A conductivity meter and cell combination, suitable for checking distilled water quality, shall be readable in ohms or mhos, and have a range of up to 2.5 megohm cm resistivity (conductivity down to 0.4 micromhos/cm) ± 1 percent. The conductivity meter may be either line/bench or battery/portable operated.

m) A drying oven shall be a gravity or mechnical convection unit with a selectable temperature control from room temperature to 180°C or higher.

n) A desiccator shall be a glass or plastic model, depending upon the particular application.

o) A hot plate may be a large or small unit and shall have a selectable temperature control for safe heating of laboratory reagents.

p) A refrigerator used for storage of organics and flammable materials shall be an "explosion proof" type. For storage of organics and flammable materials when refrigeration is not required, an explosion proof cabinet shall be provided. A refrigerator for the general storage of aqueous reagents and samples may be a standard kitchen type domestic refrigerator.

q) Glassware which is used for purposes that may subject it to damage from heat or chemicals shall be of borosilicate glass. All volumetric glassware shall be Class A, denoting that it meets Federal Specifications and need not be calibrated before use.

r) A stirred water bath shall provide from ambient temperature up to 100°C (with gable lid).

(Source: Amended at 7 Ill. Reg. 13523, effective September 28, 1983.)

Section 183.225 General Laboratory Practices

a) All prepackaged kit methods, other than the DPD Colorimetric Test Kit, are considered alternative analytical techniques and may be substituted only if approved in accordance with 40 CFR 141.27 (1982).

b) A laboratory utilizing visual comparison devices shall calibrate the standards incorporated into such devices at least every six months. These calibrations shall be documented. Preparation of temporary and permanent type visual standards shall be in accordance with the Color-Visual Comparison Method, "Standard Methods for the Examination of Water and Wastewater," 14th Edition, American Public Health Association, (Washington, D.C., 1976), pp. 64–66 and the Turbidity-Visual Methods, "Standard Methods for the Examination of Water and Wastewater," 14th Edition, American Public Health Association, (Washington, D.C., 1976), pp. 135–137. By comparing standards and plotting such a comparison on graph paper, a correction factor shall be derived and applied to all future results obtained on the now calibrated apparatus until it is recalibrated.

c) Prior to use, all glassware shall be washed in a warm detergent solution and thoroughly rinsed, first in tap water and then in distilled or deionized water. This cleaning procedure is sufficient for most analytical needs, but the procedures specified for individual parameters shall be referred to for more elaborate precautions to be taken against contamination of glassware. A separate set of glassware shall be maintained for the nitrate, mercury, and lead procedures due to the potentiality for contamination from the laboratory environment. All glassware used in organic chemical analyses shall have a final rinse with nanograde acetone or its equivalent and shall be air dried in an area free of organic contamination.

d) Distilled or deionized water shall have resistivity values of at least 0.5 megohm cm (conductivity less than 2.0 micromhos/cm at 25°C. Laboratories are advised to request a list of quality specifications for any water purchased. The quality of the distilled or deionized water shall be maintained by protecting it from the atmosphere. Quality checks of the distilled or deionized water shall be made at least once each shift and documented.

e) Reagents used for chemical analyses shall be of a quality at least equal to the grade recommended in the applicable analytical procedure reference.

f) Other than the specific requirements set forth, in these rules and regulations, laboratory safety practices are not considered an aspect of laboratory certification. However, certification officers may point out, on an informal basis, potential safety problems observed during on-site visits.

(Source: Amended at 7 Ill. Reg. 13523, effective September 28, 1983.)

Section 183.230   Methodology and Required Equipment

Minimum equipment requirements, methodology, and references for individual parameters shall be as provided in Appendix A of this Part 183.

Section 183.235   Sample Collecting, Handling and Preservation

The standards for container types, preservatives, and holding time to be met for each individual parameter[a] are shown in Table 1.

Section 183.240 Quality Control

a) A written description of the curent laboratory quality control program shall be maintained and made available to analysts in an area of the laboratory where analytical work takes places. A record of analytical quality control tests and quality control checks on materials and equipment shall be prepared and retained for 5 years.

b) A laboratory manual containing complete written instructions for each parameter for which the laboratory is certified shall be maintained and made available to analysts in an area of the laboratory where analytical work takes place.

c) A laboratory shall analyze unknown performance evaluation samples provided by the Agency so that results proving satisfactory precision and accuracy, as specified in Section 183.140, are submitted to the Agency once per year for the parameters for which the laboratory is certified. When performance evaluation sample results indicate technical error, the Agency will provide appropriate technical assistance, and followup performance evaluation samples shall be analyzed by the laboratory.

d) A current service contract shall be in effect on all analytical balances.

e) Standardized Class "S" weights shall be available at the laboratory to make periodic checks on balances.

f) At least one thermometer certified by the National Bureau of Standards (or one of equivalent accuracy) shall be available to check thermometers in ovens, etc.

g) Color standards or their equivalent shall be available to verify wavelength settings on spectrophotometers.

h) Chemicals shall be dated upon receipt of shipment and replaced as needed or, if earlier, before shelf life has been exceeded.

i) A laboratory should conduct analyses on known reference samples once per quarter for the parameters measured.

j) The following quality control procedures shall be utilized by the laboratory for inorganic parameters:

1) After a standard reagent curve composed of a minimum of a reagent blank and three standards has been prepared, subsequent standard curves shall be verified by use of at least a regent blank and one standard at or near the maximum allowable concentration. Daily checks must be within ± 10 percent of original curve; and

2) If 20 or more samples per day are analyzed, working standard curve shall be verified by running an additional standard at or

Table 1 Standards For Sample Collecting, Handling, and Preservation (Section 183.235)

Parameter[a]	Preservative[b]	Container[c]	Maximum holding time[d]
Arsenic	Conc HNO$_3$ to pH less than 2[f]	P or G	6 months
Barium	Conc HNO$_3$ to pH less than 2	P or G	6 months
Cadmium	Conc HNO$_3$ to pH less than 2	P or G	6 months
Chromium	Conc HNO$_3$ to pH less than 2	P or G	6 months
Lead	Conc HNO$_3$ to pH less than 2	P or G	6 months
Mercury	Add 20 ml per liter of sample of a solution of 2.5% potassium dichromate in 1:1 HNO$_3$	G P	38 days 14 days
Nitrate	Conc H$_2$SO$_4$ to pH less than 2	P or G	14 days
Selenium	Conc HNO$_3$ to pH less than 2	P or G	6 months
Silver	Conc HNO$_3$ to pH less than 2	P or G	6 months
Fluoride	None	P or G	1 month
Chlorinated hydrocarbons	Refrigerate at 4°C as soon as possible after collection	G with foil or Teflon-lined cap	14 days[e]
Chlorophenoxys	Refrigerate at 4°C as soon as possible after collection	G with foil or Teflon-lined cap	7 days[e]
Cyanide	Add NaOH to pH greater than 12 refrigerate & keep in dark	P or G	24 hours
Trihalomethanes	0.008% NA$_2$S$_2$O$_3$ Refrigerate at 4°C as soon as possible after collection	G with foil or Teflon-lined cap	14 days
Alkalinity	Refrigerate at 4°C as soon as possible after collection	P or G	14 days

Table 1 *(Continued)*

Parameter[a]	Preservative[b]	Container[c]	Maximum holding time[d]
Calcium	Conc HNO_3 to pH less than 2	P or G	6 months
Copper	Conc HNO_3 to pH less than 2	P or G	6 months
Hydrogen ion (pH)	None	P or G	2 hours
Iron	Conc HNO_3 to pH less than 2	P or G	6 months
Manganese	Conc HNO_3 to pH less than 2	P or G	6 months
Sodium	Conc HNO_3 to pH less than 2	P or G	6 months
Total dissolved (filterable) residue	Refrigerate at 4°C as soon as possible after collection	P or G	14 days
Zinc	Conc HNO_3 to pH less than 2	P or G	6 months

[a]If a laboratory has no control over these factors the laboratory director must reject any samples not meeting these criteria and so notify the authority requesting the analyses.

[b]The following procedure shall be utilized if the concentrated acid specified for preservation cannot be used because of shipping restrictions: (1) the sample shall be initially preserved by icing and immediately shipped to the laboratory; (2) upon receipt in the laboratory, the sample shall be acidified with the concentrated acid specified for preservation to pH less than 2; and (3) at the time of analysis the sample container shall be thoroughly rinsed with a 1:1 solution of the same type of acid and water, with the washings being added to the sample.

[c]P = Plastic, hard or soft; G = Glass, hard or soft.

[d]In all cases, samples should be analyzed as soon after collection as possible.

[e]Well-stoppered and regrigerated extracts can be held up to 30 days.

[f]Nitric acid is a negative interference if arsenic is determined by the spectrophotometric method.

(Source: Amended at 7 Ill. Reg. 13523, effective September 28, 1983.)

near the maximum allowable concentration every 20 samples. Daily checks must be within ± 10 percent of original curve.

k) The following quality control procedures shall be utilized by the laboratory for organic parameters:

1) For each day on which pesticide or phenoxyacid analyses are initiated, or trihalomethane regant water is prepared, a laboratory method blank shall be analyzed with the same procedures used to analyze samples;

2) A minimum of three calibration standards shall be analyzed each day, except that a minimum of one calibration standard per day is sufficient if the laboratory can demonstrate that the instrument response is linear through the origin and the response of the standard is within ± 15 percent of previous calibrations;

3) A field blank for trihalomethanes shall be analyzed with each sample set and resampling shall be done if reportable levels of trihalomethanes are found to have contaminated the field blank;

4) Analysis of 10 percent of all samples for trihalomethanes shall be done in duplicate, with a continuing record of results and subsequent actions maintained;

5) A known trihalomethane laboratory control standard shall be analyzed each day, so that if errors exceed 20 percent of the true value all trihalomethane results since the previous successful test are to be considered suspect;

6) Each time the trihalomethane analytical system undergoes a major modification or prolonged period of inactivity, the precision of the system shall be demonstrated by the analysis of replicate laboratory control standards;

7) Laboratories that analyze for trihalomethanes by liquid/liquid extraction shall demonstrate that raw source waters do not contain interferents under the chromatographic conditions selected; and

8) If a mass spectrometer detector is used for trihalomethane analysis, the mass spectrometer performance tests described in Section 183.220(k) using p-bromofluorobenzene shall be conducted once during each 8-hour work shift, with records of satisfactory performance and corrective action maintained.

l) The following quality control procedures shall be utilized by the laboratory for both inorganic and organic parameters:

1) At least one duplicate sample shall be run every 10 samples, or with each set of samples, to verify precision of the method;

2) Standard deviation shall be calculated and documented, as de-

scribed in "Handbook for Analytical Quality Control in Water and Wastewater Laboratories," (EPA 600/4-79-019), 1979, U.S. Environmental Protection Agency, Office of Research and Development, Cincinnati, Ohio 45268, for all measurements conducted; and

3) Quality control charts or a tabulation of mean and standard deviation shall be used to document acceptability of data, as described in "Handbook for Analytical Quality Control in Water and Wastewater Laboratories," (EPA 600/4-79-019), 1979, U.S. Environmental Protection Agency, Office of Research and Development, Cincinnati, Ohio 45268, on a daily basis.

(Source: Amended at 7 Ill. Reg. 13523, effective September 28, 1983.)

Section 183.245   Record Maintenance

Records of chemical analyses shall be kept by the laboratory for not less than one year. Since public water supplies are required by 35 Ill. Adm. Code 607.106 (prior to codification Rule 310(C) of the Illinois Pollution Control Board Rules and Regulations, Chapter 6: Public Water Supply) to maintain records of chemical analyses for not less than 10 years, laboratories which maintain records for less than 10 years may wish to give records of analyses performed to the appropriate public water supplies instead of destroying such records. The disposal of all records subject to the Local Records Act (Ill. Rev. Stat. 1981, ch. 116, pars. 43.101 et seq.) must be in accordance with the provisions of that Act. Enforcement data, which includes all raw data, calculations, quality control data and reports, shall be kept for three years. Actual laboratory reports may be kept. However, data, with the exception of compliance check samples, as detailed in 40 CFR 141.33(b), may be transferred to tabular summaries which shall include the following information:

a) Date, place, and time of sampling;

b) Name of person who collected the sample;

c) Identification of the sample origin, such as routine distribution system sample, check sample, raw or process water sample, or other special purpose sample;

d) Date of receipt of sample;

e) Records necessary to establish chain-of-custody of the sample;

f) Date of sample analysis;

g) Name of the persons and designation of the laboratory responsible for performing the analysis;

h) Designation of the analytical techniques or method used; and

i) Results of the analysis.

Section 183.250   Free Chlorine Residual and Turbidity

a) Free and total chlorine residual measurements do not need to be done in certified laboratories, but may be performed by any persons if such persons adhere to the following standards in their analyses:

 1) Samples shall not be preserved for later analysis. All analyses shall be made as soon as practicable, but no later than one hour after sample collection;

 2) Plastic or glass containers shall be used for sample collection;

 3) A DPD Colorimetric Test Kit, or a spectrophotometer, or a photometer shall be available; and

 4) The DPD Colorimetric Method specified in "Standard Methods for the Examination of Water and Wastewater," 13th Edition, American Public Health Association, (New York, New York, 1971), pp. 129–132 shall be utilized.

b) Turbidity measurements do not need to be done in certified laboratories, but may be performed by any persons approved by the Agency in accordance with Technical Policy Statement 309(B)(2) of the Illinois Environmental Protection Agency, Division of Public Water Supplies, if such persons adhere to the following standards in their analyses:

 1) Samples shall not be preserved for later analysis. All analyses shall be made as soon as practicable, but no later than one hour after sample collection;

 2) Plastic or glass containers shall be used for sample collection;

 3) A nephelometer shall be available;

 4) The Nephelometric Method specified in "Standard Methods for the Examination of Water and Wastewater", 13th Edition, American Public Health Association, (New York, New York, 1971), pp. 350–353 or in "Methods for Chemical Analysis of Water and Wastes," United States Environmental Protection Agency, Office of Technology Transfer, Washington, D.C. 20460, (1974), pp. 295–298, shall be utilized

Section 183.255 Action Response to Laboratory Results

When laboratory results indicate that a maximum allowable concentration of any parameter has been exceeded by a public water supply, the requesting facility shall be notified as soon as possible, but in any event within 48 hours, of the unsatisfactory sample result.

(Source: Amended at 7 Ill. Reg. 13523, effective September 28, 1983.)

SUBPART C: MICROBIOLOGICAL ANALYSES OF PUBLIC WATER SUPPLY SAMPLES

Section 183.305 Scope and Applicability

This Subpart C establishes standards applicable to environmental laboratories involved in microbiological analyses of samples of water from public water supplies and their sources.

Section 183.310 Personnel

a) The laboratory director shall be a person holding a minimum of a bachelor's degree in natural or physical sciences with at least 24 semester hours in chemistry or microbiology or both, and shall have had a minimum of three years experience in an environmental laboratory. The laboratory director shall be either a full-time employee or a consultant.

b) The laboratory supervisor shall be a person holding a minimum of a bachelor's degree in microbiology, biology, chemistry, or a closely related field. In addition, the laboratory supervisor shall have had a minimum of one year of bench experience in an environmental laboratory in the area of analytical responsibility and shall demonstrate ability to properly perform representative test procedures under his or her supervision while under observation by the certification officer. However, only the requirements specified in Section 183.310(c) shall be required for a laboratory supervisor employed by water or sewage treatment plants that serve communities with a population of 30,000 or less. A laboratory supervisor shall be a full-time employee.

c) An analyst is a person who performs microbiological analyses on waters, has a minimum of a high school diploma in academic or laboratory oriented vocational courses, and has had a minimum of three months experience in a microbiological analytical laboratory. In addition, an analyst shall demonstrate ability to properly perform representative test procedures with which he or she is involved while under observation by the certification officer, and shall have satisfactory results in the split water sample program. Analysts shall be under the direct supervision of the laboratory supervisor.

d) Support personnel are persons who have had a minimum of 30 days on-the-job training in areas of responsibility. Support personnel shall be under the supervision of the laboratory supervisor and shall demonstrate ability to properly perform representative test procedures with which he or she is involved while under observation by the certification officer, if requested to do so.

e) A person who, as of July 1, 1979, is serving in an environmental laboratory in any capacity as defined in Section 183.310(a–c) and does not meet the educational requirements or experience requirements or both for said position may be recommended to continue to serve in said position by the certification officer. In recommending that an existing laboratory director, laboratory supervisor, or analyst continue to serve in that position, the certification officer shall take into account the following factors:

   1) Length of experience as an offset for not meeting educational requirements:
   2) Extent of education as an offset for not meeting experience requirements; and
   3) For analysts, demonstration of ability to properly perform representative test procedures with which he or she is involved while under observation by the certification officer.

Section 183.315   Physical Facilities

The laboratory's physical facilities shall meet the following specifications:

   a) A minimum of 150 square feet of floor space shall be provided for each analyst.
   b) Floors shall be covered with asphalt tile, vinyl, concrete, or other impervious, washable surface; which can be easily maintained.
   c) Ample floor space shall be available for stationary equipment such as autoclaves, incubators, and hot-air sterilization ovens. Storage space that is free of dust and insects shall be provided for the protection of glassware, media, and portable equipment.
   d) Laboratories analyzing potable waters, non-potable source and rec-

reation waters, and sewage by microbiological methods shall have at least two separate rooms (a room for potable water, non-potable source and recreation waters; and a room for sewage).

e) A separate area for preparation and sterilization of media, glassware, and equipment shall be provided. Laboratories of water or sewage treatment plants that serve a population of 30,000 or less may carry out these activities in the same room(s) as used for microbiology, provided all activities of this nature are carried on in a special area of the room(s).

f) Walls and ceilings shall be covered with waterproof paint, enamel, ceramic tile, or other surface material that provides a smooth finish which is easily cleaned and disinfected.

g) A minimum of 6 linear feet of useable bench space, free of equipment, shall be provided for each analyst.

h) Bench tops shall be stainless steel, epoxy plastic, or other smooth impervious material which is inert, corrosion resistant, has a minimum number of seams, and is level.

i) Laboratory lighting shall be even and provide a minimum of 100 footcandle light intensity at all working surfaces.

j) The laboratory shall include a sink with hot and cold running water. All water supply outlets shall be protected by approved vacuum breakers.

k) Laboratories shall be well ventilated and free of dusts, drafts, and extreme temperature changes. Central air-conditioning is recommended to reduce contamination, permit more stable operation of incubators, and decrease moisture problems with media and analytical balances. The temperature within the laboratory shall be maintained at between 60° and 80°F.

l) An adequate electrical supply for operation of instruments and mechanical needs shall be provided. The certification officer may require verification from an offical inspector or other qualified person that the laboratory meets local and national electrical codes.

m) All electrical outlets shall be properly grounded.

n) Instruments shall be properly grounded with an internal or external regulated power supply available to each instrument.

o) All plumbing shall meet local and state plumbing codes. The certification officer may require verification from an official inspector or other qualified person that the laboratory meets such codes.

p) The laboratory shall include a vacuum source if the analyses performed so require.

q) The laboratory shall be located in an area sufficiently free from noise and vibrations to prevent interference with its functions.

r) The laboratory shall have a readily available source of laboratory pure water.

Section 183.320   Laboratory Equipment

Only those instruments that are needed to analyze for the parameters for which the laboratory is being certified are required, but those instruments shall meet the following minimum specifications. A laboratory doing all the analyses described in Section 183.335 shall have, or have access to, all of the equipment listed in this Section with the minimum specifications cited.

    a) A top loading or trip pan balance shall be clean, not corroded, and provided with appropriate weights of good quality.

        1) A torsion or trip pan balance used for weighing materials of 2 grams or more shall detect 100 mg of weight accurately at a 150 gram load.

        2) An analytical balance used for weighing quantities of less than 2 grams shall be sensitive to 0.1 mg at a 10 gram load.

    b) A magnetic stirrer shall be of variable speed, 120 volts, and use a Teflon-coated stirring bar. The magnetic stirrer may be equipped with a heating element.

    c) A pH meter shall have an accuracy of at least ± 0.05 units and a scale readability of at least ± 0.1 units. The pH meter may be either line/bench or battery/portable operated.

    d) A conductivity meter and cell combination, suitable for checking distilled water quality, shall be readable in ohms or mhos, and have a range of up to 2.5 megohm cm resistivity (conductivity down to 0.4 micromhos/cm) ± 1 percent. The conductivity meter may be either line/bench or battery/portable operated.

    e) An autoclave shall be horizontal chambered and shall meet all of the following specifications:

        1) When observed during the operational cycle or when time-temperature charts are read, the autoclave shall be in good operating condition;

        2) An operating safety valve shall be included;

        3) Separate temperature and pressure gauges shall be located on the exhaust side;

        4) The autoclave shall reach and maintain a temperature of 121°C during the sterilization cycle, and no more than 45 minutes shall be required for a complete cycle of carbohydrate media; and

5) Depressurization shall not produce gas bubbles in fermentation media.

f) A hot-air sterilization oven shall operate at a minimum of 175°C, shall be equipped with a thermometer inserted through the top porthole or be equipped with a temperature recording device, and shall be equipped with a thermostatic control that will not allow the temperature to deviate by more than ±5° from the temperature setting.

g) An incubation unit shall maintain internal temperature of 35° ± 0.5°C or 44.5° ± 0.2°C and shall be of the following type: air or water jacketed incubator, incubator room, waterbath, or aluminum block incubator. Incubation units of the aluminum block type shall have culture dishes and tubes that are snug fitting in the block.

h) An ultraviolet sterilizer shall be free from radiation leaks and shall be UV efficiency tested quarterly as described in "Microbiological Methods for Monitoring the Environment," U.S. Environmental Protection Agency, Environmental Monitoring and Support Laboratory, Environmental Research Center, Cincinnati, Ohio 45268 (EPA 600/8-78-017), December 1978. Proper eye protection shall be available for users of the ultraviolet sterilizer. The ultraviolet sterilizer shall not be used as a substitute for an autoclave.

i) A hot plate may be a large or small unit and shall have a selectable temperature control for safe heating of laboratory media and reagents.

j) A refrigerator shall maintain a temperature of between 1° and 4.4°C and shall be equipped with a thermometer located on the top shelf. The thermometer shall be graduated in at least 1°C increments and the thermometer bulb shall be immersed in liquid.

k) An agar-tempering water bath shall be of appropriate size for holding melted medium and shall be thermostatically controlled at 45° ± 1°C.

l) The following standards shall apply to temperature monitoring devices:

1) Glass or metal thermometers shall be graduated in no greater than 0.5°C units for use in 35°C incubators.

2) Glass or metal thermometers shall be graduated in no greater than 0.1° or 0.2°C units for use in 44.5°C waterbaths or aluminum block type incubators.

3) Continuous temperature recording devices shall be sensitive to at least 0.5°C when used on 35°C incubators, and shall be sensitive to at least 0.2°C when used for 44.5°C waterbaths or aluminum block type incubators.

4) An NBS certified thermometer, or one of equivalent accuracy, shall be available for calibration use and shall be accompanied with its certification papers and procedures for use. Unless otherwise specified in this Subpart B, all thermometers and temperature recording devices shall be calibrated against such certified thermometer to within $\pm 1.0°C$ ($\pm 1.8°F$).

5) Each laboratory shall have a maximum registering thermometer in the range of 200° to 400°F (90° to 200°C) graduated in increments no greater than 2°F (1°C).

6) Each laboratory shall use separate thermometers for determining the temperatures of waterbaths, ovens, autoclaves, samples, refrigerators, storage areas, etc.

7) The liquid column of glass thermometers shall have no separations.

m) Optical counting equipment shall include a low power magnification device of the dissecting or stereomicroscope type with a magnification power of 10 to 15 diameters, and an external fluorescent light source for sheen discernment.

n) A mechanical hand tally shall be available for counting colonies on membrane filters or agar pour plates.

o) Where metal loops are used, innoculation equipment shall have loops of 22 to 24 gauge Nichrome, chromel, or platinum-iridium wire; with loop diameters of at least 3 mm.

p) Membrane filter equipment shall be non-leaking, uncorroded, and made of stainless steel, glass, or autoclavable plastic. Metal plating on membrane filter equipment shall not be worn so as to expose base metal.

q) Membrane filters shall be white, grid marked, 47 mm diameter, with 0.45 micron pore size, and made from cellulose ester materials. Another pore size may be used if the manufacturer gives performance data equal to or better than the 0.45 micron membrane filter. Membrane filters shall be autoclavable or presterilized.

r) Absorbent pads shall be of uniform thickness to permit 1.8 to 2.2 ml media absorption and shall be autoclavable or presterilized. Filter paper shall be free from growth inhibitory substances.

s) Forceps used to handle membrane filters and absorbent pads shall have a round tip without corrugations.

(Source: Amended at 7 Ill. Reg. 13523, effective September 28, 1983.)

Section 183.325   Laboratory Glassware, Plastic Ware, and Metal Utensils

The following standards shall apply to glassware, plastic ware, and metal utensils used in the laboratory:

a) Except for disposable plastic ware, items shall be resistant to effects of corrosion, high temperature, and vigorous cleaning operations. Metal utensils made of stainless steel are preferred. Plastic items shall be of clear, inert, non-toxic material and shall retain accurate graduations or calibration marks after repeated autoclaving. Glassware which is used for purposes that may subject it to damage from heat or chemicals shall be of borosilicate glass. All glassware shall be free of chips, cracks, or excessive etching. All volumetric glassware shall be Class A, denoting that it meets Federal Specifications and need not be calibrated before use.

b) Media preparation utensils shall be of borosilicate glass or stainless steel, and shall be clean and free from foreign residues or dried medium.

c) Pipets shall meet the specifications set forth in "Standard Methods for the Examination of Water and Wastewater," 14th Edition, American Public Health Association, (Washington, D.C., 1976), p. 882. Containers for glass pipets shall be of either stainless steel or aluminum. Pipets used for measuring 10 ml samples or less shall be sterile and of glass or plastic. Opened packages of sterile disposable pipets shall be securely resealed between uses.

d) Sterile graduated cylinders with legible graduation marks shall be used for measurement of samples larger than 10 ml, except that membrane filter funnels marked to within an accuracy of ±2.5% may be used in lieu thereof.

e) Culture dishes shall be sterile and shall be of the tight or loose-lid plastic, or loose-lid glass type. In addition, culture dishes shall be of 100 mm × 15 mm or 60 mm × 15 mm size; and shall be clear, flat bottomed, and free from bubbles or scratches or both. Containers for culture dishes shall be of aluminum or stainless steel; or culture dishes shall be wrapped in heavy aluminum foil or char resistant paper. Open packages of sterile disposal culture dishes shall be securely resealed between uses.

f) Culture tubes shall be of borosilicate glass or other corrosion resistant glass, and shall be of sufficient size to contain culture medium, as well as the sample portions employed, without being more than three-fourths full. Culture tube closures shall be snug fitting stainless steel or plastic caps, or loose fitting aluminum caps, or plastic screw caps with non-toxic liners.

g) Dilution bottles shall be of borosilicate glass or other corrosion resistant glass, and shall be free of chips and cracks at the lip. A graduation level shall be distinctly marked on the side of dilution bottles at 99 ml. Dilution bottle closures shall be plastic screw caps with leakproof liners and shall not produce toxic substances during the sterilization process.

h) Sample bottles shall be sterile, of plastic or hard glass, wide mouthed, and of at least 120 ml capacity. Sample bottle closures shall be glass stoppers or screw caps (metal or plastic), capable of withstanding repeated sterilization, with leakproof liners, and shall not produce toxic substances during the sterilization process. Metal caps with exposed bare metal on the inside shall not be used.

Section 183.330   General Laboratory Practices

a) The following standards shall apply to sterlization procedures:
   1) Autoclaving of the following items shall be carried out at 121° ± 1°C for the durations specified below:

Item	Minimum duration of autoclaving
Membrane filters and pads	10 minutes
Carbohydrate-containing media (lauryl tryptose, brilliant green lactose bile broth, etc.)	12–15 minutes
Contaminated materials and discarded tests	30 minutes
Membrane filter assemblies (wrapped), sample collection bottles (empty), and individual glassware items	30 minutes
Rinse water volumes of 500 ml to 1000 ml	45 minutes
Rinse water volumes in excess of 1000 ml	Time adjusted for volume; check for sterility
Dilution water blanks	30 minutes

2) The maximum elapsed time for exposure of carbohydrate-containing media to any heat (from the time of closing the loaded autoclave to unloading) shall be 45 minutes.

3) Membrane filter assemblies shall be sterilized between each sample filtration series. A filtration series ends when 30 minutes or more have elapsed between sample filtrations. A UV sterilizer or boiling water may be used on membrane filter assemblies for at least 2 minutes to prevent bacterial carry-over between sample filtrations, but shall not be used as a substitute for autoclaving between sample filtration series.

4) Dried glassware to be sterilized in a hot-air sterilizing oven shall be kept at $175° \pm 5°$ for at least 2 hours.

b) Laboratory pure water, which may be distilled, deionized, or other processed water, shall meet the standards set forth in Section 183.345. Only water determined to be laboratory pure water shall be used for performing bacteriological analyses.

c) Rinse and dilution water shall be prepared in the following manner:

1) Prepare a phosphate buffer solution of potassium dihydrogen phosphate (KH_2PO_4) with laboratory pure water as specified in "Standard Methods for the Examination of Pure Water and Wastewater," 14 Edition, American Public Health Association, (Washington, D.C., 1976), p. 892.

2) The phosphate buffer solution shall be autoclaved or filter sterilized, labeled, dated, and stored at 1° to 4.4°C.

3) The stored phosphate buffer solution shall be free of turbidity.

4) Rinse and dilution water shall be prepared by adding 1.25 ml of stock phosphate buffer solution per liter of laboratory pure water, and shall have a final pH of 7.2 ± 0.2.

5) When preparing rinse and dilution water, laboratories analyzing non-potable waters may use magnesium sulfate as specified in "Standard Methods for the Examination of Water and Wastewater," 14th Edition, American Public Health Association, (Washington, D.C., 1976), p. 892, or magnesium chloride as specified in "Microbiological Methods for Monitoring the Environment", U.S. Environmental Protection Agency, (EPA 600/8-78-017), December 1978, in addition to the stock phosphate buffer solution.

d) The following minimum standards shall be met for storing and preparing media:

1) Laboratories shall use commercial dehydrated media for routine bacteriological procedures as quality control measures.

2) All media shall be prepared according to the media specifi-

cations of "Standard Methods for the Examination of Water and Wastewater," 14th Edition, American Public Health Association, (Washington, D.C., 1976), p. 892–902.

3) Dehydrated media containers shall be kept tightly closed and stored in a cool, dry location. Discolored or caked dehydrated media shall not be used.

4) All water used shall be laboratory pure water.

5) Dissolution of the media shall be completed before dispensing to culture tubes or bottles.

6) Membrane filter broths and agar media shall be heated in a boiling water both until completely dissolved.

7) Membrane filter broths shall be stored and refrigerated no longer than 96 hours prior to use. Membrane filter agar media shall be stored in a refrigerator, and used within two weeks after preparation.

8) Most probable number (MPN) media, when prepared in tubes with loose-fitting caps, shall be used within one week after preparation. If MPN media are refrigerated after sterilization, they shall be incubated overnight at 35°C to confirm usability. Tubes of MPN media showing growth, or gas bubbles shall be discarded.

9) MPN media in screw cap containers may be held up to three months, provided the media are stored in the dark and evaporation does not exceed 0.5 ml per 10 ml total volume.

10) Ampuled media such as M-Endo broth and M-FC broth may be used in emergencies and in those laboratories analyzing fewer than 30 microbiological samples from public water supplies per month, provided the ampuled media has been prepared in a microbiological water laboratory certified by the regulatory agency having responsibility for laboratory certification in the States where ampuled media is manufactured.

(*Source*: *Amended at 7 Ill. Reg. 13523, effective September 28, 1983.*)

Section 183.335 Methodology

a) The following methodology, as specified in the listed references, shall be followed for individual parameters:

b) The membrane filter procedure is preferred for the analysis of potable waters, because it permits analysis of large sample volumes in reduced analysis time. The membranes should show good colony

Type of water	Parameter	Methodology	Reference[a] (page no.)
Potable	Total coliforms	Standard total coliform MPN tests[h]	916–919
Potable	Total coliforms	Standard total coliform membrane filter procedure	928–935
Non-potable	Fecal coliforms	Fecal coliform MPN procedure	922
Non-potable	Fecal streptococcal	Multiple-tube technic	943–944
Non-potable	Fecal coliforms	Fecal coliform membrane filter procedure	937–939
Non-potable	Fecal streptococcal	Membrane filter technic	944–945
Potable and non-potable	Bacterial total count	Standard plate count	908–913

[a]"Standard Methods for the Examination of Water and Wastewater," 14th Edition, American Public Health Association, (Washington, D.C., 1976).
[b]Excluding the gram-stain technic.

development over the entire surface. The golden green metallic sheen colonies should be counted and recorded as the coliform density per 100 ml of water sample.

c) The following requirements for reporting any problems with membrane filter results shall be observed:

1) Confluent growth, with or without discrete sheen colonies, covering the entire filtration area of the membrane shall be reported as "confluent growth per 100 ml, with (or without) coliforms," and a new sample requested.

2) If the total number of bacterial colonies cannot be accurately

counted because the colonies on the membrane are too numerous (usually greater than 200 total colonies), not sufficiently distinct, or both, results shall be reported as "TNTC (too numerous to count) per 100 ml, with (or without) coliforms," and a new sample requested.

3) If the membrane exhibits confluent growth and the number of bacterial colonies cannot be accurately counted (TNTC), a new sample shall be requested. When the new sample is analyzed, the sample volumes filtered shall be adjusted to apply the membrane filter procedure; otherwise, the MPN procedure shall be used.

4) If the laboratory has elected to use the MPN test on water supplies that have a continued history of confluent growth and bacterial colonies that cannot be accurately counted, all presumptive tubes with heavy growth without gas production shall be submitted to the confirmed MPN test to check for the suppression of coliforms. A count shall be adjusted based upon confirmation and a new sample requested. This procedure shall be carried out on one sample from each problem water supply once every three months.

Section 183.340   Sample Collecting, Handling and Preservation

When the laboratory has been delegated responsibility for sample collecting, handling, and preservation, there shall be strict adherence to correct sampling procedures, complete identification of the sample, and prompt transfer of the sample to the laboratory as specified in "Standard Methods for the Examination of Water and Wastewater," 14th Edition, American Public Health Association, (Washington, D.C., 1976), pp. 904–907. In addition, the following standards for sample collecting, handling, and preservation of potable water samples shall be met:

a) In order for the sample to be representative of the potable water system, the sampling program shall include examination of the finished water at selected sites that systematically cover the distribution network.

b) Minimum sampling frequency shall be as specified in 35 Ill. Adm. Code 605.102 (prior to codification Table III of the Illinois Pollution Control Board Rules and Regulations, Chapter 6: Public Water Supply).

c) Water shall be sampled from cold water taps that are free of aerators, strainers, hose attachments, and water purification devices. Prior to sampling, a steady flow of water shall be maintained from the tap for 2 to 3 minutes to clear the service line.

d) The sample bottle shall be filled allowing at least one-quarter inch of air space from the top to provide space for mixing. A minimum sample volume of 100 ml shall be collected.

e) The sample report form shall be completed immediately after collecting the sample and shall contain complete information as specified in the "Sample Collector's Handbook," Illinois Environmental Protection Agency, (October 1978), pp. IA-6 through IA-11.

f) Sample bottles shall be of at least 120 ml capacity, of sterile plastic or hard glass, wide mouthed with glass stopper or screw cap (metal or plastic), and capable of withstanding repeated sterilization. Metal caps with exposed bare metal on the inside shall not be used. When samples are to be collected from chlorinated water supplies, sodium thiosulfate shall be added to the sample bottles in an amount sufficient to provide an approximate concentration of 100 mg per liter of sample prior to sterilization of the sample bottles. As an example, 0.1 ml of a 10 percent sodium thiosulfate solution is required for a 120 ml sample bottle.

g) The following information shall be added to the sample report form when the sample is delivered to the laboratory:
   1) Date and time of sample arrival; and
   2) Name or initials of the person receiving the sample for the laboratory.

h) Records necessary to establish chain-of-custody of the samples shall be maintained.

i) Samples delivered by collectors to the laboratory shall be analyzed on the day of arrival and no later than 48 hours after collection (preferably within 30 hours after collection).

j) Where it is necessary to send water samples by mail, bus, United Parcel Service, courier service, or private shipper, elapsed time between sampling and analyses should not exceed 30 hours. Without exception, samples arriving more than 48 hours after collection shall be refused and a new sample requested.

k) Samples of potable water for standard plate count analysis shall be refrigerated and delivered to the laboratory within 6 hours after collection.

Section 183.345   Standards for Laboratory Pure Water

The following standards shall apply to all laboratory pure water:
  a) Laboratory pure water shall have these characteristics:

Property	Value
pH	5.5–7.5
Conductivity	Less than 5.0 micromhos/cm (resistivity greater than 0.2 megohm cm) $\pm$ 1 percent at 25°C
Trace metals:	
Individual metals	Less than or equal to 0.05 mg/l
Total metals	Less than or equal to 1 mg/l
Test for bactericidal properties of distilled water	Ratio of 0.8 to 3.0
Free chlorine residual	None
Standard plate count	Less than 1,000/ml

  b) Laboratory pure water shall be analyzed annually by the test for bacteriological quality of distilled water as specified in "Standard Methods for the Examination of Water and Wastewater," 14th Edition, American Public Health Association, (Washington, D.C., 1976), pp. 888–891. Only satisfactorily tested water shall be used in preparing media, reagents, rinse, and dilution water. If the water tested does not meet the requirements, corrective action shall be taken and the water retested.
  c) Laboratory pure water shall be analyzed monthly for conductance, pH, chlorine residual, and standard plate count. Standard plate counts shall be performed as specified in "Standard Methods for the Examination of Water and Wastewater," 14th Edition, American Public Health Association, (Washington, D.C., 1976), pp. 908–913. If the water tested exceeds requirements for these properties, corrective action shall be taken and the water retested.
  d) Laboratory pure water shall not be in contact with heavy metals, and shall be analyzed initially and annually thereafter for trace metals (especially Pb, Cd, Cr, Cu, Ni, and Zn). If the water tested

exceeds requirements for trace metals, corrective action shall be taken and the water retested.

e) The following quality control tests for standard plate count shall be utilized:

1) Sterility controls shall be poured for each bottle of sterile, melted, tempered medium used.

2) Sterility of pipets and petri dishes shall be determined.

3) Microbial density of the air duing plating procedures shall be determined for each series of samples plated. When 15 or more colonies appear on an exposed plate after a 15 minute exposure period and 48 hours of incubation at 35°C, corrective action shall be taken.

(Source: Amended at 7 Ill. Reg. 13523, effective September 28, 1983.)

Section 183.350   General Quality Control Procedures

a) A written description of the current laboratory quality control program shall be maintained and made available to analysts in an area of the laboratory where analytical work takes place. A record of analytical quality control tests and quality control checks on media, materials, and equipment shall be prepared and retained for 5 years.

b) A laboratory manual containing complete written instructions for each parameter for which the laboratory is certified shall be maintained and made available to analysts in an area of the laboratory where analytical work takes place.

c) The following minimum requirements shall apply to analytical quality control tests for general laboratory practices and methodology:

1) At least 10 sheen or borderline sheen colonies shall be verified from each membrane containing 10 or more such colonies. (A positive sample for total coliform consists of one or more verified positive colonies by membrane filtration.) All sheen or borderline sheen colonies up to 10 on each membrane shall be verified. Counts shall be adjusted based on verification. The verification procedure shall be conducted by transferring growth from colonies into lauryl tryptose broth (LTB) tubes and then transferring growth from gas-positive LTB cultures to brilliant green lactose bile broth (BGLB) tubes. Colonies shall not be transferred exclusively to BGLB because of the lower recovery of stressed coliforms in this more selective medium. However, colonies may be transferred to LTB and BGLB simultaneously.

If negative, LTB tubes shall be reincubated a second day and confirmed if gas is produced.

2) A start and finish membrane filtration control test of rinse water, medium, and supplies shall be conducted for EACH FILTRATION SERIES. If sterile controls indicate contamination, all data on samples affected shall be rejected and a request made for immediate resampling of those waters involved in the laboratory error.

3) The MPN test shall be carried to completion, except for gram staining, on 10 percent of positive confirmed samples. (A positive sample for total coliform consists of one or more positive confirmed tubes by MPN.) If no positive tubes result from the potable water sample, the completed test except for gram staining shall be performed quarterly on at least one positive source water.

4) When quality control samples are available, each approved analyst shall analyze at least one per year for the parameters measured.

5) When unknown performance evaluation samples are available, each approved analyst shall analyze at least one per year for the parameters measured. When performance evaluation sample results indicate technical error, the Agency will provide appropriate technical assistance, and followup performance evaluation samples shall be analyzed by the laboratory.

6) Each approved analyst shall monthly verify fecal coliform analyses by picking at least 10 isolated colonies from membranes containing typical blue colonies and transferring to lauryl sulfate broth. The tubes shall be incubated at $35° \pm 0.5°C$ for 24 and 48 hours, and read for gas production. Growth from positive tubes shall be transferred to EC broth and incubated at $44.5° \pm 0.2°C$ for 24 hours. Gas production in EC broth verifies fecal coliform organisms.

7) Each approved analyst shall monthly verify analyses for fecal streptococci by picking at least 10 isolated pink to red colonies and transferring to brain heart infusion (BHI) agar and broth. The catalase test shall be performed on 24 hour cultures that have been incubated at $35° \pm 0.5°C$, with catalase negative cultures (possible fecal streptococci) transferred to 40 percent bile BHI broth and incubated at $35° \pm 0.5°C$. Also, catalase negative cultures shall be transferred to BHI broth and incubated at $45° \pm 0.5°C$. Growth at both temperatures verifies fecal streptococci.

8) If there is more than one analyst in the laboratory, at least once per month each analyst shall perform parallel analyses on at least one positive sample in order to compare performance between analysts.
9) The standards for laboratory pure water speciifed in Section 183.345 shall be met.

(Source: Amended at 7 Ill. Reg. 13523, effective September 28, 1983.)

Section 183.355   Quality Control for Media, Equipment, and Supplies

The following minimum requirements shall apply to quality control checks of laboratory media, equipment, and supplies:

a) The pH meter(s) shall be clean and calibrated each use period with pH 4 and pH 7, or pH 7 and pH 10 standard buffers. Each buffer aliquot shall be used only once. Commercial buffer solutions shall be dated on initial use.
b) Balances shall be calibrated at least annually using standardized Class "S" or "S-1" weights and rechecked as required.
c) Glass thermometers or continuous temperature recording devices for incubators shall be checked at least annually for accuracy and metal thermometers shall be checked at least quarterly for accuracy against an NBS certified thermometer, or one of equivalent accuracy.
d) Temperature in incubation equipment shall be recorded continuously by a temperature recording device or recorded twice daily (at times separated by at least 4 hours) from in-place thermometers immersed in liquid and placed on shelves. Temperature readings from walk-in incubators with a continuous temperature reading device shall be supplemented by readings from in-place thermometers placed on various shelves other than where the recorder probe is located.
e) Date, time, duration, and temperature of autoclaving shall be recorded continuously or recorded for each sterilization cycle. A list of materials sterilized in each cycle shall als be maintained and shall be initiated by the person(s) involved.
f) Hot air oven(s) shall be equipped with a thermometer registering up to at least 180°C, or with a temperature recording device. Date, time, duration, and temperature shall be recorded for each sterilization cycle. A list of materials sterilized in each cycle shall also be maintained an shall be initialed by the person(s) involved.

g) Only membrane filters recommended for water analysis by the manufacturer shall be utilized. Manufacturer data sheets containing information as to lot number, ink toxicity, recovery, retention, and absence of growth promoting substances for membrane filters shall be entered into the laboratory's record system.

h) Washing processes shall provide clean glassware with no stains or spotting. With initial use of a detergent or washing product and annually thereafter, the rinsing process with distilled or deionized water shall be demonstrated to provide glassware free of toxic material based on the Inhibitory Residue Test as specified in "Standard Methods for the Examination of Water and Wastewater," 14th Edition, American Public Health Association, (Washington, D.C., 1976), p. 885.

i) Each batch of clean, dried glassware or plastic ware shall be tested for residual alkaline or acid residue using bromythymol blue indicator. If the results of the indicator test are not within the desired color range of dark green to light blue, corrective action shall be taken by rinsing, then air drying and retesting.

j) At least one bottle per batch of sterilized sample bottles shall be checked for sterility by adding approximately 25 ml of sterile non-selective broth media to each bottle. The bottle shall be capped and rotated so that the broth comes in contact with all surfaces and shall be incubated at 35° ± 0.5°C for 24 hours prior to checking for growth. Prepared sample bottles from each batch shall not be used unless satisfactory results are obtained from the tested bottle.

k) At least one bottle per batch of sterilized sample bottles prepared with sodium thiosulfate shall be checked for sufficient amount of the dechlorinating reagent by properly collecting a potable sample at the laboratory tap, then checking for residual chlorine. Corrective action shall be taken if there is any residual chlorine, and bottles from the batch checked shall not be used until corrective action has been completed.

l) Current service contracts or in-house protocols shall be maintained on balances, autoclaves, hot-air sterilization ovens, water stills, deionizers, reverse osmosis apparatus, water baths, incubators, etc. Service records on such equipment shall include the date, name of the servicing person, and a description of the service provided.

m) Records shall be available for inspection on all batches of sterilized media showing lot numbers, date, sterilization time and temperatures, final pH, and name of the person(s) responsible for all or any part of the recorded data.

n) Positive and negative cultures, or a natural water of known pollution, shall be used on each new lot of medium to determine performance compared to a previous acceptable lot of medium.

o) Lot numbers of membrane filters and date of receipt shall be recorded.

p) Heat sensitive tapes, spore strips, or ampules shall be used weekly along with a maximum registering thermometer to verify sterilization temperatures within autoclaves and hot-air sterilizing ovens. A complete record of the results of heat sensitive tapes, spore strips or ampules, and maximum registering thermometers shall be maintained; and shall include the date, materials sterilized, and name of the person(s) involved.

q) When media dispensing apparatus is used, the media preparer shall check the accuracy of dispensing with a graduated cylinder at the start of each volume change and periodically throughout extended runs.

r) The refrigerator temperature shall be determined daily and the unit cleaned at least monthly. Outdated materials in the refrigerator and freezer compartments shall be discarded.

s) Ultraviolet sterilization lamps shall be tested quarterly by exposing agar spread plates containing 200 to 250 microorganisms to the light or two minutes. If such irradiation does not reduce the count of control plates by 99 percent, the lamps shall be replaced. Cleaning of ultraviolet sterilization lamps shall be done at least monthly by disconnecting the unit and cleaning the lamps with a soft cloth moistened with ethanol.

t) Water baths shall be cleaned at least monthly. The use of distilled or deionized water for water baths is recommended.

u) It is recommended that microscopes be covered when not in use, and that lens paper be used to clean optics and stage after every use.

v) Media shall be used on a first in, first out basis. Records shall be kept of the kind, amount, date received, and date opened for bottles of media. Bottles of media shall be used within 6 months after opening, except that media sorted in a desiccator may be used up to one year after opening. It is recommended that media be ordered in quantities to last no longer than one year, and that media be ordered in quarter pound multiples rather than one pound bottles in order to keep the supply sealed and protected as long as possible.

(Source: Amended at 7 Ill. Reg. 13523, effective September 28, 1983.)

Section 183.360 Data Handling

a) All records shall be initialed or signed by the person or persons responsible for recording all or any part of the data, or performing the various tests.

b) Either each unit shall be responsible for maintaining its own records, or all records shall be maintained in a general laboratory log book.

c) The laboratory shall record arrival time and date received in the laboratory, time and date of analysis, direct count, membrane filtration verified count, MPN completed count, analyst's name, and other special information on each sample report form.

d) A careful check shall be made to verify that each result is entered accurately from the bench sheet onto the sample report form. The sample report form shall be initialed or signed by the person who verified the entry of information from the bench sheet.

Section 183.365 Record Maintenance

a) A copy of the sample report form shall be maintained by the laboratory for at least five years. If results are entered into a computer storage system, a printout of the data shall be returned to the laboratory for verification with bench sheets.

b) Records of bacteriological analyses shall be kept for at least five years. Actual laboratory reports may be kept. However, data may be transferred to tabular summaries which shall include the following information:
   1) Date, place, and time of sampling;
   2) Name of person who collected the sample;
   3) Identification of the sample origin, such as routine distribution sample, resample, construction sample, raw or process water sample, surface or ground water sample, or other special purpose sample;
   4) Date and time of receipt in the laboratory;
   5) Records necessary to establish chain-of-custody of the sample;
   6) Date and time of sample analysis;
   7) Name of the persons and designation of the laboratory responsible for performing the analysis;
   8) Designation of the analytical techniques or methods used; and
   9) Results of the analysis.

c) The disposal of all records subject to the Local Records Act (Ill.

Rev. Stat. 1981, ch. 116, pars. 43.101 et seq.) must be in accordance with the provisions of that Act.

Section 183.370 Action Response to Laboratory Results

For laboratory results concerning samples from public water supplies and their sources, presumptive positive microbiological test results are to be reported to the requesting facility as preliminary without waiting for membrane filtration verification or MPN completion. After membrane filtration verification or MPN completion or both, the adjusted counts shall be reported.

(Source: Amended at 7 Ill. Reg. 13523, effective September 28, 1983.)

SUBPART D: RADIOCHEMICAL ANALYSES OF PUBLIC WATER SUPPLY SAMPLES

Section 183.405 Scope and Applicability

This Subpart D establishes standards applicable to environmental laboratories involved in radiochemical analyses of samples of water from public water supplies and their sources.

Section 183.410 Personnel

a) The laboratory director shall be a person holding a minimum of a bachelor's degree in natural or physical sciences with at least 24 semester hours in chemistry or microbiology or both, and shall have had a minimum of three years experience in an environmental laboratory. The laboratory director shall be either a full-time employee or a consultant.

b) A senior analyst is a full-time employee holding a minimum of a bachelor's degree in chemistry and having had at least one year of experience in low level radiation measurements and in the radiochemical procedures performed by the laboratory. Senior analysts shall be responsible for all radiochemical procedures performed in the laboratory.

c) An analyst is a person holding a high school diploma or its equivalent and having had a minimum of six months of training or experience or both in routine radiochemistry. Analysts shall be under direct supervision and shall perform only routine procedures which require a minimal exercise of independent judgement.

d) An analyst trainee is a person holding a high school diploma or its equivalent. During the period of training, analyst trainees shall work under the direct supervision of a senior analyst or an analyst, but shall not exercise independent judgement.

Section 183.415   Physical Facilities

The laboratory's physical facilities shall meet the following specifications:

a) A minimum of 150 square feet of floor space shall be provided for each analyst.

b) A minimum of 15 liner feet of usable bench space shall be provided for each analyst.

c) In areas where radioactive standards are prepared, bench tops shall be of an impervious material which may be covered with disposable absorbent paper, or impervious trays lined with absorbent paper shall be available.

d) The laboratory shall include a sink with hot and cold running water. All water supply outlets shall be protected by approved vacuum breakers.

e) An adequate electrical supply for operation of instruments and mechanical needs shall be provided. The certification officer may require verification from an official inspector or other qualified person that the laboratory meets local and national electrical codes.

f) All electrical outlets shall be properly grounded.

g) Instruments shall be properly grounded with an internal or external regulated power supply available to each instrument.

h) All plumbing shall meet local and state plumbing codes. The certification officer may require verification from an official inspector or other qualified person that the laboratory meets such codes.

i) A natural gas, LP gas, or propane gas supply shall be available.

j) The laboratory shall include a vacuum source.

k) A source of distilled water or deionized water or both shall be readily available.

l) The laboratory shall include at least one fume hood.

m) Counting instruments shall be located in a room separate from all other analytical activities. The temperature of such room shall be maintained between 60°F (16°C) and 80°F (27°C) and shall not vary under normal operating conditions by more than 3°C.

Section 183.420   Laboratory Equipment

Only those instruments that are needed to analyze for the parameters for which the laboratory is being certified are required, but those instruments shall meet the following minimum specifications. A laboratory doing all the analyses described in Section 183.430 shall have, or have access to, all of the equipment listed in this Section with the minimum specifications cited.

a) An analytical balance shall have a precision of ±0.05 mg and a scale readability of 0.1 mg.

b) A pH meter shall have an accuracy of at least ±0.5 units and a scale readability of at least ±0.1 units. The pH meter may be either line/bench or battery/portable operated.

c) A specific ion meter shall have an accuracy and scale readability of at least ±1 mV, and shall have expanded scale millivolt capability. The specific ion meter may be either line/bench or battery/portable operated.

d) A conductivity meter and cell combination, suitable for checking distilled water quality, shall be readable in ohms or mhos, and have a range of up to 2.5 megohm cm (conductivity down to 0.4 micromhos/cm) ±1 percent. The conductivity meter may be either line/bench or battery/portable operated.

e) A drying oven shall be of the gravity convection type. A drying lamp shall be of the infrared type.

f) A desiccator shall be a glass or plastic model, depending upon the particular application.

g) A hot plate may be a large or small unit and shall have a selectable temperature control for safe heating of laboratory reagents.

h) Glassware which is used for purposes that may subject it to damage from heat or chemicals shall be of borosilicate glass. All volumetric glassware shall be Class A, denoting that it meets Federal Specifications and need not be calibrated before use.

i) A muffle furnace shall be automatically controlled with a chamber capacity of at least 2200 cubic centimeters. The maximum operating temperature of the muffle furnace shall be at least 1100°C intermittent and 1000°C continuous.

j) A centrifuge shall be a table model with maximum speed of at least 3000 RPM and 4 × 50 ml capacity.

k) A fluorometer shall be capable of detecting 0.0005 micrograms of uranium.

l) A liquid scintillation system shall have a sensitivity that meets or exceeds the standards specified in 40 CFR 141.25(c) (1982).

m) A gas-flow proportional counting system shall have a detector of the "thin window" type. A minimum shielding equivalent to 5 cm of lead shall surround the detector. A cosmic (guard) detector shall be operated in anticoincidence with the main detector. The system shall be such that the sensitivity will meet or exceed the standards specified in 40 CFR 141.25(c) (1982)

n) A scintillation system designed for alpha counting and used for the measurement of gross alpha activities or radium-226 shall include a Mylar disc coated with a phosphor (silver-activated zinc sulfide) which is placed either directly on the sample or on the face of a photo-multiplier tube and is enclosed in a light-tight container. The system shall also include appropriate electronics (high voltage supply, amplifier, timer, and scaler).

o) The scintillation cell system for the specific measurement of radium-226 by the radon emanation method shall be designed to accept scintillation flasks ("Lucas cells"). The system shall include a light-tight enclosure capable of accepting the scintillation flasks, a detector (phototube), and the appropriate electronics (high voltage supply, amplifier, timers, and scalers). The flasks (cells) required for this measurement shall be either purchased from commercial suppliers or constructed according to the specifications published in Lucas, H. F., "Improved Low-Level Alpha Scintillation Counter for Radon," Rev. Sci. Instrum., 28:680 (1967).

p) A gamma spectrometer system shall include either a sodium iodide (NaI(Tl)) crystal, solid state lithium drifted germanium (Ge(Li)) detector, a pure germanium detector, or a gamma-X photon detector connected to a multichannel analyzer.

   1) If a sodium iodide detector is used, the crystal shall be either a 7.5 cm $\times$ 7.5 cm cylindrical crystal, or, preferably, a 10 cm $\times$ 10 cm crystal. A minimum shielding equivalent to 10 cm of iron shall surround the detector. It is recommended that the distance from the center of the detector to any part of the shield be at least 30 cm. The multichannel analyzer, in addition to appropriate electronics, shall contain a memory of not less than 200 channels and at least one readout device.

   2) If a solid state lithium drifted germanium detector, a pure germanium detector, or a gamma-X photon detector is used, a minimum shielding equivalent to 10 cm of iron shall surround the detector. The multichannel analyzer, in addition to appropriate electronics, shall contain a memory of not less than 2000 channels and at least one readout device.

(Source: Amended at 7 Ill. Reg. 13523, effective September 28, 1983).

Section 183.425 General Laboratory Practices

a) Prior to use, all glassware shall be washed in a warm detergent solution and thoroughly rinsed, first in tap water and then in distilled or deionized water. This cleaning procedure is sufficient for most analytical needs, but the procedures specified for individual parameters shall be referred to for more elaborate precautions to be taken against contamination of glassware.

b) Distilled or deionized water shall have resistivity values of at least 0.5 megohm cm (conductivity less than 2.0 micromhos/cm) at 25°C.

c) When commercially available, "analytical reagent grade" or higher quality chemicals shall be used for all procedures.

d) An enclosed, properly labeled area shall be available for the safe storage of radioactive material.

e) There shall be a designated area within the laboratory for preparation of radioactive standards and samples. Appropriate precautions shall be taken in this area to insure against radioactive contamination. Provision shall be made for safe storage and disposal of radioactive wastes, and for monitoring the work area.

Section 183.430 Methodology and Required Equipment

a) The minimum equipment requirements, methodology, and references for individual parameters are shown in Table 2.

b) When the identification and measurement of radionuclides other than those listed in Section 183.430(a) is required, the following references are to be followed, except in cases where alternative analytical techniques have been approved in accordance with 40 CFR 141.27 (1982):

1) H. L. Kreiger and S. Gold, "Procedures for Radiochemical Analysis of Nuclear Reactor Aqueous Solutions," EPA-R4-73-014, U.S. Environmental Protection Agency, Cincinnati, Ohio, (May 1973); or

2) John H. Harley, ed., "HASL Procedure Manual," USAEC Report HASL 300, ERDA Health and Safety Laboratory, (New York, New York, 1973).

c) For the purpose of monitoring radioactivity concentrations in drinking

Table 2   Methodology,[a] Equipment Requirements, References for Individual Parameters

Parameter	Methodology	Reference (page number)			Major equipment required (or its equivalent)[b]
		SM[c]	ASTM[d]	EPA[e]	
Gross alpha	Proportional accounting or alpha scintillation	598–604	–	1–3	A or B
Gross beta	Proportional counting	598–604	–	1–3	A
Strontium-89, -90	Proportional counting	604–611	–	29–33	A
Radium-226	Scintillation	617–628	–	16–23	D
Radium-228	Proportional counting[f]	–	–	–	A
Total radium	Precipitation	611–616	–	13–15	A
Cesium-134	Gamma spectrometry or proportional accounting	–	636–640	4–5	A or C
Tritium	Liquid scintillation	629–	–	34–37	E
Uranium	Fluorometry	– 681	675–	–	F

[a] Adopted from 40 CFR 141.25 (1982). All other procedures are considered alternative analytical techniques and may be substituted only if approved in accordance with 40 CFR 141.27 (1982).
[b] A = Low background proportional system; B = Alpha scintillation system; C = Gamma spectrometer (NaI(Tl) or Ge(Li)); D = Scintillation cell (radon) system; E = Liquid scintillation system; F = Fluorometer.
[c] "Standard Methods for the Examination of Water and Wastewater," 13th Edition, American Public Health Association, (New York, New York, 1971).
[d] "1975 Annual Book of ASTM Standards, Water and Atmospheric Analysis," Part 31, American Society for Testing and Materials, Philadelphia, Pennsylvania, (1975).
[e] "Interim Radiochemical Methodology for Drinking Water," EPA-600/4-75-008,

water, the required sensitivity of the radioanalysis is defined in terms of a detection limit. The detection limit shall be that concentration which can be counted with a precision of ± 100 percent at 2 times the standard deviation of the net counting rate. The standards for detection limits of radioanalyses are as follows:

1) To determine compliance with maximum allowable concentration levels for radium-226, and radium-228 systems the detection limit shall not exceed 1 pCi/l.

2) To determine the concentration of gross alpha activity (including radium-226, but excluding radon and uranium) the detection limit shall not exceed 3 pCi/l.

3) To determine compliance with maximum allowable concentration levels for beta particle and photon radioactivity from man-made radionuclides the detection limits shall not exceed the following concentrations:

Parameter	Detection Limit
Tritium	1000 pCi/l
Strontium-89	10 pCi/l
Strontium-90	2 pCi/l
Iodine-131	1 pCi/l
Cesium-134	10 pCi/l
Gross beta	4 pCi/l
Other radionuclides[a]	1/10 of applicable limit

[a]As calculated from "Maximum Permissible Body Burdens and Maximum Permissible Concentration of Radionuclides in Air or Water for Occupational Exposure," National Bureau of Standards Handbook 69 as amended August, 1963, U.S. Department of Commerce.

(Source: Amended at 7 Ill. Reg. 13523, effective September 28, 1983).

Environmental Monitoring and Support Laboratory, Environmental Research Center, Cincinnati, Ohio 45268, (1975).

[f] "Prescribed Procedures for Measurement of Radioactivity in Drinking Water," EPA-600/4-80-032. Environmental Monitoring and Support Laboratory, Office of Research and Development, Cincinnati, Ohio 45268, (1980), pages 49–57 (Method 904.0). Alternatively, "A Procedure for the Determination of a 228 Ra," (1981), by I.B. Brooks and R.L. Blanchard (available from the U.S. Environmental Protection Agency, Environmental Monitoring and Support Laboratory, Office of Research and Development, Cincinnati, Ohio 45269) may be utilized.

Section 183.435   Sample Collecting, Handling and Preservation

The following requirements for container types and preservation shall be met for each individual parameter[a]:

Parameter	Preservative[b]	Container[c]
Gross alpha	Conc HCl or HNO_3 to pH less than 2[d]	P or G
Gross beta	Conc HCl or HNO_3 to pH less than 2[d]	P or G
Strontium-89	Conc HCl or HNO_3 to pH less than 2	P or G
Strontium-90	Conc HCl or HNO_3 to pH less than 2	P or G
Radium-226	Conc HCl or HNO_3 to pH less than 2	P or G
Radium-228	Conc HCl or HNO_3 to pH less than 2	P or G
Cesium-134	Conc HCl to pH less than 2	P or G
Iodine-131	NONE	P or G
Tritium	NONE	P or G
Uranium	Conc HCl or HNO_3 to pH less than 2	P or G
Photon emitters	Conc HCl or HNO_3 to pH less than 2	P or G

[a]If a laboratory has no control over these factors, the laboratory director must reject any samples not meeting these criteria and so notify the authority requesting the analyses.

[b]Preservative shall be added to the sample at the time of collection, unless suspended solids are to be measured or unless the concentrated acid specified for preservation cannot be added because of shipping restrictions. If it is necessary to ship the sample unpreserved to the laboratory or storage area, acidification may be delayed up to 5 days. After acidification, samples shall be preserved for a minimum of 16 hours before analysis.

[c]P = Plastic, hard or soft; G = Glass, hard or soft.

[d]If HCl is used to acidify samples to be analyzed for gross alpha or gross beta activity, the acid salts shall be converted to nitrate salts before transfer of samples to planchets.

(Source: Amended at 7 Ill. Reg. 13523, effective September 28, 1983.)

Section 183.440 Quality Control

a) A written description of the current laboratory quality control program shall be maintained and made available to analysts in an area of the laboratory where analytical work takes place. A record of analytical quality control tests and quality control checks on materials and equipment shall be prepared and retained for 5 years.

b) A laboratory manual containing complete written instructions for each parameter for which the laboratory is certified shall be maintained and made available to analysts in an area of the laboratory where analytical work takes place.

c) The laboratory shall participate at least twice per year in those U.S. Environmental Protection Agency intercomparison studies that include parameters for which the laboratory is or desires to be certified. Analytical results shall be within control limits as specified by the U.S. Environmental Protection Agency.

d) The laboratory shall participate at least once per year in an appropriate unknown performance study administered by the U.S. Environmental Protection Agency. Analytical results shall be within control limits established by the U.S. Environmental Protection Agency for each parameter for which the laboratory is or desires to be certified.

e) Operating manuals and calibration protocols for counting instruments shall be available to laboratory personnel.

f) Calibration data and maintenance records on all radiation instruments shall be maintained in a permanent record.

g) The following quality control procedures shall be utilized by the laboratory on a daily basis:

1) To verify internal laboratory precision for a specific analysis, 10 percent or more duplicate analyses shall be performed. If the difference between duplicate analyses exceeds two times the standard deviation of the specific analysis as described in "Environmental Radioactivity Intercomparison Studies Program FY1977," EPA-600/4-77-001, U.S. Environmental Protection Agency, (1977), prior measurements are suspect, calculations and procedures shall be examined, and samples shall be reanalyzed when necessary.

2) When 20 or more specific analyses are performed each day, a performance standard and a background sample shall be measured with each 20 samples. If less than 20 specific analyses are performed each day, a performance standard and a background sample shall be measured along with the samples.

3) Quality control performance charts or records shall be maintained.

h) A current service contract shall be in effect on all analytical balances. Either an electronics technician shall be available or a current service contract shall be in effect for maintenance on all radiation instruments.

i) Standardized Class "S" weights shall be available at the laboratory to make periodic checks on balances.

j) Chemicals shall be dated upon receipt of shipment and replaced as needed or, if earlier, before shelf line has been exceeded.

(Source: Amended at 7 Ill. Reg. 13523, effective September 28, 1983.)

Section 183.445   Record Maintenance

a) Records of radiochemical analyses shall be kept by the laboratory for at least three years. This includes raw data, calculations, quality control data, and reports. Actual laboratory reports may be kept. However, data, with the exception of compliance check samples, as detailed in 40 CFR 141.33(b), may be transferred to tabular summaries which shall include the following information:

1) Date, place, and time of sampling;

2) Name of person who collected the sample;

3) Identification of the sample origin, such as routine distribution sample, check sample, raw or process water sample, surface or ground water sample, or other special purpose sample;

4) Date of receipt of sample;

5) Date of sample analysis;

6) Name of the persons and designation of the laboratory responsible for performing the analysis;

7) Designation of the analytical techniques or methods used; and

8) Results of the analysis.

b) The disposal of all records subject to the Local Records Act (Ill. Rev. Stat. 1981, ch. 116, pars. 43.101 et seq.) must be in accordance with the provisions of that Act.

Section 183.450   Action Response to Laboratory Results

When laboratory results indicate that a maximum allowable concentration of any parameter has been exceeded by a public water supply, the requesting facility shall be notified as soon as possible, but in any event within 48 hours, of the unsatisfactory sample result.

(Source: Amended at 7 Ill. Reg. 13523, effective September 28, 1983.)

Table 2 Methodology and Required Equipment for Chemical Analyses of Public Water Supply Samples

Parameter	Methodology (unfiltered sample)[n]	EPA[a]	SM[b]	USGS[c]	ASTM[d]	Other approved methods
Arsenic	Atomic absorption; furnace technique	206.2	—	—	—	—
	Atomic absorption; gaseous hydride	206.3	301-A-VII	I-1062-78	D2972-78B	—
	Spectrophotometric; silver diethyldithiocarbamate	206.4	404-A or 404-B (4)	—	D2972-78A	—
Barium	Atomic absorption; direct aspiration	208.1	301-A-IV	—	—	—
	Atomic absorption; furnace technique	208.2	—	—	—	—
Cadmium	Atomic absorption; direct aspiration	213.1	301-A-II or -III	—	D3557-78A or -78B	—
	Atomic absorption; furnace technique	213.2	—	—	—	—
Chromium	Atomic absorption; direct aspiration	218.1	301-A-II or -III	—	D3559.78A or -78B	—
	Atomic absorption; furnace technique	218.2	—	—	—	—
Lead	Atomic absorption; direct aspiration	239.1	301-A-II or -III	—	D3559.78A or -78B	—
	Atomic absorption; furnace technique	239.2	—	—	—	—

Table 2 Continued.

Parameter	Methodology (unfiltered sample)[n]	EPA[a]	SM[b]	USGS[c]	ASTM[d]	Other approved methods
Mercury	Manual cold vapor technique	245.1	301-A-VI	—	D3223-79	—
	Automated cold vapor technique	245.2	—	—	—	—
Nitrate	Brucine colorimetric	352.1	419-D	—	D992-71	—
	Spectrophotometric; cadmium reduction	353.3	419-C	—	D3867-79B	—
	Automated hydrazine reduction	353.1	—	—	—	—
	Automated cadmium reduction	353.2	605	—	D3867-79A	—
Selenium	Atomic absorption; furnace technique	270.2	—	—	—	—
	Atomic absorption spectrophotometry; hydride generation	270.3	301-A-VII	I-1667-78	D3859-79	—
Silver	Atomic absorption; direct aspiration	272.1	301-A-II	—	—	—
	Atomic absorption; furnace technique	272.2	—	—	—	—
Fluoride	Potentimetric ion selective electrode	340.2	414-B	—	D1179-72B	—
	Colorimetric method with preliminary distillation	340.1	414-A or -C	—	D1179-72A	—

	Automated complexone method (alizarin fluoride blue)	340.3	603	—	—	380-75WE[e]
	Automated electrode method	—	—	—	—	129-71W[f]
	Colorimetric erichrome cyanine R method	—	—	—	I-3325-78	—
Alkalinity	Electrometric titration (only to pH 4.5) manual or automated, or equivalent automated methods	310.1	—	—	—	—
		310.2	403	—	—	—
Calcium	Atomic absorption; direct aspiration	215.1	301-A-II	—	—	—
	Atomic absorption; furnace technique	215.2	—	—	—	—
	EDTA titrimetric	—	306-C	—	—	—
Copper	Atomic absorption; direct aspiration	220.1	301-A-II	—	—	—
	Atomic absorption; furnace technique	220.2	—	—	—	—
Cyanide	Colorimetric	—	308-B or -C	—	—	—
	Colorimetric with preliminary distillation[g]	335.2	413-D	—	—	—
Hydrogen ion (pH)	Electrometric measurement	150.1	424	—	—	—
Iron	Atomic absorption; direct aspiration	236.1	301-A-II	—	—	—
	Atomic absorption; furnace technique	236.2	—	—	—	—
	Colorimetric	—	310-A	—	—	—

Table 2 Continued.

Parameter	Methodology (unfiltered sample)[n]	EPA[a]	SM[b]	USGS[c]	ASTM[d]	Other approved methods
Manganese	Atomic absorption; direct aspiration	243.1	301-A-II	—	—	—
	Atomic absorption; furnace technique	243.2	—	—	—	—
Sodium	Atomic absorption; direct aspiration	273.1	—	—	—	—
	Flame photometric	—	320-A	—	—	—
Total dissolved (filterable) residue	Glass fiber filtration, 180°C	160.1	208-B	—	—	—
Zinc	Atomic absorption; direct aspiration	289.1	301-A-11	—	—	—
	Atomic absorption; furnace technique	289.2	—	—	—	—
Chlorinated hydrocarbons:	Gas chromatography[h,i]	—	509-A	—	D3086-79	—
Aldrin						
Chlordane						
DDT						
Dieldrin						
Endrin						
Heptachlor						
Heptachlor Epoxide						
Lindane						
Methoxychlor						
Toxaphene						

Chlorophenoxys: 2,4-D 2,4.5-TP	Gas chromatography[h,i]	509-B	D3478-79	—
Trihalomethanes				
Purge and trap	—	—	—	501.1[j]
Liquid/liquid extraction	—	—	—	501.2[k]
Gas chromatography/mass spectrometry	—	—	—	501.3[l]
Corrosivity				
Langelier Index	—	203	—	C400-77[m]
Aggressive Index	—	208-B	—	—
Total filterable residue	160.1	212	—	—
Temperature	—	306-C	D1126-67B	—
Calcium hardness	215.2	403	D1067-70B	—
Alkalinity	310.1	424	D1293-78A or -B	—
pH	150.1			
Chloride: potentiometric method	—	408-C	—	—
Sulfate; turbidimetric method	375.4	427-C	—	—

[a] "Methods of Chemical Analysis of Water and Wastes," U.S. Environmental Protection Agency, Environmental Monitoring and Support Laboratory, Cincinnati, Ohio 45268 (EPA 600/4-79-020), March 1979. Available from ORD Publications, CERI, USEPA, Cincinnati, Ohio 45268. For approved analytical procedures for metals, the technique applicable to total metals must be used.

[b] "Standard Methods for the Examination of Water and Wastewater," 14th Edition, American Public Health Association, (Washington, D.C., 1976).

[c] Techniques of Water-Resources Investigation of the United States Geological Survey, Chapter A-1, "Methods for Determination of Inorganic Substances in Water and Fluoride Sediments," Book 5, 1979, Stock #024-001-03177-9. Available from Superintendent of Documents, U.S. Government Printing Office, Washington, D.C. 20402.

[d] 1982 Annual Book of ASTM Standards, part 31, Water, American Society for Testing and Materials, 1916 Race Street, Philadelphia, Pennsylvania 19103.

[e] "Automated Electrode Method," Industrial Method #380-75WE, Technicon Industrial Systems, Tarrytown, New York, February 1976.

Table 2 Continued.

f"Fluoride in Water and Wastewater," Industrial Method 129-71W, Technicon Industrial Systems, Tarrytown, New York 10591, December 1972.

gAutomated distillation may be substituted. Samples exceeding the maximum allowable concentration levels contained in 35 Ill. Adm. Code 604.202 (prior to codification Table I of the Illinois Pollution Control Board Rules and Regulations. Chapter 6; Public Water Supply) must be done by reference method.

h"Methods for Organochlorine Pesticides and Chlorophenoxy Acid Herbicides in Drinking Water and Raw Source Water," (1978). Available from ORD Publications, CERI, USEPA, Cincinnati, Ohio 45268.

i"Gas Chromatographic Methods of Analysis of Organic Substances in Water," Techniques of Water-Resources Investigation of the United States Geological Survey, Chapter A-3, "Methods for Analysis of Organic Substances in Water," Book 5, 1972, Stock #2401-1227. Available from Superintendent of Documents, U.S. Government Printing Office, Washington, D.C. 20402.

jThe Analysis of Trihalomethanes in Finished Water by Purge and Trap Method," 44 Federal Register 68672-68682. (November 29, 1979). Available from U.S. Environmental Protection Agency, Environmental Monitoring and Support Laboratory, Cincinnati, Ohio 45268.

k"The Analysis of Trihalomethanes in Drinking Water by Liquid/Liquid Extraction," 44 Federal Register 68683-68689, (November 29, 1979). Available from U.S. Environmental Protection Agency, Environmental Monitoring and Support Laboratory, Cincinnati, Ohio 45268.

l"Measurement of Trihalomethanes in Drinking Water by Gas Chromatography/Mass Spectometry and Selected Ion Monitoring," (1982), U.S. Environmental Protection Agency, Environmental Monitoring and Support Laboratory, Cincinnati, Ohio 45268.

m"AWWA Standard for Asbestos-Cement Pipe, 4 in. through 24 in. for Water and Other Liquids," (1977), AWWA C400-77, Revision of C400-75, AWWA, Denver, Colorado.

nAll other methods are considered alternative analytical techniques and may be substituted only if approved in accordance with 40 CFR 141.27 (1982).

(*Source: Amended at 7 Ill. Reg. 13523, effective September 28, 1983.*)

Supplement to Table 2 Methodology and Required Equipment for
Chemical Analyses of Public Water Supply Samples

Parameter	Methodology	Reference EPA[a]
Arsenic	Inductively Coupled Plasma	200.7[o]
Barium	Inductively Coupled Plasma	200.7[o]
Cadmium	Inductively Coupled Plasma	200.7[o]
Chromium	Inductively Coupled Plasma	200.7[o]
Lead	Inductively Coupled Plasma	200.7[o]
Silver	Inductively Coupled Plasma	200.7[o]
Copper	Inductively Coupled Plasma	200.7[o]
Iron	Inductively Coupled Plasma	200.7[o]
Manganese	Inductively Coupled Plasma	200.7[o]
Zinc	Inductively Coupled Plasma	200.7[o]
Sodium	Inductively Coupled Plasma	200.7[o]
Calcium	Inductively Coupled Plasma	200.7[o]
Nitrate	Ion Chromatography	300.0[p]
Chloride	Ion Chromatography	300.0[p]
Sulfate	Ion Chromatography	300.0[p]

[o]"Inductively Coupled Plasma—Atomic Emission Spectrometric Method for
Trace Analysis of Water and Wastes—Method 200.7," available from EPA
Environmental Monitoring and Support Laboratory, Cincinnati Ohio, 45268.
[p]"The Determination of Inorganic Anions in Water by Ion Chromatography—
Method 300.0," additional information on this method is available from the
EPA Environmental Monitoring and Support Laboratory, Cincinnati Ohio,
45268.

APPENDIX E

FEDERAL REGISTER (FDA): GOOD LABORATORY PRACTICE REGULATIONS FOR NONCLINICAL LABORATORY STUDIES

Department of Health, Education, and Welfare, Food and Drug Administration [Docket No. 76N-0400]

AGENCY: Food and Drug Administration.

ACTION: Final Rule.

SUMMARY: The agency is issuing final regulations regarding good laboratory practice in the conduct of nonclinical laboratory studies. The action is based on investigatory findings by the agency that some studies submitted in support of the safety or regulated products have not been conducted in accord with acceptable practice, and that accordingly data from such studies have not always been of a quality and integrity to assure product safety in accord with the Federal Food, Drug, and Cosmetic Act and other applicable laws. Conformity with these rules is intended to assure the high quality of nonclinical laboratory testing required to evaluate the safety of regulated products.

FOR FURTHER INFORMATION CONTACT:
Paul D. Lepore, Bureau of Veterinary Medicine (HFV-102), Food and Drug Administration, Department of Health, Education, and Welfare, 5600 Fishers Lane, Rockville, MD 20857, (301-443-4313).

SUPPLEMENTARY INFORMATION: The Food and Drug Administration (FDA) is establishing regulations in a new Part 58 (proposed as Part 3e) in Title 21 (21 CFR Part 58) regarding good laboratory practice. These constitute the first of a series of regulations concerning investigational requirements which are being developed as a result of the FDA Bioresearch Monitoring Program. Proposed regulations, providing interested persons 120 days to submit comments, were published in the Federal Register of November 19, 1976 (41 FR 51206). In addition, public hearings were held on February 15 and 16, 1977 for the presentation of oral testimony on the proposal. Twenty-two oral presentations were given (transcripts are on file with the Hearing Clerk, Food and Drug Administration), and 174 written comments were received. The comments have been categorized and include the following: manufacturers of regulated products (64), associations (40), medical centers (20), private testing or consulting laboratories (18), educational institutions (15), government agencies (8), individuals (8), and an airport director (1).

In the proposal, regulations were designated as a new Part 3e. This final rule incorporates them into a new Part 58 (21 CFR Part 58). The following redesignation table correlates the new sections with those proposed, and, in most instances, reference to the new sections will be used hereinafter.

New Section	Old Section
Subpart A	
58.1	3e.1
58.3	3e.3
58.10	3e.10
58.15	3e.15
Subpart B	
58.29	3e.29
58.31	....
58.33	3e.31
58.35	3e.33
Subpart C	
58.41	3e.41
58.43	3e.43
58.45	3e.45
58.47	3e.47
58.49	3e.49
58.51	3e.51
58.53	3e.53

Subpart D

58.61	3e.61
58.63	3e.63

Subpart E

58.81	3e.81
58.83	3e.83
58.90	3e.90

Subpart F

58.105	3e.105
58.107	3e.107
58.113	3e.113
Deleted	3e.115

Subpart G

58.120	3e.120
58.130	3e.130

Subpart J

58.185	3e.185
58.190	3e.190
58.195	3e.195

Subpart K

58.200	3e.200
58.202	3e.202
58.204	3e.204
58.206	3e.206
58.210	3e.210
58.213	3e.213
58.215	3e.215
58.217	3e.217
58.219	3e.219

As a part of the overall bioresearch monitoring program that was described in the proposal, a pilot inspection program was carried out to assess the current status of laboratory practice of nonclinical testing facilities to aid in evaluating the relevance of the proposed regulations, and to identify any unanticipated difficulties in implementing an agency-wide monitoring and compliance program for the testing facilities.

The pilot inspection program began in December of 1976 and covered a representative sample of testing facilities. The results of these inspections have been evaluated, and the results of the analysis have been made available to the public as OPE Study 42, "Results of the Nonclinical Toxicology Laboratory Good Laboratory Practices Pilot Compliance Program." Notice of availability of this report was published in the Federal Register of October 28, 1977 (42 FR 56799).

TABLE OF CONTENTS FOR PREAMBLE

Testing Facilities Operation

- Standard Operating Procedures (paragraphs 130 through 145).
- Reagents and Solutions (paragraphs 146 through 149).
- Animal Care (paragraphs 150 through 167).

Test and Control Articles

- Test and Control Article Characterization (paragraphs 168 through 182).
- Test and Control Article Handling (paragraphs 183 through 184.
- Mixtures of Articles with Carriers (paragraphs 185 through 192).

Protocol for and Conduct of a Nonclinical Laboratory Study

- Protocol (paragraphs 193 through 204).
- Conduct of a Nonclinical Laboratory Study Results (paragraphs 205 through 209).

Records and Reports

- Reporting of Nonclinical Laboratory Study Results (paragraphs 210 through 216).
- Storage and Retrieval of Records and Data (paragraphs 217 through 223).
- Retention of Records (paragraphs 224 through 230).

Disqualification of Testing Facilities

- Purpose (paragraph 231).
- Grounds for Disqualification (paragraphs 232 through 233).
- Notice of and Opportunity for Hearing on Proposed Disqualification (paragraphs 234 through 238).
- Final Order on Disqualification (paragraphs 239 through 240).
- Actions upon Disqualification (paragraphs 241 through 242).
- Public Disclosure of Information upon Disqualification (paragraphs 243 through 246).
- Alternative or Additional Actions to Disqualification (paragraph 247).
- Suspension or Termination of a Testing Facility by a Sponsor (paragraphs 248 through 250).
- Reinstatement of a Disqualified Testing Facility (paragraphs 251 through 252).

Conforming Amendments

GENERAL ISSUES

1. Many of the written responses to the proposal were in two parts: a discussion of broad issues and a critique of the regulations by section and paragraph. Over a thousand individual items have been considered.

2. Thirty-two comments requested republication of the proposed regulations as guidelines.

The Commissioner of Food and Drugs advises that publishing guidelines rather than regulations was considered and rejected before publication of the proposal. The question was considered again in preparation of this order, and again rejected. The seriousness of problems encountered in testing facilities demands the use of an approach that will achieve compliance directly and promptly. Only by specifying the requirements for compliance in detailed, enforceable regulations can the Commissioner be assured of the quality and integrity of the data submitted to the agency in support of an application for a research or marketing permit.

3. Some comments objected to the incorporation by reference of other laws, recommendations, and guidelines as being either redundant or without the authority conferred by rulemaking procedures as required by the Administrative Procedure Act. It was also asserted that such incorporation could lead to confusion.

The Commissioner agrees that these regulations should not duplicate regulations and requirements subject to the purview of other agencies. Therefore, reference to animal care provisions of the Animal Welfare Act of 1970 (Pub. L. 91-570) and recommendations contained in Department of Health, Education, and Welfare (HEW) Publication No. (NIH) 74-23 have been deleted from §§ 58.43(a) and 58.90(a) (21 CFR 58.43(a) and 58.90(a)). Also, all provisions that referred to regulations of the Occupational Safety and Health Administration or were concerned with the health and safety of employees have been revised or deleted, i.e., 21 CFR 58.33(a) (by deletion of proposed 21 CFR 3e.31(a)(11)), 21 CFR 58.53(b), 21 CFR 58.81 (by deletion of proposed 21 CFR 3e.81(b)(10)), and 21 CFR 58.120(a) (by deletion of proposed 21 CFR 3e.120(a)(17)). Reference to the regulations of the Nuclear Regulatory Commission has been removed from § 58.49; and proposed § 3e.115, dealing with the handling of carcinogenic substances, has been deleted. In addition, the Commissioner has deleted reference to the various animal care guidelines cited in the proposal.

4. Some comments said the regulations should not be retroactive to

previous studies or those ongoing and should include reasonable transitional provisions for their implementation.

To give nonclinical laboratory facilities adequate time to implement required changes in their organization and physical plant, a period of 180 days after publication in the Federal Register is provided for these regulations to become fully effective. The regulations are not retroactive. All studies initiated after the effective date shall be subject to the regulations. The remaining portions of studies in progress on the effective date of the regulations shall be conducted in accordance with these regulations.

5. A number of comments challenged the general legal authority of FDA to issue good laboratory practice regulations. Other comments challenged the legal authority to require record retention or quality assurance units, or to specify the content of required records or location of storage.

The Commissioner finds that the authority cited in the preamble to the proposal (41 FR 51219; Nov. 19, 1976) provides a sound legal basis for the regulations. Although many matters covered in these regulations are not explicitly mentioned in any of the laws administered by the Commissioner, the Supreme Court has recognized, in *Weinberger* v. *Bentex Pharmaceuticals, Inc.*, 412 U.S. 645, 653 (1973), that FDA has authority that "is implicit in the regulatory scheme, not spelled out in *haec verba*" in the statute. As stated in *Morrow* v. *Clayton*; 326.F.2d 36, 44 (10th Cir. 1963).

However, it is a fundamental principle of administrative law that the powers of an administrative agency are not limited to those expressly granted by the statutes, but include, also, all of the powers that may be fairly implied therefrom.

See *Mourning* v. *Family Publications Service*, Inc., 411 U.S. 356 (1973); see also *National Petroleum Refiners Association* v. *F.T.C.*, 482.F.2d 672 (D.C. Cir. 1973). The Commissioner concludes that there is ample authority for the promulgation of good laboratory practice regulations. No comment presented any explanation or information to the contrary, let alone a cogent argument that FDA lacks legal authority under existing statutes. The standards prescribed represent amplification of the legal requirements regarding evidence of safety necessary to approve an application for a research or marketing permit and parallel, to a great extent, steps that FDA has found have been taken by members of the regulated industry to improve nonclinical laboratory operations.

6. One comment argued that the opinion of the Court of Appeals in *American Pharmaceutical Association* v. *Weinberger*, 530 F.2d 1054 (D.C. Cir. 1976), should be read to limit FDA's authority to issue regulations under section 701(a) of the act (21 U.S.C. 371(a)).

The Commissioner disagrees with the argument advanced in the comment. As discussed in the preamble to the proposed regulation, the agency's authority to issue regulations under section 701(a) of the act has been upheld by the courts. (See *Weinberger* v. *Hynson, Westcott & Dunning, Inc.*, 412 U.S. 609 (1973); see also *National Confectioners Association* v. *Califano*, No. 76-1617 (D.C. Cir. Jan. 20, 1978); *Upjohn Co.* v. *Finch*, 422 F.2d 944 (6th Cir. 1970); *Pharmaceutical Manufacturers Association* v. *Richardson*, 318 F. Supp. 301 (D. Del. 1970).) The question is not FDA's authority to issue regulations under section 701(a) of the act per se, but whether regulations issued under section 701(a) of the act appropriately implement other sections of the act. As articulated in the original proposal, and as discussed in the previous two paragraphs, the Commissioner has determined that these regulations are essential to enforcement of the agency's responsibilities under sections 406, 408, 409, 502, 503, 505, 506, 507, 510, 512, 513, 514, 515, 516, 518, 519, 520, 706, and 801 of the Federal Food, Drug, and Cosmetic Act, as well as the responsibilities of FDA under sections 351 and 354-360F of the Public Health Service Act.

7. A number of comments said various sections of the act did not specify the submission of safety data or did not deal with "applications for research or marketing permits."

The Commissioner has reviewed the comments and finds that the comments are based on a misunderstanding of the phrase, "applications for research or marketing permits." This concept is discussed in relation to § 58.3(e) below. Each cited provision contains authority for FDA either to require submission of, or to use, nonclinical safety data to justify a decision to approve the distribution of a regulated product.

8. A number of comments said the cost of implementing the proposed regulations would be prohibitive to smaller testing laboratories and would, at the least, result in a substantial increase in the cost of product testing.

The Commissioner agrees that implementation of these regulations will increase the cost of nonclinical laboratory testing. The Commissioner finds, however, that such costs are justified on the basis of the resultant increase in the assurance of the quality and integrity of the safety data submitted to the agency. The agency has previously concluded (see the Federal Register of November 19, 1976 (41 FR 51220)) that this document does not contain regulations requiring preparation of an inflation impact statement under Executive Orders 11821 and 11929, Office of Management and Budget Circular A-107 and the guidelines issued by the Department of Health, Education, and Welfare. For a notice on the availability of the agency's economic impact assessment regarding rules for good laboratory practice for nonclinical laboratory studies, see the Federal Register of February 7, 1978 (43 FR 5071). The revisions in this final rule, along with the findings of the pilot program, which showed that many of the inspected facilities were already substantially in compliance with the proposed regulations, should allay some of the concerns of small facilities regarding cost or feasibility of compliance.

9. Many comments suggested changes in language, grammar, terminology, punctuation; sentence structure, and other editorial changes to clarify or improve upon the requirements as stated in the regulations or to eliminate redundancies or inconsistencies. Comments that raised significant policy questions, suggested changes in the substance of the regulation, or otherwise required, in the Commissioner's opinion, a specific response, are discussed individually below. Many of the suggested changes, however, were editorial and stylistic and do not warrant a detailed discussion.

The Commissioner has reviewed each of these numerous editorial and language changes to determine whether it offered an improvement in clarity or definition, eliminated an obvious error or redundancy, promoted consistency with other portions of the regulations, or otherwise identified textual problems that had not been previously noted by FDA. Where the proposed alternative language or other changes suggested by the comments were superior to the proposal, they were adopted in substance or verbatim. Where they did not offer any improvement, the Commissioner declined to accept them.

GENERAL PROVISIONS

Scope

10. Numerous comments addressed the stated scope of the proposed regulations (§ 58.1). Six comments said the proposed scope was vague. Ten comments said the scope should be limited to long-term animal toxicity studies. Twenty-two comments indicated that the scope should be limited to animal safety studies to be submitted to FDA. Individual comments recommended limiting the scope to studies performed on marketed products, studies performed on animals and other biological test systems or studies submitted in support of a color additive petition, food additive petition, investigational new drug application, new drug application, or new animal drug application.

In the preamble to the proposed regulations, the Commissioner set forth the reasons for the broad terminology employed in the statement of scope, stating "these regulations are intended to ensure, as far as possible, the quality and integrity of test data that are submitted to FDA and become the basis for regulatory decisions made by the Agency." In the proposed rule (41 FR 51210), the Commissioner specifically invited comments on which laboratories and/or studies should be subject to the regulations, and further, on whether the scope of the regulations should be defined in terms of the type of testing facility rather than the type of study performed. Based on the review of the comments, the Commissioner has chosen to describe the scope of the regulations in language that is only slightly changed form the proposal. Further clarification of scope is achieved by the specific definition of the key terms, "nonclinical laboratory study" and "application for research or marketing permit" in § 58.3. Taken together, these provisions eliminate any vagueness in the scope of these regulations.

The Commissioner has rejected the request to narrow the scope by listing in the regulation specific types of studies covered. Any such list, if it included all types of studies used by the agency to assess the safety of all the products it regulates, would be cumbersome and might exclude specific types of studies that could become important to future safety decisions. the Commissioner emphasizes that this decision does not mean, however, that the scope of the regulations is unlimited. The scope of the GLP regulations is limited in several ways.

First, they apply only to nonclinical laboratory studies that are submitted or are conducted for submission to the agency in support of a research or marketing permit for a regulated product. Language has been added that provides that the scope includes studies "intended" to support applications for research or marketing permits. This language was included in the preamble to the proposed regulation (41 FR 51209), and the Commissioner has added the language to the regulation because it helps to make clear in advance when a study should comply with the regulation and when a study should be listed on a testing facility's master schedule sheet as a nonclinical laboratory study subject to these regulations (§ 58.35(b)(1)). Tests never intended to be submitted to the agency in support of (i.e., as the basis for) the approval of a research or marketing permit, such as exploratory safety studies and range-finding experiments, are not included even though they may be required to be submitted as part of an application or petition.

Second, the definition of "nonclinical laboratory study" (§ 58.3(d)) makes it very clear that studies utilizing human subjects, clinical studies, or field trials in animals are not included.

Third, the scope of coverage is now limited to safety studies, i.e., those which can be used to predict adverse effects of, and to establish safe use characteristics for a regulated product. "Functionality studies" have been excluded in the final rule.

Fourth, the definition of "test system" (§ 58.3(i)) taken together with the definition of "nonclinical laboratory study" makes it clear that the scope of coverage is confined to studies performed on animals, plants, microorganisms or subparts thereof.

Products regulated by the agency, for which safety data may be required, cover a wide range of diverse items that pose quite different types of risk.

Examples include implantable medical devices; indirect food additives which may occur in food in very small quantities; direct food additives which may be consumed on a daily basis in larger quantities; human drugs intended for prescription or over-the-counter use; animal drugs intended for use in pets and other companion animals of social importance, drugs used in food-producing animals (drug residues can become a part of food); radiation products used in the diagnosis and/or treatment of a disease or condition; radiation products (e.g., microwave ovens and television sets) widely used by the public; vaccines; and blood components and derivatives.

The guarantee of the safety of each of these product classes requires conducting a broad spectrum of safety tests, all of which should be subject to the same standards. Therefore, the Commissioner rejects the proposal to limit the scope of these regulations to long-term animal toxicity studies. Median lethal dose (LD_{50}) and other short-term tests are covered by the regulations because they may serve as part of the basis for approval of, for example, use of an indirect food additive or an investigational new drug in man.

In vitro biological tests are included insofar as such tests have a bearing on product safety, even though they are not now used in agency decisions, because they may in the future become important indicators of safety. Examples of such tests include short-term mutagenicity tests as well as various other tissue culture and organ tests.

Also included in the scope of these regulations are studies of safety of regulated products on target animals, acute toxicity studies on a final product formulation, studies of test articles that are completed in 14 days or less, studies conducted on test articles used in "minor food producing species of animals," and studies on test articles which are not widely used.

11. Several comments closely related to the concerns expressed in paragraph 10 of this preamble requested that further language be added to the regulation exempting certain specific types of studies from coverage.

The Commissioner has reviewed the requests and has chosen not to change the language of the regulation itself to exclude specific study types other than those already mentioned (e.g., studies utilizing human subjects). The regulations apply to any study conducted to provide safety data in support of an application for a research or marketing permit for an FDA-regulated product, and a specific type of study which may be important in the overall safety evaluation of one type of regulated product may not be important in evaluating another. The Commissioner believes it useful to identify in this preamble further examples of studies that are—or are not—within the scope of the GLP regulations.

Examples of studies that are not within the scope of these GLP regulations include:

a. Clinical tests performed solely in conjunction with product efficacy.

b. Chemical assays for quality control.

c. Stability tests on finished dosage forms and products.

d. Tests for conformance to pharamcopeial standards.

e. Pharmacological and effectiveness studies.

f. Studies to develop new methodologies for toxicology experimentation.

g. Exploratory studies on viruses and cell biology.

h. Studies to develop methods of synthesis, analysis, mode of action, and formulation of test articles.

i. Studies relating to stability, identity, strength, quality, and purity of test articles and/or control articles that are covered by good manufacturing practice regulations.

Further examples of types of tests not covered included:

a. Food additives: Tests of functionality and/or appropriateness of the product for its intended use; tests of extractability of polymeric materials that contact food; and all chemical tests used to derive the specifications of the marketed product.

b. Human and animal drugs: Basic research; preliminary exploratory studies; pharmacology experiments; studies done to determine the physical and chemical characteristics of the test article independent of any test system; and clinical investigations.

c. Medical devices: All studies done on products that do not come in contact with or are not implanted in man.

d. Diagnostic products: Essentially all are excluded.

e. Radiation products: Chemical and physical tests.

f. Biological products: All tests conducted for the release of licensed biologicals described in part 601 (21 CFR Part 601) of this chapter.

These examples do not represent all the exclusions from the regulations, but provide guidance in applying the agency's safety considerations to specific situations. The defined scope of the regulations is necessarily broad to encompass the wide range of types of safety tests, types of testing facilities and regulated products for which proper safety decisions are important.

12. More than 20 comments sought the addition of specific language exempting various classes of FDA-regulated products, such as medical devices, from coverage by the regulations.

The Commissioner has generally elected not to permit exemptions based on broad categories of regulated products because no compelling reasons have been presented that would support the contention that assurance of safety is less desirable for one class of regulated products than for another. Proper safety decisions are important for all these products; accordingly, the processes by which such safety data are collected should be subjected to identical standards of quality and integrity.

13. Several comments said that the animal care provisions should apply only to these nonclinical studies using laboratory animals and should not apply to nonclinical studies which involve large animals.

It is clear that the animal care provisions are directed toward the use of laboratory animals, and therefore certain of these provisions may not apply to studies not involving laboratory animals, such as tissue residue and metabolism studies conducted in cattle. Although these studies do fall within the definition of a nonclinical laboratory study, the animals used in such a study are not generally kept in a laboratory setting. Because the husbandry requirements for laboratory animals differ greatly from those for large animals, the agency does not require that large animals be reared and maintained under the same conditions as laboratory animals. The regulations are revised to include terms such as "when applicable" and "as required" in those provisions for which a wide latitude of acceptable husbandry practice exists.

14. Three comments said the regulations should apply to all studies whether submitted in support of or as a challenge to an "application for a research or marketing permit."

The Commissioner agrees, in principle, that all nonclinical studies should be performed in a manner designed to ensure the quality and integrity of the data. FDA is requiring that, at the time a study is submitted, there be included with the study either a statement that the study was conducted in compliance with Part 58 requirements or, if the study was not conducted in compliance with those requirements, a statement that describes in detail all deviations. This requirement means that, at the time a study not conducted in compliance with the requirements is submitted, the agency may evaluate the effects of the noncompliance and take one of the following actions: (1) Determine that the noncompliance did not affect the validity of the study and accept it, or (2) determine that the noncompliance may have affected the validity of the study and require that the study be validated by the person submitting it, or (3) reject the study completely. The standard of review applied to studies that contain data adverse to a product is no different. That is, a study that failed to comply with these regulations might, nonetheless, contain valid and significant data significant data demonstrating a safety hazard. Thus, FDA is not proposing a double standard, but is, rather, seeking to address those studies that present the most serious regulatory problems.

The preamble to the proposed regulation (41 FR 51215) discussed this issue as follows:

Valid data and information in an otherwise unacceptable study which are adverse to the product, however, may serve as the basis for regulatory action.

This disparity in treatment merely reflects the fact that a technically bad study can never establish the absence of a safety risk but may establish the presence of a previously unsuspected hazard. It reflects current agency policy; even in situations where the scientific quality of an investigational drug study is not in question, FDA may receive data but not use it in support of a decision to approve testing or commercial distribution because of ethical improprieties in the conduct of the study. (See 21 CFR 312.20).

A positive finding of toxicity in the test system in a study not conducted in compliance with the good laboratory practive regulations, may provide a reasonable lower bound on the true toxicity of the substance. The agency must be free to conclude that scientifically valid results from such a study, while admittedly imprecise as to incidence or severity of the untoward effect, cannot be overlooked in arriving at a decision concerning the toxic potential of the product. The treatment of studies conducted by a disqualified testing facility is discussed in paragraph 231a, below.

15. Exemptions from coverage by these regulations were requested for various types of facilities. Requests were received that they not apply to academic, medical, clinical, and not-for-profit institutions.

The public health purpose of these regulations applies to all laboratory studies on which FDA relies in evaluating the safety of regulated products, regardless of the nature of the facilities in which the studies are conducted. The Commissioner finds that granting an exemption based on type of facility would frustrate the intent of the good laboratory practice regulations. Many other comments urged that such exemptions not be considered because the standards applied to nonclinical testing should be uniform. Many of the requests for exemption were based on the idea that academic or not-for-profit institutions conduct primarily basic research and ought, therefore, to be specifically excluded. Insofar as academic institutions are concerned, the Commissioner notes that such institutions conduct significant amounts of commercial testing pursuant to contracts. He also notes that significant levels of noncompliance with GLP requirements have been found in such institutions. Moreover, as noted in paragraph 11, basic research on drugs is outside the scope of these regulations. In short, no justification has been presented to warrant granting an exemption to such a facility, and any such exemption from the regulations by the type of facility collecting safety data would not provide equal application of the principles of good laboratory practice. Product safety decisions are equally important whether data are collected by the largest commercial nonclinical laboratory facility or by the smallest nonprofit facility. Therefore, the data collected in all types of facilities should be subjected to the same standards of quality and integrity. The results of the pilot program show that the proposed regulations represent achievable standards.

16. Exemption of or different standards for studies conducted outside the United States were requested.

These regulations are designed to protect the public health of the American people by assuring the scientific integrity and validity of laboratory studies that the agency relies on in evaluating the safety of regulated products. The same assurance is needed, whether the studies relied on are foreign or domestic in origin. The Commissioner notes that FDA clearly may refuse to accept studies from any nonclinical testing facility, foreign or domestic, that does not follow the requirements set forth in these regulations. To exempt from the requirements imposed on studies conducted in domestic testing facilities a nonclinical study conducted in a testing facility outside the United States that is submitted to FDA in support of an application for a research or marketing permit or to impose different standards for such studies, would only have the effect of discriminating against U.S. Firms. Although inspection of a foreign facility may not be made without the consent of that facility, FDA will refuse to accept any studies submitted by any facility that does not consent to inspection. These same conditions apply to other FDA regulations, e.g., the current good manufacturing practice regulations (21 CFR Part 210), a program of inspection of foreign facilities for compliance with those regulations has been conducted by FDA for several years. A similar inspection program of foreign laboratory facilities conducting studies within the scope of this regulation will be conducted; several foreign laboratories were inspected during the pilot program, and mechanisms for such inspections are being worked out with representatives of the responsible regulatory authorities in foreign countries.

Definitions

The Commissioner received hundreds of comments regarding definitions (§ 58.3). General comments are listed immediately below; comments regarding specific definitions follow in numerical order.

17. Several comments asked that commonly used terms such as "batch," "area," "laboratory," "pathologist," "quality data," "data integrity," "supervisor," and "management" be defined or clarified.

The Commissioner finds that, with the exception of "batch," the terms set out above do not require individual definitions. The term "pathologist" is used in its ordinary sense, as are the terms "supervisor" and "management" and the phrases "quality data" and "data integrity." As a general rule, the regulation defines separately only those words which will be used in a sense which differs from that given in currently accepted dictionaries or words whose meaning will be limited by the regulation. A new definition has been added for the term "batch" because it is used in these regulations in a context different from other agency regulations, e.g., the good manufacturing practice regulations. "Batch" in these regulations means a specific quantity of a test or control article that has been characterized according to § 58.105(a).

18. Several comments on § 58.3(b) questioned the applicability of the term "test substance" to medical devices, radiation products in vitro diagnostic products, and botanical materials.

The Commissioner has reviewed the comments carefully and finds that many of the comments submitted regarding the term "test substance" argued that the term, as defined, did not accurately reflect the scope intended to be covered. Because the term "substance," in common usage, refers to chemical compounds and biological derivatives of more or less defined composition, and because the term is not commonly understood to include devices or electronic products, the Commissioner has changed the term "test substance" to "test article." The term "article" is intended to include all regulated products which may be the subject of an application for a research or marketing permit as defined in § 58.3(e).

The Commissioner has deleted the reference to botanical materials because all botanical materials subject to FDA jurisdiction are adequately encompassed by the other articles specifically mentioned in the definition.

19. Clarification of the term "control substance" (§ 58.3(c)) was requested. Several comments asked whether the term was to include carrier substances and solvents and vehicles. Other comments suggested this term could be confused with the same term used by the Drug Enforcement Administration.

The term is changed to "control article" to parallel the revised definition for test article. This change avoids any potential conflict with definitions used by other agencies. The term is intended to define those materials given to control groups of test systems for establishing a basis of comparison. The Commissioner recognizes that for certain nonclinical laboratory studies, no control groups are used, and therefore this definition would not apply. For example, testing the safety of implantable pacemakers in animals would require either no control animals or animals that have only been "sham-operated." The definition includes carrier materials when such carrier materials are given to control groups within test system and likewise for administered vehicles and solvents. The term also applies to articles used as positive controls.

20. Many comments on § 58.3(d) addressed the definition of the term "nonclinical laboratory study." A great many, if not the majority, of the comments sought to change the definition by adding language excluding certain specific tests, products, or types of laboratories.

The Commissioner notes that many of these comments overlap with or are identical to comments submitted in response to § 58.1 (Scope). To the extent that the comments and issues are the same, they have been dealt with in the discussion of § 58.1, above. Other comments are dealt with specifically below.

21. Many comments stated that the proposed language which included studies intended to assess the functionality and/or effectiveness of a test article should be deleted. One comment stated that efficacy testing in nonclinical tests is, by definition, preliminary and should be excluded to be consistent with the scope defined in § 58.1. Other comments stated that the language was too broad and too ambiguous and could be interpreted to include many studies which were not safety studies at all.

The Commissioner has considered these comments and agrees that the language related to functionality and/or effectiveness is too broad. He has, therefore, deleted the sentence.

22. Several comments requested that the last sentence of § 58.3(d) be modified by deleting the proposed examples of tests.

The Commissioner finds that the examples included in the proposal tended to confuse rather than to clarify. The examples, therefore, have been deleted.

23. Section 58.3(e), which defines the various types of submissions to FDA, was criticized for use of the term "application for research or marketing permit." Several comments said the term was misleading because not all products are regulated through the use of "permits."

The Commissioner believes the term is appropriate for the purpose of these regulations. As stated in the proposal, this definition includes all the various requirements for submission of scientific data and information to the agency under its regulatory jurisdiction, even though in certain cases no permission is technically required from FDA for the conduct of a proposed activity with a particular product, i.e., carrying out research or continuing marketing of a product. The term is intended solely as a shorthand way of referring to the separate categories of data (identified in the proposal) that are now, or in the near future will become, subject to requirements for submission to the agency.

24. One comment stated that proposed § 3e.3(e)(14) should be deleted because the language was overly broad and because it contradicted the intent expressed in the preamble to limit GLP regulations to safety studies.

The Commissioner notes that the preamble to the proposal (41 FR 51209) stated that studies conducted to determine whether a drug product conforms to applicable compendial and license standards were excluded from the regulation. Safety data submitted to obtain the initial licensing of a biological product are covered by these regulations in § 58.3(e)(13). Once a biological is licensed, however it becomes subject to testing procedures similar to compendial testing procedures. The Commissioner finds that postlicensing testing of biologicals is conducted more for quality control purposes than for establishing the basic safety of the biologic product and has, accordingly, deleted postlicensing testing from the definition of research and marketing permit.

25. Several comments stated that in vitro diagnostic tests (proposed § 3e.3(e)(15)) should not be included because in vitro diagnostic products do not come in contact with patients and do not, therefore, require preliminary animal safety testing.

Because in vitro diagnostic products do not require any nonclinical laboratory tests for agency approval, the Commissioner agrees that in vitro diagnostic products need not be included in the definition "application for a research or marketing permit." Proposed § 3e.3(d)(15) has, therefore, been deleted from the final regulation.

26. Several comments objected to the inclusion of medical devices in § 58.3(e) (16), (17), and (18), stating that medical devices were not "test substances," that medical devices should not be included because the rules for data submission for such devices were as yet undefined, and that inclusion of medical devices would be unduly restrictive. These comments suggested either total or partial exclusion from coverage under the good laboratory practice regulations.

For reasons stated previously, the Commissioner does not agree that medical devices, as a category, should be excluded. Implantable devices may be composed of polymeric materials that contain components capable of leaching from the device into the body of the recipient or may themselves be adversely affected by body constitutents. In either case, safety studies would be necessary to demonstrate that components of the device did not cause harm or that the body constituents did not promote breakdown or malfunction of the device.

27. Comments also requested deletion of all terms relating to radiation products in § 58.3(e) (20), (21), and (22), stating that to include such products would restrict experimentation unduly, and arguing that radiation products were not "test substances."

The Commissioner rejects these comments. The quality and integrity of the safety data are no less important for radiation products than they are for other agency-regulated products. He does not agree that including radiation products will unduly restrict experimentation. The remaining argument is covered in the discussion of "test article" above. A new paragraph § 58.3(e)(19) is added to cover data and information regarding an electronic product submitted as part of the procedure for obtaining an exemption from notification of a radiation safety defect or failure of compliance with a radiation performance standard, described in Subpart D of Part 1003 (21 CFR Part 1003).

28. Many comments stated that the term "sponsor" in § 58.3(f) was too broadly defined. For example, two comments stated that the definition, as written, would cover a company which provides a grant to a university, a fact which, if true, would inhibit giving grants. One comment said that the definition is so broad that it could be interpreted to apply to stockholders.

The Commissioner advises that a person providing a grant may be a sponsor. In the area of nonclinical laboratory studies, most grantor ultimately submit the data to the agency. The Commissioner does not agree that because the definition of "sponsor" includes grantors it will inhibit the giving of grants. No data were submitted to support this argument. The Commissioner further advises that the definition does not include stockholders.

29. Other comments on § 58.3(f) asked whether the regulation allowed for multiple sponsors and whether government agencies could be sponsors.

"Person," as defined in § 58.3(h), includes government agencies, partnerships, and other establishments such as associations. Therefore, a government agency can clearly be a sponsor. In addition, the Commissioner advises that the definition does not preclude joint sponsorship of a study.

30. Several comments asked that the definition of "testing facility" in § 58.3(g) be revised to indicate clearly that a facility conducting a study subject to the regulations should be subject only to the extent that the facility is involved with and responsible for the study.

The Commissioner concludes that no revision to the definition is necessary. The definition clearly does indicate that a facility is covered by the regulations only to the extent that the facility is conducting or has conducted nonclinical laboratory studies.

31. Numerous comments addressed the definition of "test system" in § 58.3(i). Eighteen comments stated that the definition, as written, could be interpreted to require testing of beakers and test tubes. Two comments pointed out that the "test system" is not the container being tested for extractables but rather it is the animal, microorganism, or cellular components used to test the extractables for safety.

The Commissioner has carefully reviewed the proposed definition in light of the comments and has made a number of changes. The terms "cellular and subcellular" have been replaced for clarity with "subparts therefore" which refers to animals, plants, and microorganisms. The revised definition now reads: " 'Test system' means any animal, plant, microorganism, or subparts therefore, to which the test or control article is administered or added for study. 'Test system' also includes appropriate groups and components of the system not treated with the test or control articles." The revisions should make the definition clearly consistent with § 58.3(d) ("nonclinical laboratory study"), which states that studies to determine physical or chemical characteristics of a test article or to determine potential utility are not included. Therefore, testing of beakers and test tubes, which fall into the category of physical and chemical tests, is excluded.

32. Section 58.3(j), which defines "specimen," drew several comments. These included requests for precise definition of the terms "material" and "tissue" and requests for a clearer definition of the term "specimen."

The Commissioner is modifying the term "specimen" to include any material derived from a test system for examination or analysis. Under these circumstances, blood, serum, plasma, urine, tissues, and tissue fractions are all included if they are intended for further examination or analysis. The definition includes all materials that yield data related to the safety decision on a regulated product.

33. Many comments were received on the definition of "raw data" in § 58.3(k). Included were requests to clarify the term "certified" and to state whether carbons, photocopies, and written reports of dictated material could be classified as "raw data." Other issues concerned whether financial information and first drafts of reports were "raw data."

The Commissioner concludes that the proposed definition should be clarified. The word "exact" is substituted for the word "certified." "Certified" connotes a legal document that requires notarization; "exact" has not such connotation and more precisely reflects the Commissioner's intention. The definition is further clarified by inserting, after the first sentence, a new sentence which reads: "In the event that exact transcripts of raw data have been prepared (e.g., tapes which have been transcribed verbatim, dated, and verified accurate by signature), the exact copy or exact transcript may be substituted for the original source as raw data." This clarification will permit data collection by tape recorders without requiring the retention of the original tapes. Carbons and photocopies satisfy the regulations, provided they are exact and legible copies of the original information. Neither financial information nor first drafts of reports are raw data within the meaning of the term.

34. Several comments said only recorded data contributing substantially to the study should be retained and, similarly, only computer printouts contributing substantially should be retained. Several comments requested clarification of the method for storing machine-generated data and definition of "on line data recording system."

Because the parenthetical example ("derived from on-line data recording systems") served more to confuse than to clarify, it has been deleted. However, an "on line data recording system" pertains to an instrument that can feed data directly into a computer that analyzes and stores the information. The product of this activity usually consists of a memory unit plus a computer program for extracting the information from the unit. Hard-copy computer printouts are unnecessary, provided the computer memory and program are accompanied by a procedure that precludes tampering with the stored information.

The Commissioner cannot agree that only those portions of the data that contribute substantially to the study need to be retained. Such an approach would require a judgment to be made which, if in error, could lead to improper or incorrect study reconstruction. The purpose of retaining the raw data is to permit the quality assurance unit and agency investigators to reconstruct each phase of a nonclinical laboratory study. Discarding essential records would frustrate this purpose. Raw data may be stored in separate areas provided the archival indexes give the data location.

35. Many comments addressed "quality assurance unit" in § 58.3(l).

The Commissioner has reviewed these comments and concludes that they are more concerned with the concept of the quality assurance unit than with the definition. The comments are therefore dealt with in detail in that section of the preamble concerned with § 58.35 of the regulations. (See paragraphs 75 through 92 below.)

36. Several comments addressed "study director" in § 58.3(m). These comments requested clarification, permission to have more than one study director per study, and that the term "implementation" be changed to "conduct."

The Commissioner has revised the definition to read: " 'Study Director' means the individual responsible for the overall conduct of a nonclinical laboratory study." The revision is intended to emphasize that the study director is responsible for the entire study, as well as being responsible for the interpretation, analysis documentation, and reporting of results.

The Commissioner concludes that the other comments received on the definition of "study director" addressed the concept rather than the definition, and these comments are dealt with under the discussion of § 58.33 (see paragraphs 59 through 74, below).

Applicability to Studies Performed Under Grants and Contracts

37. Two comments requested revision of § 58.10 to specify clearly that the sponsor is ultimately responsible for data validity, even if the data are obtained by a sponsor from a grantee or contractor.

The Commissioner concludes that no revision of § 58.10 is necessary. All persons involved in a nonclinical laboratory study are responsible for part or all of the study, depending upon the extent of their participation. Although a sponsor who submits studies to FDA bears the responsibility for the work performed by a subcontractor or grantee, that fact in no way relieves a grantee or subcontractor from individual responsibility for the portion of the study performed for the sponsor. Indeed, the purpose of the requirement that the sponsor notify a grantee or subcontractor that the work being performed is part of a nonclinical laboratory study which must be conducted in compliance with the good laboratory practice regulations is to assure that all parties submitting data are aware of their responsibilities under the regulation.

38. Several comments requested exemption for certain specialized services which are not commonly available, e.g., ototoxicity studies with diuretics. The comments stated that these specialized services would probably not be available to them if the stringent requirements of the regulations had to be met by the service organization.

The Commissioner concludes that certain specialized services cannot be exempted from these regulations. The specialized services may contribute in large measure to the agency decision to approve a research or marketing permit. If the studies are intended to provide safety data in support of an application for a research or marketing permit, their conduct falls within the scope of these regulations.

Inspection of a Testing Facility

39. Comments on the inspection provisions (§ 58.15) expressed concern regarding the competence and scientific qualifications of FDA investigators.

The agency has endeavored, through a specialized training program, to assure that FDA investigators are competent to perform good laboratory practice inspections. The EILP program is new, and training and evaluation will continue to improve it. The results of the pilot inspection program and the manner in which it was conducted should provide added assurance to testing facility management regarding the competence of FDA investigators. The quality of the program is not, however, dependent on the competence or training of any single individual. Inspection of findings are always subject to supervisory review within the agency, and no official action may be taken without concurrence of a number of qualified persons.

40. Several comments stated that agency inspection should be limited to those facilities under current FDA legal authority.

The scope of the regulations and the definition of a "nonclinical laboratory study" define those studies covered by the regulations. The agency intends to inspect all facilities which are conducting such studies. Many of these facilities are subject to inspection under express statutory authority vested in FDA. As noted in the preamble to the proposal (41 FR 51220):

Inspections of many, perhaps most, testing facilities will not be conditioned upon consent. Under section 704(a) of the act, FDA may inspect establishments including consulting laboratories, in which certain drugs and devices are processed or held, and may examine research data that would be subject to reporting and inspection pursuant to section 505 (i) or (j) or 507 (d) or (g) of the act. In addition, any establishment registered under section 510(h) of the Act is subject to inspection under section 704 of the act. Thus, most manufacturing firms that conduct in-house non-clinical laboratory studies on drugs and devices, and those contract laboratories working for such firms, would be subject to FDA inspection whether or not they consented.

Facilities that are not subject to statutory inspection provisions will be asked to consent to FDA inspection. The absence of any statutory authorization does not bar FDA from asking permission to conduct an inspection, and the agency should not bar itself from seeking permission. Thus, the proposal in the comment is not accepted.

41. Several comments requested that FDA make its enforcement strategy known as promised in the preamble to the proposal.

The enforcement strategy was discussed in the preamble to the proposal (41 FR 51216) and is amplified in the compliance program which implements this regulation. The compliance program is publicly available and may be obtained by sending a written request to the agency official whose name and address appear at the beginning of this preamble as the contact for further information.

42. Two comments on § 58.15 as proposed requested that the requirement that the testing facility permit inspection by the sponsor be deleted. The comments argued that the rights and obligations of a sponsor and its laboratory are a matter of contract between them alone, and not a proper subject for government regulation.

The Commissioner has considered this issue, is persuaded that the comments are correct, and has deleted the phrase "the sponsor of a nonclinical laboratory study." At the same time, however, the Commissioner reemphasizes that, because a sponsor is responsible for the data he or she submits to the agency, the sponsor may well wish to assure that the right to inspect a testing facility is included in any contract.

43. Other comments suggested that the sponsor should accompany the FDA investigator during an inspection of a contract testing facility and that FDA access to data should require the sponsor's consent.

The Commissioner disagrees with these comments. An agency investigator may be inspecting the results of studies from several sponsors during an inspection. The logistics required to notify and arrange for several sponsors to accompany an investigator, or to obtain sponsor consent to information release, would be unworkable. FDA's practice of unannounced inspections has proved to be an effective and efficient use of scarce resources. Because of resource limitations, FDA cannot inspect each facility as often as it would like to, and the Commissioner finds that the possibility of unannounced FDA inspections at any time motivates compliance.

44. Many comments were concerned that trade secret information obtained during inspection would be released by FDA.

The Commissioner notes that trade secrets obtained as a result of an inspection are fully protected under the provisions of section 301(j) of the act (21 U.S.C. 331(j)), as well as 18 U.S.C. 1905 and the Freedom of Information Act (5 U.S.C. 552(b)(4)) and the FDA's implementing regulations (21 CFR 20.61). Interested parties may refer to the agency's public information regulations (21 CFR Part 20), which govern agency release of documents.

45. One comment requested that the results of government laboratory inspections be made public.

The Commissioner notes that no distinctions will be made between government or nongovernment laboratories. The results of an inspection of testing facilities will be available after all required followup regulatory action has been completed.

46. The phrase "and specimens" has been added to § 58.15(a). The Commissioner finds that examination of specimens may be required to enable the agency, where necessary, to reconstruct a study from the study records.

47. Many comments stated that the inspection of records should not extend to certain records compiled by the quality assurance unit.

The Commissioner agrees and has exempted from routine inspection those records of the quality assurance unit which state findings, note problems, make recommendations, or evaluate actions taken following recommendations. These exemptions from inspection are discussed in greater detail under the discussion of § 58.35.

48. A new paragraph (b) has been added to § 58.15. This paragraph is similar to proposed § 58.200 and reiterates that a determination that a nonclinical laboratory study will not be considered in support of an application for a research or marketing permit does not relieve an applicant from any obligation under any applicable statute or regulation (e.g., 21 CFR Parts 312, 314, 514, etc.) to submit the results to FDA. If a testing facility refuses inspection of a study, FDA will refuse to consider the study in support of an application for a research or marketing permit. This refusal,

however, does not relieve the sponsor from any other application regulatory requirement that the study be submitted.

ORGANIZATION AND PERSONNEL

Personnel

49. A number of comments addressed the definition of training, education, and experience in § 58.29. Several comments considered such references too vague; several others suggested that appropriate qualifications be established by professional peer groups.

It would be inappropriate, if not impossible, for FDA to specify exactly what scientific disciplines, education, training, or expertise best suit a specific nonclinical laboratory study. These factors, which vary from study to study, are left to the discretion of responsible management and study directors. They are responsible for personnel selection and for the quality and integrity of the data these personnel will collect, analyze, document, and report. The Commissioner urges, however, that management and study directors carefully consider personnel qualifications as they relate to a particular study. The agency has uncovered instances, discussed in the preamble of the proposal (41 FR 51207), in which the conduct of a study by inadequately trained personnel resulted in invalid data. Although the Commissioner recognizes the value of certification by professional peer groups, he does not agree that the concept is appropriate for regulatory personnel.

50. Several comments said the study director should be given responsibility for assurance of qualifications of personnel.

The Commissioner agrees that, generally, the study director will be responsible for ensuring that personnel selected to conduct a nonclinical laboratory study meet necessary educational, training, and experience requirements. The Commissioner notes however, that management also has selection and hiring responsibilities and privileges.

51. One comment stated that the requirement of § 58.29 that each individual engaged in the conduct of a study have sufficient training or experience to enable the individual to perform the assigned function should be limited to those personnel engaged in supervision and collection and analysis of data.

The Commissioner disagrees. These factors are important and should be considered for personnel other than supervisors or those engaged in collection and analysis of data. The approach suggested by the comment would ignore the fact that specific expertise is required, for example, by animal caretakers, physical science technicians, and by persons using pesticides near animal-holding areas. While the degree of education, training, and experience necessary for these positions will be quite different from the qualifications necessary for supervisors or scientific staff, the need for sufficient training or experience is no less important.

52. One comment pointed out the appropriateness of changing the term "person" to "individual" in § 58.29(a).

Because the term "person" as defined in § 58.3(h) includes partnerships, corporations, etc., the Commissioner agrees that "individual" is the proper term and has so amended § 58.29(a).

53. Seventeen comments questioned the use of, or objected to reference to, the term "curriculum vitae" for nontechnical personnel such as animal caretakers, as required in proposed § 58.29(b).

Another comment asserted that the requirement infringed on management's prerogatives without specifying how any such infringement occurred. One comment stated that the requirement that such records be retained after termination of employment was unnecessarily cumbersome.

The Commissioner does not agree that the requirement infringes on management's prerogatives. However, the Commissioner agrees with the remaining comments and has revised the section. "Curriculum vitae" has been changed to "summaries of training and experience plus job descriptions." Reference to the maintenance of records of terminated employees is deleted from this section because the requirement is redundant to the record retention requirements set forth in § 58.195(e).

54. Ten comments said the wording of § 58.29(c), relating to "sufficient numbers of personnel" and to "timely" conduct of the study, was vague.

The Commissioner purposely left the paragraph broad in context and coverage because differences in types of studies preclude any specific approach to defining numbers of personnel. The precise number of personnel required for a specific study, as well as for all ongoing studies, is a management decision. FDA experience, however, indicates that a shortage of qualified personnel can lead to inadequate or incomplete monitoring of a study and to delayed preparation and analysis of results, and the numbers of personnel conducting a study should be sufficient to avoid these problems.

55. Ten comments requested deletion of § 58.29(d) or clarification of the language regarding employee health habits, stating that the section was too vague and that an employer was responsible for health habits only at work. One comment submitted alternative language.

The Commissioner adopts with modifications the alternate language. The paragraph now requires only that personnel take necessary personal sanitation and health precautions to avoid contamination of test and control articles and test systems.

56. Several comments asked that the term "laboratory" in § 58.29(e), as applied to protective clothing, be deleted because it is too restrictive. Other comments suggested that the requirement that clothing be changed as often as necessary to prevent contamination be eased by changing "prevent" to "help prevent." Four related comments requested modification to reflect only "contamination affecting validity of studies."

The Commissioner agrees to the elimination of "laboratory" as applied to clothing. The provision of specialized clothing is, however, an established and well-known procedure for preventing contamination in a variety of situations. The Commissioner disagrees with any suggested modification of this section which weakens the intent of the regulation. The objective is to prevent contamination of the test system.

57. A number of comments addressed several aspects of § 58.29(f) regarding personal illnesses, personal health records, types of illnesses, and records of illnesses. Comments said disclosure of medical records was an invasion of privacy and of little relevance to the proper conduct of a nonclinical laboratory study.

The Commissioner agrees that documentation of personal illnesses may constitute an unwarranted invasion of privacy, and this requirement is deleted. The Commissioner disagrees with the requests for deletion of the entire paragraph, noting the relationship between personnel health and possible contamination of test systems. Revised § 58.29(f) requires individuals with illnesses that may adversely affect the quality and integrity of nonclinical laboratory studies to be excluded from direct contact with test and control articles and test systems. All personnel should be instructed to report such medical conditions to their immediate supervisor, who should protect test systems from personnel reporting as ill.

Testing Facility Management

58. Many comments on the responsibilities of the study director objected that some of the responsibilities assigned to the study director were more properly assigned to management.

The Commissioner agrees that several of the responsibilities previously assigned to the study director should be assigned to the testing facility management. For clarification, a new § 58.31 is added to the regulations. It is management's responsibility to assure that for each study there is a study director and an independent quality assurance unit, as required by the regulations. It is also management's responsibility to ensure that any deviations from the regulations which are reported by the quality assurance unit are, in turn, reported to the study director and that corrective actions are both taken and documented. Designating management responsibilities in this manner merely clarifies the fact that the study director should be viewed as the chief scientist in charge of a study. Duties which are more administrative than scientific are the responsibility of management; however, management may delegate appropriate administrative duties to the study director.

Study Director

59. More than 50 comments addressed the scope of responsibilities proposed for the study director. Many comments stated that these responsibilities were much too broad for one person.

In the proposal, the Commissioner advanced the concept of a single fixed point of responsibility for overall conduct of each nonclinical laboratory study. Experience has demonstrated that if responsibility for proper study conduct is not assigned to one person, there is a potential for the issuance of conflicting instructions and improper protocol implementation. The study director is charged with the technical direction of a study, including interpretation, analysis, documentation, and reporting of results. As discussed in paragraph 58, several of the responsibilities proposed for the study director have been transferred to testing facility management. This transfer should allay concerns regarding the magnitude of the responsibilities assigned to the study director.

60. Nine comments object to the term "ultimate" as applied to the study director's responsibility.

The Commissioner agrees that "ultimate" responsibility for the study rests with facility management and/or the sponsor. Therefore, the word "ultimate" has been replaced by "overall" in § 58.33.

61. Several comments argued that more than one study director should be allowed for each study.

The Commissioner rejects these comments. As noted above, there must be a single point of responsibility for overall technical conduct of the study. The potential for conflicting instructions and confusion in study implementation is too great to diffuse the responsibility by, for example, study direction by a committee. The regulation does not, however, preclude the study director from directing more than one study.

62. Many comments stated that the requirements would interfere with management's prerogatives to organize and conduct studies as it so chooses.

The requirement that the study director be the single point of responsibility for technical conduct of the study need not interfere with normal delegation of authority by management.

63. Five comments argued that the proposed requirements for study director and quality assurance unit were duplicative.

The Commissioner has carefully reviewed the proposal and comments and has clearly separated the responsibilities in the final regulation to avoid duplication. The first sentence in § 58.33 has been revised to specify clearly that each study shall have a study director. The second sentence has been revised to amplify the concept: "The study director has overall responsibility for the technical conduct of the study, as well as for the interpretation, analysis, documentation and reporting of results and represents the single point of study control."

64. One comment suggested revising § 58.33(a) to specify that the sponsor must approve the protocol and the study director must approve any change.

The Commissioner advises that § 58.120(a)(15) requires that the sponsor approve the protocol, and § 58.120(c) requires that the study director approve any changes or revisions to the protocol. The language in § 58.33(a) has been revised to reference § 58.120.

65. Five comments objected to the proposed requirement that the study director assure that test and control articles or mixtures be appropriately tested. The comments argued that this was not a proper function of the study director.

The Commissioner agrees that this responsibility is more properly assigned to testing facility management. Therefore, the requirement has been transferred to § 58.31(d).

66. Three comments suggested that, rather than the study director assuring that test systems are appropriate, the study director should assure that the test systems are as specified by the protocol.

The Commissioner agrees that the determination of the appropriateness of the test system is a scientific decision beyond the scope of these regulations. Section 58.33(d) has been revised to state: "Test systems are as specified in the protocol."

67. Four comments argued that the scheduling of personnel, resources, facilities, and methodologies was not a proper requirement for the study director.

The Commissioner agrees that this scheduling is beyond the scope of the study director's responsibilities and has, therefore, transferred it to the responsibilities of testing facility management in § 58.31(e).

68. Two comments object to the requirement that personnel clearly understand the functions they are to perform.

The Commissioner finds that it is essential that personnel be adequately trained to assure the integrity and validity of the data. However, the Commissioner concludes that training is a proper responsibility of testing facility management and has transferred the requirement to § 58.31(f).

69. Three comments suggested deletion of the phrase "and verified" from the proposed requirement that the study director assure that all data

are accurately recorded and verified. Four comments requested definition of the term "verified."

The Commissioner disagrees with the requested deletion. Recording and verifying data are key operations in the successful completion of a study. The Commissioner intends that the study director assure that data are technically correct and accurately recorded. "Verified" is used in its ordinary sense of "confirmed" or "substantiated." The process by which verification is achieved may be determined by the study director.

70. One comment stated that proposed § 3e.31(a)(8) merely repeated proposed § 3e.31(a)(7).

The Commissioner finds that the two sections can be combined for clarity. According, § 58.33(c) now reads "unforeseen circumstances that may affect the quality and integrity of the nonclinical laboratory study are noted when they occur, and corrective action is taken and documented."

71. One comment stated that the requirement that the study director assure that responses of the test system are documented is unreasonable.

The Commissioner disagrees. Assuring that all experimental data including unforeseen responses to the test system are accurately observed and documented is a critical part of study conduct and is a responsibility properly assigned to the study director.

72. Two comments stated that the requirement that the study director assure that good laboratory practice regulations are followed either should be modified to make it more flexible or should be deleted. One comment suggested that the study director should be allowed to delegate the responsibility.

The Commissioner rejects these comments. The regulations constitute an effective means to aid study directors in achieving better control of complex studies. Responsibility for assuring compliance properly rests with the study director. While delegation of authority is always the prerogative of a manager, responsibility cannot be delegated.

73. Several comments stated that the requirement that the study director assure that study documentation is transferred to the archives is redundant to § 58.190.

The Commissioner does not agree that the sections are redundant. Section 58.190 requires that the study records be retained, and § 58.33(f) requires that the study director assure that the records are transferred for retention. The phrase "and other information to be retained" has been deleted from § 58.33(f) because the phrase is subsumed by raw data, documentation, protocols, specimens and final reports.

74. Thirteen comments questioned the proposed approach to study director replacement, specifically objecting to the requirement that justification of such replacement be documented and retained as raw data. The

comments argued that justification carries a negative connotation and that replacement of a study director is a management prerogative.

The Commissioner is persuaded that replacement of the study director should remain within the discretion of management that the requirement that justification for such replacement be documented and retained is an inappropriate subject for these regulations. Consequently, the requirement for justification for such replacement has been deleted. The requirement that the study director be replaced promptly when necessary has been transferred to § 58.31(b).

Quality Assurance Unit

75. More than 100 comments objected to part or all of § 58.35 as proposed. Many comments questioned the need for a quality assurance unit as proposed. Some comments stated that the establishment of such a unit would increase the administrative burden and costs of performing nonclinical studies to the point of forcing small facilities out of business. Others stated that the provisions would interfere with management's prerogatives to organize the facility or with the informed scientific judgment of principal investigators or study directors.

The Commissioner has retained the requirement that each testing facility have a quality assurance unit (QAU) to monitor the conduct and reporting of nonclinical laboratory studies. In view of the potential gain to management, to sponsors, and to FDA, through the added assurance of well-conducted studies, increased costs, if any, are justified. The quality assurance unit need not be a separate organizational entity composed of personnel permanently assigned to that unit. All nonclinical studies falling within the scope of this regulation must be monitored by a quality assurance unit composed of at least one person. Within this framework, management retains its organizational prerogatives Because different individuals may be responsible for quality assurance functions at different times, it is important that quality assurance unit records be centrally located, and § 58.35(e) has been modified to so require. The regulations permit a study director for a particular study to serve as a part of the quality assurance unit or as the quality assurance unit for a different study. However, for any given study a separation must exist between individuals actually engaged in the conduct of a study and those who inspect and monitor its progress. In those situations in which several different individuals are performing the quality assurance functions for different studies, each such individual must maintain that portion of the master schedule sheet which relates to the study he or she is monitoring. This means that several people may be responsible for maintaining the master schedule sheet. Because the function of the quality assurance unit is administrative rather than scientific, the Commissioner does not agree that the functions of a QAU will interfere with the study director's control of the overall technical conduct of the study. In order to emphasize this point, the following language has been added to § 58.35(a): "For any given study the quality assurance unit shall be entirely separate from and independent of the personnel engaged in the direction and conduct of that study."

76. Sixteen comments objected to the word "unit" in the term "quality assurance unit" and suggested alternate words such as "function" or "program."

The Commissioner has elected to preserve the word "unit" to conform to similar wording in other regulations such as the current good manufacturing practice regulations. The Commissioner agrees, however, with the rationale of the comments that the important objective of this section is that there be a quality assurance function operating for each nonclinical study. As indicated in paragraph 75, the exact organizational means by which this function is achieved is the prerogative of facility management and may vary from facility to facility.

77. Numerous comments addressed the composition of the quality assurance unit. Four comments sought inclusion of criteria for education, training, and experience of QAU personnel. Seven comments indicated that compliance with this section was impractical because of a shortage of people qualified to staff such a unit.

The Commissioner has not attempted to specify the qualifications of quality assurance personnel because qualifications should be determined by management and will vary according to the type of facility and the types of studies conducted by each facility. Because the function of the quality assurance unit is to assure compliance with procedural and administrative requirements rather than to oversee the technical aspects of study conduct, QAU personnel need not be limited to professional personnel and/or scientists. The Commissioner does not agree, therefore, that there exists a serious shortage of qualified people to fulfill this function.

78. Two comments indicated that the quality assurance unit should be composed of outside consultants in order to assure the independence of the function. One comment requested that quality assurance unit membership be restricted to employees of the facility.

The Commissioner notes that the quality assurance functions may be performed by outside consultants. This fact should enable small facilities or facilities conducting nonclinical laboratory studies for resubmission to the FDA on an irregular basis to meet the quality assurance requirements in a cost-effective manner. At the same time, the Commissioner does not agree that the QAU function must be performed by an outside body. The organizational separation of the QAU from the study team should be sufficient to assure objective monitoring by the QUA.

79. Four comments questioned the last sentence in § 58.35(a) as proposed, stating that it seemed to require monitoring of some studies by two QAU's—that of the sponsor and that of the contract facility.

The Commissioner has deleted from § 58.35(a) the sentence in question. The QAU of the testing facility is solely responsible for fulfilling the quality assurance functions for studies conducted within that facility. In those cases where portions of a study, e.g., feed analysis, are performed by a contract facility which, because it is not itself a nonclinical facility, does not have a QAU, the person letting the contract, and not the contract facility, is responsible for the performance of the quality assurance functions.

The Commissioner believes that the mechanism by which a sponsor is assured of the quality of nonclinical studies performed for it under contract is a matter that can be left to the contracting parties and need not be addressed in these regulations.

80. Three comments suggested that testing facilities be licensed or certified in lieu of having an ongoing quality assurance unit.

The Commissioner considered such an approach and rejected it before publishing the proposed regulations. (See 41 FR 51208-51209). No persuasive arguments for changing this decision were presented in the comments. The diversity in the size and nature of nonclinical testing facilities subject to the provisions of these regulations makes licensing or certification procedures impractical. The regulation is intended to assure the quality and validity of the data obtained by each nonclinical laboratory study, and the QAU provides a mechanism to monitor each ongoing study. Licensing a testing facility could not achieve the same result.

81. Many comments objected to the provisions of § 58.35(b)(1) which require that the quality assurance unit maintain a master schedule sheet of all nonclinical laboratory studies. Some comments believed the requirement was excessive, while others questioned the proposed format and contents of the list. One comment pointed out that not every study includes all items listed.

The Commissioner is convinced that maintenance of a master schedule sheet is essential to the proper function of the Quality Assurance Unit. Only through such a mechanism can management be assured that the facilities are adequate and that there are sufficient numbers of qualified personnel available to accomplish the protocols of all nonclinical studies being conducted at a facility at any given time.

Upon careful review of the items required to be listed, the Commissioner agrees that the requirement that animal species be identified may be deleted because the requirement that "test system" be listed adequately covers this point. He has, in addition, deleted the examples of study types because he agrees that including the information is not necessary to achieve objectives of this section. The Commissioner has further reworded this section to eliminate reference to whether the final report has been approved for submission to the sponsor because the language was strictly applicable only to studies done under contract. The revised language simply requires that the status of the final report be listed.

83. Nine comments objected that § 58.35(b)(2) required too much duplicative paper.

The Commissioner has concluded that the QAU must maintain copies of study protocols to assure that they are followed and amended in accordance with the further provisions of these regulations. The Commissioner agrees that the requirement that the QAU maintain copies of all standard operating procedures would substantially increase the volume of records needed to be retained by this unit. Because there should be many copies of standard operating procedures present throughout the facility which should be freely available to QAU members, the Commissioner has deleted the requirement that these be maintained by the QAU.

83. Fifteen comments suggested that § 58.35(b)(3) be deleted on the basis that FDA should not dictate how the QAU achieves its objectives. One comment suggested that "inspect" be changed to "audit."

The Commissioner remains convinced of the need for a formal mechanism through which the QAU maintains oversight of the conduct of a study. Such a mechanism must be based on direct observation in order that the independence of the QAU be preserved. The Commissioner has retained the word "inspect" in preference to "audit." "Inspect" more accurately conveys the intent that the QAU actually examine and observe the facilities and operations for a given study while the study is in progress, whereas "audit" could be interpreted to mean simply a detailed review of the records of a study. Because the QAU function is to observe and report the state of compliance with the regulations and to determine whether the protocol is being followed rather than to verify the results of a study, "inspect" more properly conveys the agency's intent.

84. Fourteen comments addressed the need to inspect "each phase of a study *** periodically," seeking clarification or different language. Nine of these comments called for the use of random sampling procedures in choosing studies or phases of studies to inspect in order to decrease the workload and resource requirements of the QAU.

The Commissioner does not agree that random sampling would be an adequate method of evaluation in the nonclinical laboratory setting. In situations which involve the repetition of similar or identical procedures, random sampling can provide an adequate means of quality control. Here, however, the differences among study operations and among the personnel conducting them invalidate any assumption that the conduct of one phase of one study is representative of the conduct of that phase of another or of other phases of a single study. The term "each phase" is intended to emphasize the need for repeated surveillance at different times during the conduct of a study so that each critical operation is observed at least once in the course of the study. The term "periodically" is retained to indicate the need for more than one inspection of certain repetitive continuing operations that are part of the conduct of longer term studies such as animal observations and diet preparation.

85. Many comments objected to the proposed requirement that any problems found by the QAU be brought to the attention of management and appropriate responsible scientists. Some felt that this would require that excessive resources be spent on minor problems. Others felt that notification of appropriate supervisory personnel rather than management was sufficient.

The Commissioner agrees that only those problems likely to affect the outcome of the study need to be brought to the immediate attention of personnel who are in a position to resolve those problems, and the language of § 58.35(b)(3) has been changed accordingly. The term "management" in its ordinary usage means appropriate supervisory personnel and has not, therefore, been changed.

86. More than 40 responses to proposed § 3e.33(b)(4) objected to the specific time frames required for evaluation. Several comments suggested that the paragraph be deleted. Others objected to the specific requirements, and still others stated that appropriate times for evaluations should be selected by management.

The Commissioner advises that periodic inspection is necessary and that the time periods specified are the minimum required to assure that a study is being conducted in compliance with the regulation. Should deviations be found during the periodic inspections, there may still be time to take corrective action. The Commissioner has, however, determined that inspection of studies lasting less than 6 months need only be conducted at intervals adequate to assure the integrity of the study and that specific time intervals for such studies need not be set out in this regulation. The requirement that studies lasting more than 6 months be inspected every 3 months remains unchanged. The section has been added to § 58.35(b)(3).

87. Several comments requested that the phrase "complete evaluation" in proposed § 3e.33(b)(4) be clarified.

The Commissioner has changed the term "complete evaluation" to "inspect." The function of the QAU is to inspect studies at specified intervals to maintain records required by this regulation, and to report to management and the study director deviations from the protocol and from acceptable laboratory practice. Evaluation of any reported deviations is left to the study director and to management.

88. Fifteen comments sought deletion of § 58.35(b)(4), which requires the periodic submission of status reports to management and the study director. Three comments questioned the need to note problems and corrective actions taken.

The Commissioner has retained this provision as proposed. Only through the submission of such status reports can management be assured of the continuing conformity of study conduct to the provisions of these regulations. Because § 58.35(b)(3) has been revised to require that only significant problems be reported immediately to management, the periodic status report becomes even more important as a means of informing management of minor problems and normal study progress. The status reports are needed to document problems and corrective actions taken so that management can be certain that quality is being maintained and that management intervention is not required. The timing of such reports may be determined by management.

89. Six comments objected that the term "prior" preceding "authorization" in § 58.35(b)(5) was too restrictive. The comments pointed out that unforeseen circumstances may prevent prior authorization for deviations from standard procedure and that the QAU should be concerned with the documentation of the deviation, not with whether prior authorization existed. Two comments stated that the QAU cannot assure that deviations do not occur but can determine, by inspection, whether deviations were documented.

The Commissioner is persuaded that prior authorization cannot always be obtained. For example, a fire in the facility would necessitate immediate action. The Commissioner agrees that documentation of the deviation rather than prior authorization is the important point and has deleted "prior" and added "documentation." In addition, "assure" has been changed to "determine" to respond to the comments and to reflect more accurately the Commissioner's intent. Section 58.35(b)(5) now reads: "Determine that no deviations from approved protocols or standard operating procedures were made without proper authorization and documentation."

90. Several comments objected to the wording of § 58.35(b)(6), which states that the QAU shall review the final study report. The comments stated that such review requires a scientific judgment and is not an appropriate function for the QAU to perform. One comment suggested that the requirement should be modified to allow for random sampling rather than a complete review of all studies.

The Commissioner agrees that the QAU should not attempt to evaluate the scientific merits of the final report. Therefore, he has modified the paragraph. The QAU must however ensure that the final report was derived from data obtained in accordance with the protocol. Data in the final report significantly contributing to the quality and integrity of a nonclinical laboratory study shall be reviewed. A random sampling approach is not acceptable.

90a. The Commissioner has added to § 58.35 new paragraph (b)(7) which requires that the QAU prepare and sign a statement to be included with the final report which specifies that dates inspections of the study were made and findings reported to management and the study director. This requirement clarifies the fact that QAU review should extend through the completion of the final report and provides a mechanism for documenting that the review has been completed. A conforming section has been added to the final report requirements of § 58.185 as new paragraph (a)(14).

91. Many comments argued that requiring all portions of a quality assurance inspection to be available for FDA inspection might serve to negate their value as an effective management tool for ensuring the quality of the studies during the time in which the studies are being conducted.

The Commissioner shares the concerns of the comments that general FDA access to QAU inspection reports would tend to weaken the inspection system. He believes that FDA's review of quality assurance programs is important, and he recognizes the need to maintain a degree of confidentiality if QAU inspections are to be complete and candid. Therefore, the Commissioner has decided that, as a matter of administrative policy, FDA will not request inspections and copying of either records of findings and problems or records of corrective actions recommended and taken; and §§ 58.15 and 58.35(c) have been revised to separate those records subject to regular inspection by FDA from those records not subject to such inspection. Exempt from routine FDA inspection are records of findings and problems as well as records of corrective actions recommended and taken. All other records are available. Although the Commissioner is deleting the requirement in new § 58.35(d) that testing facility management shall, upon request by an authorized employee, certify in writing that the inspections are being performed and that recommended action is being or has been taken. Upon receiving such a request, management is required to submit the certification of compliance. A person who submits a false certification is liable to prosecution under 18 U.S.C. 1001.

The one exception to FDA's policy of not seeking access to records of findings and problems or of corrective actions recommended and taken is that FDA may seek production of these reports in litigation under applicable procedural rules, as for otherwise confidential documents.

92. Many comments objected to requiring internal quality assurance audits to be available to the agency might violate the constitutional privilege against compelled self-incrimination.

The Commissioner disagrees with the comments. It is settled that the privilege against compelled self-incrimination is an individual privilege relating to personal matters; the privilege is not available to a collective entity, such as a business enterprise, or to an individual acting in a representative capacity on behalf of a collective entity. *California Bankers Ass'n* v. *Schultz*, 416 U.S. 21, 55 (1974); *Bellis* v. *United States*, 417 U.S. 85 (1974); *United States* v. *Kordel*, 397 U.S. 1, 8 (1970); *Curcio* v. *United States*, 354 U.S. 118, 122 (1957); *United States* v. *White*, 322 U.S. 694, 699 (1944); *Wilson* v. *United States*, 221 U.S. 361, 382-384 (1911); *Hale* v. *Henkel*, 201 U.S. 43, 74-75 (1906). Even for individuals, the privilege against compelled self-incrimination is inapplicable where a reporting requirement is applied to an "essential noncriminal and regulatory area of inquiry," where self-reporting is the only feasible means of securing the required information, and where the requirement is not applied to a "highly selective group inherently suspect of criminal activities" in an "area permeated with criminal statutes." *California* v. *Byers*, 402 U.S. 424, 430 (1971); *Marchetti* v. *United States*, 390 U.S. 39 (1968); *Albertson* v. *SACB*, 382 U.S. 70, 79 (1965); *Shapiro* v. *United States*, 335 U.S. 1 (1948).

Access to Professional Assistance

93. Comments on proposed § 3e.35 suggested rephrasing the statement to specify that professional assistance be authorized by the study director, that it be either in person or by telephone, that it be available within a reasonable period, and that reference to availability of a veterinary clinical pathologist be included. Other comments suggested that the concept was duplicative of the function of the study director and should be deleted.

The Commissioner proposed this requirement to assure that a scientist or other professional would be available to respond to requests for assistance or consultation from less experienced personnel. However, because management is responsible for assuring that personnel are available and that personnel clearly understand the functions they are to perform, and because the study director has overall responsibility for the technical conduct of the study, access to professional assistance is a matter best left to management's discretion. Therefore, the section is deleted from the final regulations.

FACILITIES

General

94. Many comments requested definition or clarification of the terms denoting separation (i.e., separate area, defined area, separate space, and specialized area), which are used in §§ 58.41, 58.43, 58.47, 58.49, and 58.90.

The Commissioner's intent in proposing that there be defined (and, where required, separate or specialized) areas in a testing facility was to assure the adequacy of the facility for conducting nonclinical laboratory studies. This intent is more clearly stated in the revised second sentence of § 58.41, which now reads: "It shall be designed so that there is a degree of separation that will prevent any function or activity from having an adverse effect on the study." The important point is that the facility be designed so that the quality and integrity of the study data is assured. The manner in which the separation is accomplished may be determined by testing facility management.

Adequate separation may be, in various situations, a function of such factors as intended use of the specific part of the facility, space, time, and controlled air. The broad variety of test systems, test and control articles, and the size and complexity of testing facilities preclude the establishment of specific criteria for each situation. For these reasons the Commissioner declines to include in the regulation either a definition or specific examples of methods for achieving adequate separation.

95. One comment suggested that a number of additional animal care and facility requirements be added to the regulations. The suggestions included, e.g., ambience to assure nonstressful conditions; ventilation and room access arranged to prevent cross contamination; and surveillance of animal health before and during a test or experiment.

The Commissioner concludes that no additional requirements need to be added because the regulation, as it stands, adequately covers the additions proposed by the comments. For example, ventilation and room access arranged to prevent cross contamination are addressed by the degree of separation requirement in § 58.41.

Animal Care Facilities

96. Many comments suggested that accreditation of animal care facilities by a recognized organization should provide adequate evidence that a testing facility is in compliance with § 58.43(a). One comment suggested accreditation by recognized organizations for analytical laboratories.

Although the Commissioner is aware of the value of accreditation programs, he cannot delegate FDA's responsibility for determining compliance with these regulations to an organization over which FDA has no authority. Few, if any, accreditation programs cover the same areas covered by this regulation. Furthermore, the Commissioner is unaware of any facility accreditation program which is mandatory. The agency's obligation to inspect a testing facility for overall compliance would not be altered by the fact that a facility was otherwise accredited.

97. Numerous comments objected to the requirements concerning separation of species, isolation of projects, and quarantine of animals as impractical and not necessary in all instances, e.g., separation of species in large animal studies and quarantine of all newly acquired animals. Some of the comments stated that the requirements of this section allow no latitude for judgment concerning their applicability.

The Commissioner reiterates that all requirements may not be applicable or necessary in all nonclinical laboratory studies and that the degree to which each requirement should apply in each case can be determined by informed judgment. Because of the variability of nonclinical laboratory studies, a degree of flexibility in applying the requirements of § 58.43(a) is necessary, and the language of § 58.43(a) is amended to read: "A testing facility shall have a sufficient number of animal rooms or areas, as needed, to assure proper: (1) separation of species or test systems, (2) isolation of individual projects, (3) quarantine of animals, and (4) routine or specialized housing of animals." As noted in the general discussion at the beginning of this preamble, all references to other standards ("The Animal Welfare Act") have been deleted.

98. Several comments suggested that § 58.43(b) be amended to include isolation of test systems with infectious diseases as well as isolating studies conducted with infectious or otherwise harmful test articles.

The Commissioner agrees that test systems with infectious diseases should be isolated. Proposed § 3e.49(b) provided for specialized areas for handling volatile agents and hazardous aerosols. Section 3e.49(b) also provided for special procedures for handling other biohazardous materials. Proposed § 3e.49(c) provided for special facilities or areas for handling radioactive materials.

To clarify all these requirements, the Commissioner has amended § 58.43(b) to read: "A testing facility shall have a number of animal rooms or areas separate from those described in paragraph (a) of this section to ensure isolation of studies being done with test systems or test and control articles known to be biohazardous, including volatile substances, aerosols, radioactive materials, and infectious agents." The provisions in proposed § 3e.49(b) and (c) regarding specialized areas for handling volatile agents, hazardous materials and radioactive materials are deleted from § 58.49.

99. One comment on § 58.43(c) suggested that, in addition to the area designated for the care and treatment of diseased animals, a separate area should be provided for animals with contagious diseases.

The Commissioner agrees, and the paragraph is amended to allow for an area for treatment of animals with contagious diseases, and it is to be separate from the area designated for the care and treatment of diseased animals.

100. Several comments questioned the requirement for separate areas for diseased animals, indicating that often such animals are sacrificed rather than treated.

The Commissioner does not agree that a separate area is not always needed for diseased animals. Although diseased animals may be sacrificed, this is not always the case, and it may not always be possible immediately to sacrifice diseased animals. Thus, a separate area should be available for such animals until sacrifice can be accomplished.

101. One comment requested that § 58.43(e), which deals with facility design, construction, and location to minimize disturbances that interfere with the study, should also define the acoustic and sound-insulating requirements necessary to satisfy this requirement.

The Commissioner concludes that it is impractical to attempt to define acoustic and sound insulation requirements. It would be equally impractical to attempt to define all other types of possible disturbances that might interfere with a study.

Animal Supply Facilities

102. One comment asked that § 58.45 be clarified by specifically excluding "carriers" from the storage requirements.

The term "carrier," as used in § 58.113, is the material with which the test article is mixed, e.g., feed. The Commissioner concludes that it is necessary to provide facilities for proper storage of carriers and declines, therefore, to exclude them from the storage requirements.

103. One comment requested deletion of the section, stating that it discusses items not appropriate for FDA concern.

Improper storage of feed, carriers, bedding, supplies, and equipment can adversely affect the results of a study. Therefore, the Commissioner finds these matters to be of legitimate concern to FDA and declines to delete the section.

104. Two comments stated that separate storage space need not be required as long as material is properly stored and does not interfere with the conduct of the study.

The Commissioner agrees with these comments, in principle, but is convinced that storage areas for feed and bedding should be separate from the areas housing the test system to preclude mixups and contamination of the test systems. The section has been modified by adding the words "as needed."

Facilities for Handling
Test and Control Articles

105. One comment stated that § 58.47, as worded, represented an impossible standard and suggested that use of the "designed to prevent" concept would be more realistic.

The Commissioner rejects this comment. The inherent purpose of "design" of all regulations is to prevent or require some action, and the use of the phrase "designed to prevent" would be an awkward and ambiguous modification of § 58.47.

106. Numerous comments objected to creating the number of separate or defined areas proposed by § 58.47, stating that the volume of testing would make it infeasible to create all the separate areas. One comment asked whether eight separate areas were required.

The Commissioner reiterates that the purpose of this section is to assure that there exists a degree of separation that will prevent any one function or activity from having an adverse effect on the study as a whole. Because of the wide variety of studies covered by these regulations, a degree of flexibility is appropriate in applying these requirements, and the degree to which each requirement should apply in each case may vary. To make this clear, the term "defined" has been deleted from § 58.47. Section 58.47(a) now reads: "As necessary to prevent contamination or mixups, there shall be separate areas for * * *." There is no specific requirement for eight separate areas.

Laboratory Operation Areas

107. A number of comments stated that § 58.49 required clarification, that in some instances more than one activity could be permitted in the same room, and that certain of the requirements would not be appropriate in every case.

The Commissioner agrees that the section as proposed was subject to misinterpretation. Because of the nature and scope of the types of studies subject to these regulations, it would be inappropriate to set specific uniform requirements for all studies. Therefore, the provisions are revised to make it clear that reasonable judgments regarding area and space requirements may be made on the basis that a particular function or activity will not adversely affect other studies in progress. Proposed § 58.49(b) has been revised, and the references to biohazardous materials has been added to the list of activities in § 58.49(a). (See the discussion at paragraph 98 above).

108. Two comments suggested that the wording of § 58.49(a) be changed to refer to "adequate" rather than "separate" laboratory facilities, stating that animal studies require that laboratory facilities be available on the immediate premises. One comment requested that provisions be made for the use of outside laboratory facilities.

The Commissioner concludes that the term "separate" is proper in the context of § 58.49(a). He does not agree that laboratory facilities must be available on the immediate premises of the testing facility, and finds that many laboratory functions can be conducted properly in separate buildings or by independent laboratories located outside the testing facility.

109. Two comments on § 58.49(b) stated that the requirement that space and facilities separate from the housing areas for the test systems be provided for cleaning, sterilizing, and maintaining equipment and that supplies should apply only to major equipment.

The Commissioner does not agree. The objective of the requirement is to prevent the occurrence of those adverse effects which might result to a study from the activities of cleaning, sterilizing, and maintaining. No meaningful distinctions based on "major" or "not major" equipment can be made.

110. One comment on § 58.49(b) stated that the proposed wording did not have useful application in all test systems or studies and that the section should be rewritten to focus on the intended principle and not on the way to achieve it.

The section has been revised. It now reads, "separate space shall be provided for cleaning, sterilizing, and maintaining equipment and supplies used during the course of the study." The revised wording grants flexibility in application as long as study results are not affected.

Specimen and Data Storage Facilities

111. Several comments asked whether § 58.51 applied to completed or ongoing studies. Concern was also expressed that limiting access to storage areas to authorized personnel was not feasible.

This section is amended to apply to archive storage of all raw data and specimens from completed studies. The Commissioner cannot agree, however, that limiting access of the archives to authorized personnel only is not feasible. Prudence would dictate such limited access even in the absence of a requirement. The potential for misplaced data and specimens is too great to allow unlimited access to the archives.

Administrative and Personnel Facilities

112. One comment on § 58.53(a) stated that the section was unnecessary because administrative functions had been previously defined in §§ 58.29, 58.33, and 58.35.

The Commissioner notes that this section specifies facilities rather than duties. References to OSHA regulations have been deleted.

EQUIPMENT

Equipment Design

113. Five comments on § 58.61 stated that the section was fragmented and redundant.

The Commissioner agrees with these comments and has consolidated the section into one paragraph, which reads: "Automatic, mechanical or electronic equipment used in the generation, measurement or assessment of data and equipment used for facility environmental control shall be of appropriate design and adequate capacity to function according to the protocol and shall be suitably located for operation, inspection, cleaning and maintenance." This consolidation eliminates the fragmentation and redundancy of the proposal and specifies clearly that the requirements are limited to that equipment which, if improperly designed, or inadequately cleaned and/or maintained, could adversely affect study results.

114. Two comments objected to the undefined general terms "adequate" and "appropriate" in § 58.61.

The Commissioner points out that broad terms are necessary because of the wide range of equipment used in the studies covered. Exact design and capacity requirements for each piece of equipment are clearly beyond the scope of these regulations.

115. Four comments on § 58.61 stated that how cleaning is accomplished is irrelevant and that the regulation should emphasize accomplishment rather than ease of accomplishment.

The Commissioner agrees that the primary concern is that adequate cleaning be accomplished. However, past experience has demonstrated that when equipment is not designed and located to facilitate cleaning and maintenance, it is much less likely to be adequately cleaned and maintained.

Maintenance and Calibration of Equipment

116. Five comments suggested that § 58.63(a) should allow the required functions to be performed at the time the equipment is used rather than specifying that the functions be performed regularly.

The Commissioner agrees that performing these functions at the time of use is satisfactory and is amending § 58.63(a) to provide flexibility. The second sentence of this section now reads: "Equipment used for the generation of data shall be adequately tested, calibrated and/or standardized."

117. Two comments suggested that "calibrated" should be changed to "standardized" because the word "calibrated" normally means a performance check against known standards, whereas "standardized" normally means to make uniform.

The Commissioner finds that for some equipment the term "calibrated" is more appropriate and for other equipment the term "standardized" is more appropriate. Revised § 58.63(a) allows application of either term.

118. Two comments suggested that the reference to the use of cleaning the pest control materials is misplaced in § 58.63(a).

The Commissioner agrees that this use is more appropriately addressed under "Testing Facility Operations" and the requirements have been transferred to § 58.90(i).

119. Comments requested a precise definition of the equipment for which § 58.63(b) requires written standard operating procedures.

The Commissioner advises that because of the range of study and product types covered, such a list is impractical. The language of this section is retained as proposed to encompass the total range of equipment used in conducting nonclinical studies.

120. Eleven comments questioned the appropriateness of designating a responsible individual in § 58.63(b).

The Commissioner has changed "individual" to "person" as defined in § 58.3(h) to allow for designation of an organizational unit.

121. One comment indicated the need for a clear FDA policy regarding primary calibration standards.

The Commissioner concludes that proper standards are the responsibility of management, and these are to be set forth in the standard operating procedures.

122. One comment agreed with the standard operating procedure requirements of § 58.63(b), but suggested a several year phase-in period.

The Commissioner concludes that 180 days is a sufficient time period for developing standard operating procedures. Furthermore, the Commissioner's intent to require such procedures has been known since November 1976, when the proposed regulation was published.

123. Seven comments suggested that the manufacturer's recommendations should be sufficient for standard operating procedures. Additionally, one comment pointed out that maintenance could be subcontracted and a certificate should be allowed.

The Commissioner advises that the regulation does not preclude the use of manufacturer's recommendations as part of the standard operating procedures, nor does it preclude subcontracting maintenance. The Commissioner advises, however, that if a facility decides to subcontract maintenance, that fact does not relieve the facility of the responsibility for maintenance.

124. One comment argued that the requirement that all equipment records specify remedial action to be taken is excessive, and two comments said there are too many variables to specify in advance the remedial action to be taken.

The Commissioner notes that trouble-shooting charts are available for most equipment. The remedial action taken may influence the results of the study and therefore must be documented.

125. Several comments suggested that the equipment for which standard operating procedures are required be limited by rewording in one or the following ways: "major" equipment, "equipment used in data collection," or "delicate, complex equipment."

The Commissioner has considered the comments and has modified the language of § 58.63(b) to require that standard operating procedures describe in "sufficient" detail the procedures to be used in cleaning, testing, and standardizing equipment. The Commissioner points out that § 58.81(a) (standard operating procedures) states that the written standard operating procedures are to be those which management is satisfied are adequate to ensure the quality and integrity of study data. While the Commissioner does not find it feasible to confine the requirement for standard operating procedures to "major" equipment, he does find that the regulation clearly contemplates that the required procedures need be only as detailed as deemed necessary to assure the integrity of the study data. Simple equipment, therefore, should require only brief standard operating procedures.

126. Five comments suggested that written records for nonroutine repairs should only be required where the nature of the malfunction could affect the validity and integrity of the data.

The Commissioner rejects this suggestion because it is not always possible to make this judgment ahead of time.

127. Many comments argued that the recordkeeping requirements of § 58.63(c) are excessive.

The Commissioner has concluded that the cost of maintaining records of cleaning exceeds the benefits, and this requirement is deleted. However, the requirement for maintaining records of all inspections, maintenance, testing, calibrating and/or standardizing operations is retained because these records may be necessary to reconstruct a study and to assure the validity and integrity of the data.

128. One comment proposed that a new sentence, reading as follows, be added to § 58.63(c): "Where appropriate, the written record noted above may consist of a notation temporarily fastened to the piece of equipment stating when the last specified action with respect to the equipment was taken."

The Commissioner finds that the suggested approach is not precluded by the language of the section as written, but cautions that where such an approach is used, the notations constitute records which must be retained as required by § 58.195(f).

129. One comment asked whether each client of a contract facility must receive a copy of the equipment maintenance and calibration records.

The Commissioner concludes that the regulation does not so require.

TESTING FACILITIES OPERATION

Standard Operating Procedures

130. Two comments suggested deleting § 58.81 in whole or in part. Several others said the requirements for standard operating procedures were unnecessary and burdensome.

The Commissioner does not agree. The use of standard operating procedures is necessary to ensure that all personnel associated with a nonclinical laboratory study will be familiar with and use the same procedures. These requirements will prevent the introduction of systematic error in the generation, collection, and reporting of data, and they will ensure the quality and integrity of test data that are submitted to FDA to become the basis for decisions made by the agency. The Commissioner recognizes that the requirements for standard operating procedures may place an additional burden on testing facilities, but finds that the resulting benefits should outweigh the burden. The requirements will benefit the public by producing better quality data and will benefit the testing facility by reducing the need to repeat nonclinical laboratory studies because of errors in the data.

131. A few comments suggested that responsibility for the standard operating procedures should be specified.

The Commissioner has concluded that this function should reside with the management of a facility, and the first sentence of § 58.81(a) is revised accordingly.

132. Several comments suggested that the responsibility for authorizing significant changes in established procedures be vested in someone other than management.

The Commissioner disagrees. Because standard operating procedure will often apply to more than one study in a testing facility, the Commissioner believes that significant changes to a standard operating procedure, which could affect several different studies, should be authorized by management.

133. Several comments stated that standard operating procedures should not apply to certain types of test systems, that the requirement would introduce difficulties in open-ended exploratory experimentation and electromedical equipment testing, that the approach would not lend itself to rapidly changing methodology such as mutagenicity testing, and that requiring chemical standard operating procedures for each test and procedure was not realistic.

The Commissioner agrees that routine standard operating procedures should not apply to exploratory studies involving basic research. He does not agree, however, that electromedical equipment testing should be exempt unless such testing does not fall under the definition of "nonclinical laboratory study." Standard operating procedures are feasible for studies using methods which change rapidly and for studies using any test system. In the case of chemical procedures, the Commissioner finds that it is realistic to require written standard operating procedures for each test.

134. One comment recommended that the phrase "written standard operating procedures" in § 58.81(a) be changed to "documented appropriate operating procedures." The same comment suggested that the term "ensure" in the first sentence of § 58.81(a) be changed to "maintain."

The Commissioner disagrees with both suggestions. The term "standard operating procedures" refers to routine and repetitive laboratory operations. "Appropriate operating procedures," as a phrase, implies that such procedures could be changed at will. The Commissioner also rejects the suggestion that "ensure" be changed to "maintain." The purpose of written standard operating procedures is to ensure the quality and integrity of the data generated in the course of nonclinical laboratory study. The term "maintain" assumes the procedures already in existence are sufficient to ensure the quality and integrity of the data when, in fact, they may not be sufficient.

135. One comment said that the term "adequate" in the first sentence of § 58.81(a) is a nonprecise term.

The Commissioner agrees, but finds that a testing facility may have a broad range of divergent standard operating procedures for many different studies and that it is impractical to define the adequacy of such procedures for all types of tests. A determination of the adequacy of each standard operating procedure is the responsibility of the management of the testing facility.

136. Numerous comments asked what changes or deviations from standard operating procedures should be documented in the raw data, as required in § 58.81(a). One comment said any deviation should be documented, whether authorized or not.

Every deviation or change in a standard operating procedure should be documented in the raw data. The second sentence of § 58.81(a) has been revised for clarity. It now reads: "All deviations in a study from standard operating procedures shall be authorized by the study director and shall be documented in the raw data."

137. Seven comments indicated that it is inappropriate to require that every minor deviation be documented and reported in writing to the QAU.

The Commissioner agrees that, because the QAU is no longer required to maintain copies of standard operating procedures, it is inappropriate to require that every deviation be reported in writing to the QAU. It is sufficient that all deviations from standard operating procedures be authorized by the study director and documented in the raw data. No exceptions can be made for "minor" deviations. Because any deviation or change may affect the outcome of a study, it is not possible to judge in advance whether or not a deviation is, in fact, "minor."

138. Several comments indicated that the requirement for standard operating procedures should be general in nature.

The Commissioner disagrees. In the proposal, the Commissioner cited evidence from agency investigations of certain testing facilities that had failed to maintain written standard operating procedures of the kind outlined in § 58.81(b). As a result, certain technical personnel were unaware of the proper procedures required, e.g., for care and housing of animals, administration of test and control articles, laboratory tests, necropsy and histopathology, and handling of data. The Commissioner has concluded that a specific delineation of standard operating procedures will allow for uniform performance of testing procedures by personnel and consequent improvement in the quality of the data.

139. Two comments indicated that the requirements for standard operating procedures set out in § 58.81(b) (1) through (12) largely concern animal studies and that this should be so indicated in this section.

The Commissioner agrees that many of the provisions listed in § 58.81(b) are applicable only to studies involving animals. Such is true, however, of many provisions throughout the regulations, and no special mention of the fact is required here. The Commissioner emphasizes that operations requiring standard operating procedures are not limited to those listed in § 58.81(b).

140. One comment suggested that the phrase "and control" be deleted from the first sentence of § 58.81(b)(3), which requires standard operating procedures for test and control articles, because a control article may often be a competitor's product.

The Commissioner does not agree. Where a control article is a commercially available product, its specifications and characterization may be documented by its labeling.

141. Several comments, suggested that the last sentence of proposed § 58.81(b)(3), which reads: "The testing program shall be designed to establish the identity, strength, and purity of the test and control substances, to assess stability characteristics, where possible, and to establish storage conditions and expiration dates, where appropriate" be deleted or suggested that the sentence be transferred to another section.

The Commissioner agrees. The sentence is deleted from § 58.81(b)(3), and appropriate portions of the sentence are transferred to § 58.105(a). The concepts expressed in this sentence properly belong in the section of the regulations relating to "Test and Control Article Characterization." The phrase "testing and administration" has been deleted from the first sentence of § 58.81(b)(3) for the same reason. To specify clearly the Commissioner's intent, "method of" has been added to § 58.81(b)(3) to modify "sampling." Revised § 58.81(b)(3) now reads: "Receipt, identification, storage, handling, mixing and method of sampling of the test and control articles."

142. One comment stated that § 58.81(b)(9), "Histopathology," and § 58.81(b)(8), "Preparation of specimens," were duplicative.

The Commissioner has revised § 58.81(b)(8) to read: "Collection and identification of specimens" to distinguish the requirement from § 58.81(b)(9), "Histopathology." The term "histopathology" covers the examination of specimens, not their collection and identification.

143. Eight comments recommended a rewording of the requirement in proposed § 3e.81(b)(12) that standard operating procedures be established for the preparation and validation of the final study report.

The Commissioner concludes that the requirement should be deleted because the reporting provisions of § 58.185 adequately describe the requirements for final reports. A new paragraph, § 58.81(b)(11), covering "maintenance and calibration of equipment," has been added to reflect the requirements of § 58.63(b).

144. Seven comments suggested that in § 58.81(c) the requirement that standard operating procedures be available at all times to personnel in the

immediate bench area be broadened to be within "easy access." Another comment said the location of such materials should be left to the facility's discretion.

The Commissioner has concluded that unless standard operating procedures are immediately available within the laboratory area they are not with "easy access" and may not be consulted by personnel when routine operations are being performed. The first sentence in § 58.81(c) has been edited for clarity, but the requirement remains.

145. Several comments were received regarding § 58.81(c) and the use of textbooks as standard operating procedures. One comment suggested that textbooks be considered appropriate as part of a standard operating procedure. Two comments assumed that standard operating procedures would permit the incorporation of textbooks by reference. One comment suggested that supplementary material should be written to augment textbooks. An additional comment suggested that textbooks be used in the absence of standard operation procedures.

Standard operating procedures should be set forth in writing, and textbooks may be used as supplements to written standard operating procedures. Reference to applicable procedures in scientific or manufacturer's literature may be used as a supplement to written standard operating procedures. For example, a standard operating procedure could refer to the pertinent pages of any portion(s) of a textbook or other published literature that might be pertinent to a laboratory procedure performed; these supplementary materials need not be incorporated verbatim in the standard operating procedure, but would be required to be immediately available in the laboratory area for the use of personnel. The last sentence of § 58.81(c) is revised to make this point clear. Additionally, § 58.81(d) regarding a historical file of standard operating procedures has been clarified to read: "A historical file of standard operating procedures, and all revisions thereof, including the dates of such revisions, shall be maintained."

Reagents and Solutions

146. Numerous comments on § 58.83 said that to require that the labeling of reagents and solutions in laboratory areas include the method of preparation was neither feasible nor necessary.

The Commissioner agrees and is deleting the phrase "method of preparation" from § 58.83 because the method of preparation could be too lengthy to fit readily on the label. The method of preparation of reagents and solutions should, however, be addressed by the standard operating procedures.

147. Several comments stated that the provision for the handling and use of deteriorated materials and materials of substandard quality should specify only that they not be used and should not specify or require their removal from the laboratory because their removal should be left to the discretion of the laboratory.

The Commissioner agrees, and § 58.83 has been revised accordingly.

148. One comment suggested that the phrase "used in nonclinical studies" be substituted for the phrase "in the laboratory areas" in the first sentence of § 58.83.

The Commissioner disagrees with this comment. All reagents and solutions used in a laboratory conducting a nonclinical study should be properly labeled as provided in the regulation to preclude inadvertent mixups of reagents and solutions that are used in such studies with those that are not intended for such use.

149. Two comments suggested that the phrase "Deteriorated materials and materials of substandard quality" in the second sentence of the section be changed to incorporate the terms "reagents" and "solutions."

The Commissioner agrees and is revising the second sentence of § 58.83 accordingly. Revised § 58.83 now reads: "All reagents and solutions in laboratory areas shall be labeled to indicate identity, titer or concentration, storage requirements, and expiration date. Deteriorated or outdated reagents and solutions shall not be used."

Animal Care

150. Several comments raised the issues of unnecessary animal experimentation and the humane care of animals.

The issue of using animals in laboratory experiments designed to establish the safety of regulated products has been raised many times in the course of agency rulemaking. The position of FDA has been consistent on this issue. The use of animal tests to establish the safety of FDA-regulated products is necessary to minimize the risks from use of such products by humans. The humane care of test animals is a recognized and accepted scientific and ethical responsibility and is encouraged both by various agency guidelines and the Animal Welfare Act. The good laboratory practice regulations should, in fact, encourage the humane treatment of animals used in nonclinical laboratory studies by establishing minimum requirements for the husbandry of animals during the conduct of such studies. In addition, there should occur a reduction in the amount of animal testing that has to be repeated or supplemented because the original studies were inadequate or inappropriate to establish the safety of FDA-regulated products.

151. Numerous comments objected to the incorporation by reference of guidelines and standards proposed in § 58.90(a).

As noted early in the preamble, all references to other standards such as the Animal Welfare Act of 1970 and HEW Publication No. (NIH) 74-23 have been deleted. Section 58.90(a) is revised to read: "There shall be standard operating procedures for the housing, feeding, handling and care of animals."

152. Several comments stated that the quarantine of animals required in § 58.90(b) was impossible in some cases, unnecessary under certain conditions, and would prevent the use of certain animals, such as "timed-pregnant" mice. Other comments said the paragraph could be interpreted to require a separate quarantine area or an extensive quarantine time period.

The purpose of this paragraph is to require that the health status of newly received animals be known before they are used. This requires a separate quarantine area where necessary to determine animal health status. The concept of "separate areas" has been previously discussed. In some cases, depending on such factors as the species or type (e.g., time-pregnant) of animal, or the source and the nature of the expected use of the animal, a health evaluation can be made immediately, or soon after arrival, resulting in a very short quarantine period. The regulation does not preclude this type of health evaluation if it is done in accordance with acceptable veterinary medical practice.

153. Several comments stated that quarantine is unnecessary when animals are obtained from reputable or specific pathogen-free sources.

A health evaluation is required of all newly received animals regardless of the supply source, although the source can be a factor in determining the degree or depth of health evaluation required. Seldom can the conditions under which animals are transported from their source be considered certain to preclude the possibility of exposure of the animals to disease.

154. Some comments requested deletion of § 58.90(b) because it duplicates the animal care requirements regulations.

The Commissioner rejects these comments. The agency is responsible for animal care procedures as they pertain to testing facilities conducting nonclinical laboratory studies, and the provisions are appropriately included in § 58.90(b).

155. Several comments said that the requirements of § 58.90(c) and (d) concerning the isolation of known or suspected diseased animals and keeping animals free of disease or conditions that would interfere with the conduct of the study were impractical.

For clarity, these paragraphs are revised and combined in § 58.90(c). This paragraph deals only with those diseases and conditions that might interfere with the study. This excludes a wide range of diseases and conditions and allows the consideration of such factors as etiology and whether the disease is communicable. The section does not require isolation of all animals in a shipment from a study when only one or some of the animals are diseased, and it covers only those animals that are known or suspected to be diseased.

156. Some comments suggested that specific requirements be provided for the management of diseased animals, and one comment said the veterinary staff should be able to treat diseased animals as they deem proper.

The Commissioner concludes that it is beyond the scope and purpose of these regulations to describe detailed requirements concerning the management of diseased animals and that § 58.90(c) is sufficiently explicit to exclude the use of diseased animals that would interfere with the purpose or conduct of a nonclinical laboratory study. The regulation does not prohibit the treatment of diseased animals if such treatment does not interfere with the study. If treatment will interfere with the study, the diseased animals shall be removed from the study.

157. More than 60 comments objected to or requested revision of proposed § 3e.90(e), which called for the unique identification of all animals used in nonclinical laboratory studies. Fifty-four of the comments addressed specific issues related to this concept, e.g., unique identification of mice, costs of such systems, application to suckling rodents, injury to animals from identification systems, effects of dyes or tattoos, a lack of need in single-dose or short-term experiments, and cage identification instead of animal identification with precautions being taken to prevent animal mixups.

In the absence of a proven and acceptable method of unique identification for small rodents, the Commissioner is revising § 58.90(d) to require appropriate identification for warmblooded animals, excluding suckling rodents, which require manipulations and observations over extended periods of time. Suckling rodents have been excluded from the requirements because of potential cannibalization by the mother. The same information needed to specifically identify each animal is required on the outside of housing containers or cages. Such identification should substantially reduce the possibility for animal mixup. Because of the varied nature of the tests conducted and the test systems used, the manner of identification is left to the discretion of the testing facility.

The Commissioner advises that whenever a study requires that animals be removed from and returned to their home cages, there is a potential for mixup. Thus, if a single-dose or short-term study requires such manipulations, the animals shall receive appropriate identification.

Because the requirement for unique identification has been deleted, the concerns expressed regarding cost, injury to the animals from various identification systems, and the effects of dyes or tattoos are no longer germane.

158. Two comments questioned whether the study director could in practice assure unique identification as proposed in § 3e.90(e), without direct observation.

The requirement has been deleted, along with the requirement for unique identification.

159. Two comments requested deletion of the last sentence of proposed § 3e.90(e) regarding the identification of specimens.

The Commissioner concludes that proper specimen identification is an integral part of proper study conduct, but that the requirement more properly belongs under standard operating procedures. Consequently, § 58.81(b)(8) now incorporates this provision.

160. One comment inquired whether, in the event animals of the same species in different tests were in the same room, FDA would require identification of all compounds. This, it was felt, would raise confidentiality questions for a contract testing facility.

The Commissioner advises that the use of coding to identify test or control articles is not precluded by § 58.90(e). The concluding phrase, "to avoid any intermixing of test animals," was deleted as redundant.

161. Proposed § 3e.90(g) required comparison of cage and animal identification for each transfer, procedures for verification, and written permission of the study director for location transfer. Seventeen comments objected to part or all of these requirements as vague, burdensome, unnecessary, and redundant.

The Commissioner agrees, and the paragraph is deleted. Procedures for the transfer and proper placement of animals are required as standard operating procedures in § 58.81(b)(12).

16? several comments claimed that the requirements of proposed § 3e.90(h), redesignated § 58.90(f), were redundant in view of the requirement for standard operating procedures in § 58.81. Other comments stated that the incorporation of guidelines by reference was inappropriate.

The Commissioner concludes that the requirement that animal cages, racks, and accessory equipment be cleaned is appropriately included in this section even though there is some overlap with the language of § 58.81, standard operating procedures. The reference to other agency guidelines has been deleted.

163. Three comments asserted that sanitization should not always be done, because it could in certain cases interfere with the conduct of the study.

The Commissioner agrees and points out that the language in redesignated § 58.90(f) permits cleaning and sanitization at appropriate intervals.

The section now reads: "Animal cages, racks and accessory equipment shall be cleaned and sanitized at appropriate intervals."

164. Many comments objected to proposed § 3e.90(i), redesignated § 58.90(g), which requires periodic analysis of feed and drinking water for "known interfering contaminants." Certain of these comments requested clarification or deletion, or expressed concern about the costs involved. Others argued that the use of positive and negative controls would accomplish the intent of the requirement, or that certificates of analysis from local water supply authorities and feed manufacturers should be permissible. Finally, a few comments said analysis of feed and water should only be required when there is reason to believe that a particular contaminant may have an effect on the study, and comments said the analysis requirements should be specified in the protocol.

Most of the objections raised against the analytical requirements of the section were based on misinterpretation of such requirements. The intent of the Commissioner was to require analysis for contaminants known to be capable of interfering with the nonclinical laboratory study and reasonably expected to be present in the feed or water, and not to require analysis of feed and water for all contaminants known to exist. Certain contaminants could affect study outcome by masking the effects of the test article, as was observed in recent toxicological studies of pentachlorophenol and diethylstilbestrol, in which the feeds used as carriers for the test articles were found to contain varying quantities of pentachlorophenol and estrogenic activity, respectively, that invalidated these studies by producing erratic results. The use of positive and negative controls in these examples was insufficient to compensate for the variability in contaminant content. Therefore, the Commissioner agrees with the comments that suggested that analysis of feed and water only be done when there is reason to believe that a particular contaminant may have an effect on the study, and may be present in the feed or water, and the language of both redesignated § 58.90(g) and § 58.120(a)(9) have been revised to make this clear. This clarification of the regulations should allay the concerns of those comments relating to certificates of analysis, costs, and precise definition of impurities. Acceptable contaminant limits must be specified by the protocol (§ 58.120 (a)(9)), and should be determined at the time the protocol is developed, taking into account the scientific literature, the availability of suitable analytical methodology, and the practicability of controlling the level of the contaminant.

165. One comment suggested additional requirements for, e.g., analysis of nutrients and reserve samples of feed at the testing facility.

Nutrient analysis should be addressed by the facility's standard operating procedures. Requirements for reserve samples of test or control articles/

carrier mixture (e.g., feed) are set forth in § 58.113(b). The Commissioner concludes that minimum requirements for those items are set forth in the regulation. The regulation does not preclude the setting of additional requirements by the sponsor and/or the testing facility.

166. Proposed § 3e.90(j) would have required feed to bear an expiration date. Twenty-three comments argued that this requirement is of dubious value, is beyond the current state of the art because of varied storage conditions, and that commercially available feed is not expiration dated, making the requirement impractical or impossible.

The Commissioner agrees with these comments, and this requirement is deleted.

167. Several comments argued that the requirement for weekly changes of bedding should be deleted. The comments stated that, in certain cases, weekly bedding changes are contraindicated.

The Commissioner agrees, and the phrase "at least once per week" is removed from § 58.90(h), which now reads, "Bedding * * * shall be changed as often as necessary to keep the animals dry and clean."

TEST AND CONTROL ARTICLES

Test and Control Article Characterization

168. One comment suggested that § 58.105 be deleted; another suggested that the entire subpart be condensed; and three comments suggested that the section is not generally applicable to nonclinical device studies, particularly with reference to such terms as "identity, strength, quality, and purity."

The Commissioner does not agree that the section should be deleted. Its purpose is to assure that the article being tested has been thoroughly characterized or defined and that either the sponsor or the testing facility has a thorough understanding of what is being tested. The Commissioner agrees that the subpart should be condensed and has shortened it. Section 58.105(a) is modified by the inclusion of the sentence "the identity, strength, purity, and composition or other characteristics which will appropriately define the test or control article." This addition provides for characterization of various products, including devices in terms suited to their identity or uniqueness.

169. One comment argued that the requirement that "??? contained in the test and control substances" be accounted for, as proposed in § 58.105(a), was vague.

By this provision the Commissioner intended to indicate the need to identify and characterize solvents, excipients, inert ingredients and/or impurities that might be part of the test substance. Because these materials are included by definition in the term "test article," the Commissioner has determined that the original language was unnecessary and has deleted it.

170. Three comments sought definition of the word "batch" as used in § 58.105(a).

The term "batch" is now defined in § 58.3(n).

171. Seventeen comments on § 58.105(a) stated that because some control or reference articles might be a competitor's or a supplier's product, the assay and method of synthesis might not be available or might be confidential.

The Commissioner concludes that, in those cases where a competitor's or supplier's product is used as a control article, such products will be characterized by the labeling and no further characterization is necessary.

172. One comment stated that the testing facility should not be responsible for identity, strength, quality and purity and that this responsibility should rest with the sponsor. This comment also suggested that the requirement, as written, would inhibit the conduct of blind studies.

The Commissioner concludes that it is the responsibility of testing facility management to assure that the requisite tests have been done, either by the sponsor or by the test facility (see § 58.31(d)). In those cases where a testing facility is unable to perform the characterization test or is performing blind studies, the sponsor should perform the required testing and notify testing facility management that the characterization of the test or control article has been performed. The section, as reversed, does not inhibit the conduct of blind studies: it does not require that the sponsor give the characterizing information to the testing facility, only that the sponsor notify the testing facility that the required characterization has been done.

173. One comment suggested that the requirements of § 58.105 should only apply if the integrity of the study is threatened, and another suggested that any contaminants in a test or control article should be evaluated only with respect to their impact on study validity.

The Commissioner does not agree that the requirement should be so limited. Thorough characterization of the article under test is essential because the results of the test may be compromised by possible contamination. Only by knowing the identity and quantity of the components can one predict their effect on the study. The evaluation of the impact of test and control article contaminants on the validity of the study is an important part of the thorough characterization of the test and control articles.

174. Thirteen comments suggested that characterization of the test article be permitted during the study, after its completion, or left to such time as specified in the protocol.

The Commissioner concludes that characterization of the test or control article should be determined before the initiation of the study in order to provide a means of controlling variations from batch to batch as well as to make certain that the test article meets the specifications of the protocol. As previously stated, a thorough understanding of the nature of the test article is a basic requirement for assuring the absence of contaminants that may interfere with the outcome of the study. When the stability of the test and control articles has not been determined before initiation of the study, the regulation requires periodic reanalysis of each batch of test and control articles as often as necessary while the study is in progress.

175. One comment stated that the phrase "verifying documentation" in § 58.105(a) was not clear.

The Commissioner has determined that the phrase is not needed, and § 58.105(a) is revised to delete it.

176. Seven comments suggested that stability studies required by § 58.105(b) may not always be necessary; three comments suggested that common vehicles and placebo controls, such as water, should be omitted from stability studies.

Some degree of instability may be associated with every test article that might be the subject of nonclinical laboratory study. The Commissioner concludes, therefore, that stability information must be included as part of the information upon which the agency bases a decision regarding the safety of the article. If the stability of common vehicles is generally recognized and can be documented, stability testing is not required.

177. Twelve comments suggested that the term "production" in proposed § 3e.105(c) should be deleted or changed by substitution of other terms such as "approved" or "released," stating that the use of the word was confusing. Several other comments stated that the requirement that test and control substances be derived from the smallest number of production batches consistent with their stability was not always possible or necessary.

The Commissioner agrees that the section was confusing and finds that the requirement is adequately covered by § 58.105(a). The word "batch" has been defined in § 58.3(n), and proposed § 3e.105(c) has been deleted.

178. One comment suggested that the test and control articles should be derived from a large number of batches to increase the probability that test and control articles are representative.

The Commissioner agrees that, in some cases, combining representative samples of test or control articles from various production sources or lots to form a batch may be desirable. Where this is done, however, the resulting batch, rather than the individual samples, must be characterized in accordance with § 58.105(a).

179. Eight comments on § 58.105(d) suggested that the requirement for reserve sample retention be restricted to those substances whose stability had not been previously determined. Another comment suggested that the section seems to require that a reserve sample of water be retained if water is used as the control article, and another comment suggested that the retention of a reserve sample should be left to the discretion of the sponsor.

The Commissioner does not agree that the decision to retain a reserve sample should be at the discretion of the sponsor. Maintaining a reserve sample is necessary to provide independent assurance that the test system was exposed to the test article as specified in the protocol. Reserve samples need not be reanalyzed routinely if the stability of the test or control article is well established. If, however, the results of a study raise questions as to the composition of the test or control article, retention of reserve samples allows resolution of the question. Retention of a reserve sample of water is required when it serves as the control article in a nonclinical laboratory study.

180. Eight comments on § 58.105(d) suggested that containers should be comparable rather than identical to maintain approximate ratio of mass of article to container volume.

Reserve samples should be stored in containers and under conditions that maximize their useful life. The specifications for containers are deleted from § 58.105(d), however, and are now left to the discretion of the study director.

181. Six comments said § 58.105(d) duplicated §§ 58.105(b) and 58.113 (a)(2); three said that the requirement that the reserve sample be analyzed at the time the batch is depleted, at the termination of the study, or at the expiration date may result in unnecessary testing. One comment suggested that a portion of the remaining article should be tested rather than testing the reserve sample.

The Commissioner agrees that the requirement for routine reanalysis of all test or control articles is unnecessary where stability characteristics have been well established, and this requirement has been deleted. The Commissioner does not agree that the cited sections duplicate one another. Section 58.105(b) concerns the stability of test and control articles in a carrier mixture. But § 58.105(d) concerns reserve samples of test and control articles.

182. A number of comments on proposed § 3e.105(f) sought clarification of the requirements, definition of the term "quarantine," and deletion of the requirement to reanalyze batches returned from distribution.

The Commissioner has examined the provision as proposed and has found that the intent is achieved by the provisions of § 58.107 (test and control article handling). Proposed § 3e.105(f) has, therefore, been deleted.

Test and Control Article Handling

183. One comment asserted that § 58.105 covered the specifics for handling test and control substances and that § 58.107 should be deleted.

The Commissioner disagrees with the assertion that § 58.107 repeats § 58.105. The provisions of § 58.105 apply to the characterization of test and control articles and their storage prior to use. Section 58.107 sets forth provisions for the handling and distribution of test and control articles during the course of a nonclinical laboratory study. The purpose of this section is to provide further mechanisms to assure that test and control articles meet protocol specifications throughout the course of the study, and that test article accountability is maintained.

184. Other comments argued that the language of § 58.107 should be modified and that, as written, the section was impractical.

The Commissioner does not agree that the requirements are impractical. The section has, however, been edited for clarity. Section 58.107(a) now reads, "There is proper storage." Because contamination is only one of the consequences that may result from improper handling during distribution, the Commissioner has revised § 58.107(b) to read: "Distribution is made in a manner designed to preclude the possibility of contamination, deterioration, or damage."

Mixtures of Articles with Carriers

185. Many comments stated that the requirements of § 58.113 should only apply to certain types of studies, such as long term feeding studies, or should apply only in cases where problems of instability might result from mixing the test article with a carrier.

The Commissioner does not agree. The need to know that the test system is being exposed to the amounts and types of test and control articles that are specified in the protocol is common to all types of studies. The effect of mixing on the concentration and stability of the test or control article in the mixture cannot be predicted beforehand.

186. Six comments stated that the requirement that each batch of a test or control article that is mixed with a carrier be tested for uniformity of mix, stability, and release, as proposed in § 58.113, was excessive.

The Commissioner has reviewed the reasons advanced by the comments and has deleted the "for each batch" requirement. Once the uniformity of the mixture has been established for a given set of mixing conditions, it is not necessary to establish the uniformity of each subsequent batch that is mixed according to the same specifications. Similar considerations apply to stability testing. Section 58.113(a)(1) introductory text and (a) now read: "For each test or control article that is mixed with a carrier, tests by appropriate analytical methods shall be conducted: (1) to determine the uniformity of the mixture and to determine, periodically, the concentration of the test or control article in the mixture." The sentence, "[I]f the nonclinical study is to be performed as a blind study, enough individual samples of the mixture shall be returned to the sponsor for analysis," has been deleted. The requirement for analysis of test or control article mixtures is adequately addressed by the revised language of § 58.113(a)(1). The mechanism of satisfying the requirement is left to the testing facility. Blind studies are discussed in paragraph 172 above.

187. One comment stated that the possibility of administration by other than the oral route should be considered.

The Commissioner agrees, and reference to the route of administration is removed.

188. Several comments said the acute and subacute toxicity studies are often conducted before there is extensive knowledge about a drug's stability and that in such cases the drug might be prepared daily. In addition, it was suggested that § 58.113(a)(2) allow for concurrent stability studies.

The Commissioner agrees with the comment and has revised the regulation to allow concurrent studies of stability to proceed with the ongoing nonclinical laboratory study.

189. Three comments on § 58.113 suggested that establishing expiration dates for a substance used up in a week seemed too stringent. Many comments suggested that the expiration dating requirement be eliminated entirely because batch sizes are established so that they will be used up prior to deterioration of the test article.

The Commissioner has considered the comments and has revised, as noted above, the requirement for labeling each batch of test or control article carrier mixture to permit concurrent stability testing. The Commissioner declines to eliminate entirely the requirement for listing of expiration dates. Expiration dates should be used, when known, to minimize the possibility that subpotent, unstable, or decomposed test or control article carrier mixtures will be used. New § 58.113(c) requires that, where any of the components of the test or control article carrier mixture has an expiration date, that date shall be clearly shown on the container. If more than one component has an expiration date, the earliest date shall be shown.

190. Many comments on proposed § 3e.113(a)(3) stated that the requirement for tests to determine the release of the test or control substance from the carrier needed to be clarified, might be impossible to do, and were not always necessary.

The Commissioner has reviewed the comments and the section and finds that such testing should be adequately addressed by the protocol. He has, therefore, deleted the section.

191. Eleven comments suggested that the requirement that reserve samples of each batch of test or control article-carrier mixture be retained was excessive and impractical.

The Commissioner does not agree. Maintenance of reserve samples of these mixtures is necessary for the same reasons that reserve samples of test and control articles themselves are necessary. These reasons are stated in paragraph 179 above.

192. Proposed § 3e.115 incorporated principles set forth in other regulations and has, accordingly, been deleted. (See the discussion in paragraph 3.)

PROTOCOL FOR AND CONDUCT OF A NONCLINICAL LABORATORY STUDY

Protocol

193. Several comments said the protocol requirements of § 58.102(a) were not relevant to specific test articles, e.g., electronic diagnostic instrumentation. Other comments objected to requiring a protocol for short-term studies or for routine tests described elsewhere in 21 CFR Chapter I. Additional comments proposed that specific requirements be imposed only where applicable, and one comment said the protocol should focus on what is intended rather than on how the intended result is to be achieved.

The Commissioner has previously discussed the types of tests and the conditions within the scope of Part 58. Because of the broad range of studies covered, specific sections may not apply to all studies. However, the Commissioner declines to exempt short-term studies or routine tests from these requirements. Any study which qualifies as a nonclinical laboratory study is subject to the requirements. The good laboratory practice regulations are both process-oriented and product-oriented, and are designed to ensure, insofar as possible, the quality and integrity of nonclinical laboratory data submitted to FDA in support of regulated products. The Commissioner recognizes that some of the requirements of this section have often not been traditionally included in a protocol. He has nonetheless concluded that the requirements are essential to ensure that all operations needed to fulfill the objectives of a study are performed and that the complete list of information required by this section is necessary to ensure that deviations, should they occur, are readily apparent.

194. One comment asked what was meant by "all methods" in § 58.120; one suggested deletion of the word "approved" to describe the protocol; and another suggested that reference to statistical methods in § 58.120(a) be deleted and that a new paragraph on statistical methods be added to the list of information required.

"All methods" refers to all operations necessary to achieve the objectives of the study, e.g., analytical methods, randomization procedures, etc. If such methods are from published sources, citation of the source would fulfill this requirement. If the methods are not from published sources, full descriptions would need to be included in the protocol. The word "approved" is retained to emphasize that a sponsor or testing facility should have a mechanism for evaluation and approval of initial protocols and all amendments. A new paragraph (a)(16) is provided to emphasize the need to consider statistical methodology in preparing a protocol.

195. Ten comments objected to the inclusion, in proposed § 3e.120(a)(3), of stability methodology as a protocol requirement because such methodology may not have been developed before the study was begun. Another comment suggested deletion of this requirement as not relevant to a protocol, while three comments suggested revision.

The Commissioner recognizes that stability data may not be available when a study is initiated, and this requirement is deleted from the section. The Commissioner emphasizes, however, that determination of the stability of the test and control articles is a responsibility of the study director, that determination of the stability of the articles per se is required under § 58.105(b), and that determination of the stability of the article/carrier mixes is required under § 58.113.

196. Numerous comments on proposed § 3e.120(a)(4) objected to the listing of the names of laboratory assistants and animal care personnel in the protocol because these jobs are subject to constant turnover or periodic rotation.

The Commissioner agrees that laboratory assistants and animal care personnel need not be identified in the protocol. The list of personnel required to be named is transferred to § 58.185(a)(12).

197. One comment proposed that listing the name of the sponsor and name and address of the testing facility required by § 58.120(a)(3) be restricted to studies done under contract.

The Commissioner does not agree with restricting this requirement to studies done under contract because a testing facility, though a division of the sponsor, may have a specific designation and a location different from the sponsor's, and this information is necessary to determine the exact location of the study.

198. Numerous comments on § 58.120(a)(4) objected to specifying starting and completion dates in the protocol because changing priorities may make such specification impractical. Another comment proposed deletion of the requirement for dates as not relevant to a protocol.

Changing priorities may cause changes in starting dates. For this reason the requirement calls for the proposed dates. If the actual dates differ from the proposed dates, the change should be reflected in a protocol amendment. The dates may be needed in the reconstruction of the study.

199. Ten comments on proposed § 3e.120(a)(7) objected that the proposed date for submission of the final study report to management or to the sponsor was not relevant to a protocol, and one requested a definition of the term "completion date."

The Commissioner agrees that the proposed submission date is not relevant, and the provision is deleted.

200. Numerous comments on § 58.120(a)(6) suggested requiring age of the test system only where applicable or substituting age range for age. Several objected to the requirement for justification for selection of the test system as not relevant to protocol requirements. Additional comments proposed that the requirement for justification be limited to nonroutine systems.

The Commission agrees that age of the test system may not always be critical, and § 58.120(a)(6) now requires number, body weight range, sex, source of supply, species, strain and substrain, and age of the test system only "where applicable." The Commissioner does not agree that justification for selection of the test system is not relevant to a protocol or should be limited to nonroutine systems. Such justification is an integral and essential part of every protocol and to emphasize its importance, the Commissioner is establishing a separate paragraph for this requirement, § 58.120(a)(5).

201. Several comments on § 58.120(a)(8) (proposed § 3e.120(a)(10)) objected that the method of randomization was not relevant to the protocol and suggested requiring justification for the selected method only when nonroutine methods are selected; four comments said justification of the method of randomization is unnecessary; and one comment proposed revised language regarding method of randomization.

The Commissioner finds that the method of randomization or other methods of controlling bias are relevant and are essential parts of a protocol, whether the methods used may be described as routine or nonroutine. The suggested revision is adopted in part, and § 58.120(a)(8) now reads: "A description of the experimental design, including the methods for the control of bias."

202. One comment said a description of the diet used in the study (proposed § 3e.120(a)(11), now § 58.120(a)(9)) was unnecessary unless the diet was unusual. The comment further said that the necessity for including solvents and emulsifiers was questionable because these might not be known at the time the protocol is written.

The Commissioner advises that the phrase "and/or identification" in § 58.120(a)(9) permits a commercial animal diet to be identified by its name. The need for using solvents or emulsifiers may not be known when the protocol is written; however, when this information is available and the solvents, etc., are selected, this fact should be reflected in a protocol amendment.

203. Nine comments pointed out that the degree of absorption (proposed § 3e.120(a)(14)), now § 58.120(a)(12)) is usually unknown at the time of the preparation of the protocol.

The Commissioner recognizes that absorption studies may be conducted concurrently with or as part of the nonclinical laboratory study and points out that the requirements of § 58.120(a)(12) can be fulfilled by amending the protocol.

204. Nine comments suggested deletion of the requirement that the protocol include the records to be maintained (proposed § 3e.120(a)(16), now § 58.120(a)(14)) because this duplicates the requirements under another provision of the regulation.

The Commissioner concludes that the protocol should include a plan identifying the records to be maintained and, therefore, does not agree that § 58.120(a)(14) should be deleted.

Conduct of a Nonclinical Laboratory Study

205. Several comments objected to the § 58.130(c) requirement that specimens be identified. Three comments proposed revisions to eliminate the list of specific items (test system, study, nature, date of collection) included for identification of specimens. Numerous comments objected to the identification system as overly restrictive, stating that a coding system should be permitted.

The Commissioner rejects the suggested modifications because the requirements are designed to preclude error. The specific items required to identify a specimen are the minimum necessary to prevent mixup of specimens and permit orderly storage. The Commissioner does not agree that this system is overly restrictive because it does not preclude a coding system.

206. Numerous comments objected to the requirement, in § 58.130(e), for recording data in bound books with prenumbered pages as costly, time-consuming, overly restrictive, and difficult for long-term studies. Six were concerned that much information is too voluminous to be recorded directly and that reference to other documents should be permitted to justify changes, and two comments objected to recording "dictated observations" in ink.

The Commissioner agrees that the requirement for bound books is too restrictive in view of both the variety of data recording procedures that can be used in nonclinical laboratory studies covered by this part and the many ways in which data are generated and collected for these studies. He is, therefore, revising the section. As revised, § 58.130(e) does not preclude reference to other documents if the documents are clearly identified and available. The requirements of the section can be met by maintaining the dictation media or an exact transcription.

207. Three comments proposed that § 58.103(e) be revised to reflect the three types of computer entries, i.e., direct on-line recording, input from computer readable forms, and input transcribed from recorded raw data. An additional comment suggested revised language to achieve this purpose; and two comments stated that computer printouts of interim display data need not be maintained when the data are wholly contained in subsequent iterations.

The revised wording of § 58.130(e) is equally applicable to the various forms of computer data entries. The Commissioner advises that where the data for computer input are in machine-readable form, such as marketed-sense cards, or are transcribed from recorded raw data, the machine-readable forms or the recorded raw data would constitute raw data within the definition of this part. Where input is via direct on-line recording, the magnetic media and the program would constitute raw data within the meaning of this part.

208. Three comments objected that a daily signature and date for each entry would be burdensome in studies involving daily measurements on each animal.

Section 58.130(e) does not require signing and dating of every individual item recorded. An entry can consist of several observations of several animals made by the same person.

209. Three comments suggested deletion of proposed § 3e.130(f), which required the review of all recorded data, because this duplicated the function of the study director.

The Commissioner agrees that these requirements are adequately addressed by § 58.33(b), and the paragraph is deleted.

RECORDS AND REPORTS

Reporting of Nonclinical Laboratory Study Results

210. Seven comments said the requirement that the final report include all raw data and calculations proposed in § 3e.185(a)(3) is not practical and that a recapitulation should be adequate.

The Commissioner agrees, and the requirement that all raw data be included in the final report is deleted.

211. Two comments on § 58.185(a)(3) stated that the scope of the term "method" was not clear.

The Commissioner advises that "method" does not mean that either the actual calculations or a step-by-step reiteration of the process be included. The name of the method, the description of the method, or a reference to an article or test describing the method will be sufficient.

212. Several comments on § 58.185(a)(4) stated that the final report should provide only a reference to the information on "strength, quality, and purity" rather than the actual values for those characteristics.

The Commissioner does not agree. The final report should include actual values for all characteristics required for proper identification. Because the actual values for strength, quality, and purity are not, in every case, sufficient for adequate identification, the word "quality" has been stricken and the words "and composition or other appropriate characteristics" have been added. The additional language will permit the use of any characteristic which facilitates identification of the test and control article.

213a. Several comments on § 58.185(a)(5) stated that the requirement that stability of the test and control articles be described should be narrowed.

The Commissioner finds that stability information must be submitted as part of the final report. The extent of stability testing required by these regulations is discussed at paragraphs 176, 185, 186, and 189 above.

b. Comments on proposed § 3e.185(a)(8) (now § 58.185(a)(7)) requested that the words "appropriate and necessary" be inserted following the words "procedure used," for identifying the test system.

The Commissioner is modifying § 58.185(a)(7) to require reporting such details where applicable.

214. Seven comments on § 58.185(a)(12) protested the requirement that the final report include reports of each of the individual scientists or other professionals involved in the study.

The Commissioner concludes that the individual reports are required to assure that the final results reported accurately reflect the findings of the individual scientists.

215. A number of comments on § 58.185(a)(3) objected to reporting the location of the raw data in the final report.

For the purpose of information retrieval, the Commissioner is of the opinion that the location of the raw data should be specified.

216. The Commissioner advises that the list of personnel required to be named in the final report as specified in § 58.185(a)(12) has been broadened to include all professionals. (See paragraph 196 above.)

Storage and Retrieval of Records and Data

217. Several comments requested revision and clarification of "other information" in § 58.190(a).

The phrase "and other information" is deleted because it is subsumed by the specific requirements for documentation.

218. Five comments requested clarification of the term "specimen" as used in § 58.190(b).

The term "specimen" is defined in § 58.3(j) and means any material derived from a test system for examination or analysis. This includes wet specimens, histological blocks, and slides that yield information pertinent to the outcome of the study. Such specimens are required to bear sufficient labeling to permit identification and expedient retrieval.

219. Several comments stated that the prohibition against "intermingling" of specimens was unnecessary if specimens are properly labeled and indexed.

The Commissioner agrees and finds that the storage requirements are adequate to achieve their purpose without any further prohibitions. The reference to intermingling of samples is, therefore, deleted.

220. Seven comments said proposed § 3e.190(c) was unclear or redundant and required the maintenance of unnecessary duplicative files by both the testing facility and the sponsor.

The Commissioner agrees with the comments, and the paragraph is deleted.

221. A number of comments requested that § 58.190(c) provide that more than one person be permitted to be responsible for the archives.

The Commissioner reaffirms the need for one individual to be accountable for the maintenance and security of the archives to prevent access by unauthorized personnel. Such access could lead to the loss of, or damage to, records and specimens required to be maintained by these regulations. This provision does not preclude delegation of duties to other individuals who may help maintain the archives.

222. Comments on § 58.190(e) suggested that coding of archival contents

should be allowed and objected that the section would require four-way indexing.

The paragraph is revised for clarity. As revised, the use of a coding system is permitted; however, the cross-reference indexing system is retained as a requirement.

223. Section 58.190(g) is deleted because the inspection requirements are adequately addressed by § 58.15.

Retention of Records

224. Several comments stated that the proposed record retention requirements were inconsistent with those previously established.

A new paragraph (a) is added to § 58.195 to make it clear that the record retention requirements of this section do not supersede those of any other regulations in this chapter.

225. Several comments pointed out that IND's are not "approved" and asked that the record retention requirements for IND's be clarified.

The Commissioner agrees that the record retention requirements, as they apply to both IND's and IDE's, need clarification. In addition to the fact that IND's are not, in a technical sense, "approved," the Commissioner has considered the fact that when either an IND or an IDE is submitted to the agency, the application may contain voluminous data collected over a number of years. It was not the intent of these regulations that such supporting IND or IDE data be destroyed after 2 years because not all studies submitted at the time of filing may be of interest to the agency until several years after submission. Therefore, a new sentence is added to § 58.195(b)(1), which states that the 2-year retention requirement does not apply to studies supporting notices of claimed investigational exemptions for new drugs (IND's) or applications for investigational device exemptions (IDE's). These records are governed by § 58.195(b)(2) and shall be retained for at least 5 years. This additional language clarifies both agency policy and current scientific practice which is, in most cases, to maintain such study records far longer than 5 years.

226. One comment said the variable record retention periods are unworkable, and another said records should be maintained as long as the public is exposed to a chemical.

The record retention period represents the minimum deemed appropriate. For uniformity, all records may be retained for 5 years. Longer retention periods are unnecessary because each nonclinical testing facility will be inspected every 2 years. Studies conducted at facilities that are in substantial compliance with these regulations will be presumed to be valid. When significant deviations are discovered, steps will be taken to validate individual studies before the record retention period expires.

227. Twenty-three comments on § 58.195(b)(3) objected to the record retention requirement as it applies to terminated or discontinued studies, stating that the requirement goes beyond the intent expressed in the definitions or that FDA lacks the authority to require that such studies be retained.

The Commissioner finds that such studies are frequently capable of yielding information applicable to evaluations of related compounds. In the interest of the public health, all such data derived from studies originally intended to be submitted to the agency should be available to the agency. This is particularly important when studies are terminated because of preliminary findings that the test article causes adverse effects at such low levels that any safe use of the article is precluded. The general question of FDA's authority is discussed in paragraph 5 above.

228. With respect to retention of appropriate samples, including wet specimens, several comments on § 58.195(c) requested that the regulations specifically set forth conditions of storage. Others felt that this requirement would be of doubtful value, and several were concerned that the retention period not exceed that which could adversely affect sample integrity.

The Commissioner states that it would be impractical to attempt to specify the specific storage conditions for sample retention. This should be left to the judgment of the testing facility. It is essential as a check on recorded observations that, wherever possible, samples be retained for confirmation of findings. Such samples should be retained for the minimum period specified in the regulations. The regulation clearly states that fragile samples shall be retained only so long as the quality of the preparation affords evaluation.

229. Three comments on § 58.195(e) objected to archive retention of curricula vitae and job descriptions of all personnel involved in the study.

Section 58.195(e) is revised to permit this information to be retained as part of the testing facility employment records.

230. One comment on § 58.195(f) stated that equipment records should be maintained in an independent log rather than maintained as part of each study.

The Commissioner advises that the language of the section does not preclude such an approach. Records of maintenance and calibration of equipment may be kept in a repair manual or on a tag affixed to the instrument. The reference to cleaning records is deleted.

DISQUALIFICATION OF TESTING FACILITIES

Purpose

231. Many comments were received concerning the general concept and purpose of disqualification.

The Commissioner believes that many of these comments were based, at least in part, on misunderstanding of the frequency with which disqualification might be used. The Commissioner believes disqualification is an important alternative to rejection of specific studies and legal prosecution because it can reduce by consolidation the number of FDA investigations and administrative proceedings that might be required if FDA acted only on a study-by-study basis. To clarify the agency's intent regarding the disqualification mechanism and to allay fears that this sanction might be abused, the Commissioner is revising Subpart K of the regulations to define more clearly the grounds for disqualification.

231. Section 58.200(a) has been revised to clarify the purposes of disqualification. The first purpose stated in the section is to permit FDA to exclude from consideration any completed studies conducted by a testing facility which has failed to comply with good laboratory practice requirements until it can adequately be demonstrated that the noncompliance did not occur during, or did not affect the validity of data generated by, a particular study. Thus, for studies completed before disqualification, the order of disqualification creates a rebuttable presumption that all studies previously conducted by the facility are unacceptable. Such a study may be accepted, however, upon presentation of evidence demonstrating that the noncompliance which resulted in the disqualification did not affect the particular study. The second purpose set forth in the revision of § 58.200(a) is to exclude studies completed after the date of disqualification from consideration until the facility can satisfy the Commissioner that it will conduct studies in compliance with the regulations. (See also the discussion in paragraph 241.)

Ground for Disqualification

232. Many comments argued that the disqualification provisions appeared to be overly harsh, arbitrary, and ambiguous.

To clarify the agency's intent, the Commissioner is revising the section. The primary function of the agency's regulation of nonclinical laboratory testing is to assure the quality and integrity of data used in making judgments about the safety of products regulated by the agency. The grounds for disqualification are based on those types of noncompliance that significantly impair achievement of those objectives. Proposed § 3e.202(a) through (p) is deleted, and new § 58.202(a) through (c) clarifies the policy that a testing facility may be disqualified only if the Commissioner finds all three of the following: (1) That the testing facility failed to comply with one or more of the standards set forth in Part 58 or in any other FDA regulations regarding standards for nonclinical testing facilities (e.g., any supplemental requirements in the IND or IDE regulations); (2) that the noncompliance adversely affected the validity of the data produced by the study; and (3) that other lesser regulatory actions, such as warnings or rejection of data from individual nonclinical laboratory studies, have not been or probably will not be adequate to achieve compliance. This approach will assure that the sanction will not be used in trivial situations, but will be invoked only when the violation has compromised the integrity of a study. It further requires the Commissioner to consider the availability and probable effectiveness of lesser sanctions as an alternative to disqualification. It would not, however, preclude disqualification without prior warning.

As pointed out in the preamble to the proposed regulations, the provisions for disqualification are not to be interpreted as either the exclusive or primary administrative action for noncompliance with good laboratory practice. Disqualification is designed to provide FDA with an enforcement tool that is more efficient and effective than a study-by-study review when it becomes apparent that a testing facility is not capable of producing accurate and valid test results. The disqualification of a nonclinical testing facility will be reserved for the rare case when the rejection of a particular study is an inadequate regulatory response. The testing facility and/or the sponsor of the nonclinical laboratory study may also be prosecuted for violations of Federal criminal laws, including section 301(e) of the Federal Food, Drug, and Cosmetic Act (failure to make a report required under certain other sections of the act, because a grossly erroneous or inadequate report does not fulfill the statutory obligation) and 18 U.S.C. 1001 (submission of a false report to the government). Even where the testing facility is not under a direct statutory obligation to submit information to FDA, and in fact does not send data to the agency but merely transmits them to the sponsor, the facility is likely to be aware that FDA will be the ultimate recipient. In such cases, it may be liable for aiding and abetting in the violation (18 U.S.C. 2) or for causing the violation to be made by a third party.

233. Two comments stated that the disqualification regulation seemed to apply only to private firms.

This interpretation is incorrect. The preamble to the proposed regulations makes clear the policy that the good laboratory practice regulations are to apply to any institution that generates or otherwise prepares safety data for submission to FDA. Included in that definition, to the extent that they prepare safety data to be submitted to FDA in support of petitions for regulated products, are, for example, veterinary and medical clinics, universities and State experimental stations, and State and Federal Government research laboratories. Accordingly, disqualification provisions apply equally to all facilities that prepare safety data for submission to FDA. The language regarding the intended use of sanctions is incorporated into § 58.202(c).

Notice of and Opportunity for Hearing on Proposed Disqualification

234. Several comments stated that the disqualification process, as proposed, would violate due process, deny a formal hearing, and deny a right of appeal to the courts.

The Commissioner advises, and the revisions to § 58.202 make clear, that the disqualification procedure will not be invoked for minor violations of the regulation. In addition, § 58.204 provides that a regulatory hearing may be conducted in accordance with 21 CFR Part 16. Such a hearing provides all the safeguards essential to due process. See also the FEDERAL REGISTER of 40 FR 40713 et seq. (preamble to Subpart F of 21 CFR Part 2, recodified as 21 CFR Part 16—Regulatory Hearing Before the Food and Drug Administration; section 201(y) of the act (21 U.S.C. 321(y)) (procedural requirements of an "informal hearing"); *Goldberg* v. *Kelly*, 397 U.S. 254 (1970). Judicial review of final administrative action is provided by the Administrative Procedure Act (5 U.S.C. 701 et seq.) See also § 10.45 *Court Review of final administrative action; exhaustion of administrative remedies* (21 CFR 10.45); and 40 FR 40689-40691 (preamble to procedural regulations, § 2.11 (recodified as 21 CFR 10.45)).

235. Several comments expressed the concern that any regulatory hearing conducted under 21 CFR Part 16 should provide for the confidentiality of all data on which the hearing is based.

The Commissioner advises that § 16.60(a) (21 CFR 16.60(a)) provides adequate safeguards when required to maintain the confidentiality of commercial information.

236. One comment stated that if notice for such a hearing should be mailed to a facility, more than 3 days should be allowed for a facility to be able to prepare itself to come to a meeting.

The Commissioner finds that the provisions of § 16.22 (21 CFR 16.22) provide adequate flexibility for any party responding to a notice of opportunity for a hearing. See also the comments addressed to 21 CFR 52.204, set out in the preamble to the proposed regulations on obligations of sponsors and monitors, published in the FEDERAL REGISTER of September 27, 1977 (42 FR 49619).

237. One comment suggested that § 58.204 include a provision specifying that a sponsor be allowed to intervene in the hearing process when a notice of opportunity for a hearing has issued to a testing facility that is performing studies under contract for the sponsor.

Inasmuch as the disqualification process in such a case is directed at the testing facility rather than the sponsor and inasmuch as the alleged violations involved would be those of the testing facility, the Commissioner finds that intervention by a sponsor (or, in many cases, multiple sponsors) would serve no useful purpose. As noted in the preamble to the proposed regulation (41 FR 51218), a sponsor who wishes to contest a finding that a particular study or studies is or are inadequate will be provided an opportunity to do so by the procedures for denying or withdrawing the approval of an application for a research or marketing permit.

238. Concern was also expressed that a reasonable time be provided to allow a sponsor to conduct a new test prior to termination or withdrawal.

The Commissioner emphasizes that in those cases in which a safety decision has been based on data that have subsequently been called into question, protection of the public requires that proceedings be instituted without delay. As previously noted, opportunity to contest a finding that a particular study is so inadequate that it will not support a claim of safety of a product will be provided by procedures set forth in other regulations, e.g., withdrawal of an NDA.

Final Order on Disqualification

239. Several comments stated that § 58.206 should provide specifically for appeal to the Federal courts following a final decision to disqualify by the Commissioner.

The Commissioner notes that the provisions of 21 CFR 16.120 and 10.45 adequately address this point. These regulations clearly state the provisions that apply to court review of final administrative action.

240. One comment suggested that § 58.206(b) be modified to require that sponsors be notified, when applicable, at the time of issuance of a final order to a testing facility.

The Commissioner advises that such notification, which is discretionary, is expressly provided for in § 58.213(b). Additionally, § 58.206(a) and (b) are revised to reflect the requirement that the Commissioner must make the findings required by § 58.202 before a final order disqualifying a nonclinical testing facility shall issue.

Actions upon Disqualification

241. Several comments objected to the retroactive provisions of § 58.210(a), which state that once a testing facility has been disqualified, each application for a research or marketing permit, whether approved or not, that contains or relies upon any nonclinical laboratory study conducted by the disqualified testing facility may be examined to determine whether these studies were or would be essential to a decision.

The Commissioner advises that calling into question studies performed by a subsequently disqualified testing facility does not represent a departure from prior FDA policy in other areas. FDA must make additional inquiries to establish safety any time a question is raised about data previously submitted, regardless of whether a disqualification procedure exists. Section 58.210(a) allows the person relying on the study in question to establish that the study was not affected by the circumstances that led to disqualification. The safety of the public would not be adequately protected were no such validation required when serious questions are raised regarding the adequacy of data upon which regulatory decisions are based.

Section 58.210 is revised by the addition of paragraph (b), which states that no nonclinical laboratory study begun after a facility has been disqualified will be considered in support of any application for a research or marketing permit unless the facility has been reinstated under § 58.219. This addition makes it clear that, in such a case, no subsequent information can be submitted for purposes of subsequent validation. If the facility is reinstated, however, the study might be acceptable to FDA. This provision does not relieve the applicant from any other requirement under FDA regulations that all data and information regarding clinical experience with the article in question be submitted to the agency.

242. Many comments regarding § 58.210 were based on the assumption that the disqualification process might be invoked for a minor violation of the good laboratory practice regulation and stated that calling studies into question based on a minor violation was unreasonable.

As previously discussed, § 58.202 is revised to make it clear that the disqualification process will be reserved for those situations in which lesser sanctions, e.g., rejection of individual studies, will not suffice. Because disqualification will be reserved for use in serious situations, the Commissioner finds that calling into question all studies done before or after disqualification is warranted.

Public Disclosure of Information upon Disqualification

243. Several comments said that proprietary or trade secret documents should not be released. Others urged that disqualification records not be disclosed.

The Commissioner advises that release of all such documents is governed by the provisions of the Freedom of Information Act (5 U.S.C. 552) and 21 CFR Part 20 and need not be separately dealt with in this regulation. Interested parties are referred specifically to Part 20—Public Information (21 CFR Part 20). Section 20.61 (21 CFR 20.61) deals with trade secrets and commercial information and § 20.64 (21 CFR 20.64) deals with investigatory records. The preamble to the public information regulations (39 FR 44602 et seq.) (since recodified as Part 20) discusses these issues at length.

244. One comment on § 58.213 stated that no notification of other government departments or agencies should issue until completion of the judicial process.

The Commissioner disagrees and finds that withholding notification until completion of the administrative process by the agency provides an adequate opportunity for a testing facility to be heard prior to the issuance of any such notification.

245. Another comment stated that because FDA is a Federal agency, notification of State agencies is outside FDA's jurisdiction

The Commissioner points out that section 705(b) of the act (21 U.S.C. 375(b)) provides for dissemination of information regarding food, drugs, or devices in situations involving imminent danger to health or gross deception of the consumer. In addition, the Commissioner emphasizes that he proposes to notify the States only in those situations for which adequate cause has been established and for which a final order has been issued. Section 58.213(a) is amended to make it clear that such notification shall state that it is given because of the relationship between the testing facility and the person notified and that the Food and Drug Administration is not advising or recommending that any action be taken by the person notified. Additionally, § 58.213 is modified to make it clear that notification of disqualification may be sent by the Commissioner not only to other Federal agencies but to any other person known to have professional relations with the disqualified testing facility. This includes sponsors of studies being performed by the facility.

246. A comment suggested that the scope of notification should be limited to those nonclinical laboratory studies upon which the decision to disqualify was based.

The language of § 58.213 makes it clear that notification may be given at the discretion of the Commissioner whenever he believes that such disclosure would further the public interest or would promote compliance with the good laboratory practice regulations. The Commissioner finds that, given the expressed purpose of notification, further limitation would be inappropriate.

Alternative or Additional Actions to Disqualification

247. One comment on § 58.215 suggested that informal procedures be used prior to the institution of more formal procedures.

The Commissioner notes that this approach was discussed in the preamble to the proposed regulation at 41 FR 51218. Because such informal procedures have, in the past, doubled the time and expense of all involved parties without discernible benefit, the Commissioner has decided not to provide for informal procedures in these regulations.

Suspension or Termination of a Testing Facility by a Sponsor

248. Many comments on § 58.217 said that the section seemed to be an attempt on the part of FDA to provide legal grounds for the unilateral breaking of contracts between private parties.

The Commissioner finds that the section, as written, was subject to a great deal of misunderstanding. Therefore, the section is revised. The Commissioner advises that nothing in Part 58 is intended to infringe upon or alter the private contractural arrangements between a sponsor and a nonclinical testing facility. A sponsor may terminate a testing facility for reasons of its own whether or not FDA has begun any action to disqualify that facility. Where a sponsor has independent grounds for suspending or terminating studies performed for that sponsor by the facility under contract, the fact that FDA has not itself disqualified the facility may not be raised by the contract facility as a defense against the sponsor.

249. Several comments said notification within 5 days was impractical.

The Commissioner agrees, and the time period is extended to 15 working days.

250. A number of comments said the notification requirement provided a sponsor with an unfair opportunity to impugn a contract facility that would have no opportunity for response.

The Commissioner emphasizes that termination of a nonclinical testing facility by a sponsor should be subject to the contract between the two parties. A nonclinical testing facility, as a party to the contract, may protect itself from unjust termination by the terms of its contract with the sponsor. Remedies for both parties to such a contract may be spelled out in the contract and are governed by principles of contract law. The Commissioner further emphasizes that the requirement that a sponsor notify FDA when it has terminated or suspended a testing facility applies only to those cases in which an application for a research or marketing permit has been submitted. Where no application has been submitted, no notification is required.

Reinstatement of a Disqualified Testing Facility

251. One comment on § 58.219 expressed concern that when read with § 58.210, it was confusing.

The Commissioner finds that the addition of § 58.210(b) substantially clarifies the status of studies conducted before, during, and after disqualification and that further amendment is unnecessary.

252. A typographical error in the last sentence of § 58.219 has been corrected. The last sentence now reads: "A determination that a testing facility has been reinstated is disclosable to the public under Part 20 of this Chapter."

CONFORMING AMENDMENTS

253. The Commissioner is adding to or revising provisions in the regulations regarding food and color additives, new drugs for investigational use, new drug applications, OTC drug products, antibiotic drugs, new animal drug applications, biological product licenses, and performance standards for electronic products to incorporate appropriate implementing provisions for, and cross references to, Part 58, which is being added by this document. Each of the regulations requires the submission of data which may include nonclinical laboratory studies. The regulations are being revised to require, with respect to each nonclinical laboratory study contained as part of the submitted information, either a statement that the study was conducted in compliance with the good laboratory practice regulations set forth in Part 58 of this chapter, or, if the study was not conducted in compliance with such regulations, a statement that describes in detail all differences between the practices used in the study and those required in the regulations. The revisions highlight the fact that although studies not conducted in compliance with the regulations may continue to be submitted to FDA, the burden

of establishing that the noncompliance did not affect the quality of the data submitted is on the person submitting the noncomplying study.

Therefore, under the Federal Food, Drug, and Cosmetic Act (secs. 406, 408, 409, 502, 503, 505, 506, 507, 510, 512–516, 518–520, 701(a), 706, and 801, 52 Stat. 1049–1053 as amended, 1055, 1058 as amended, 55 Stat. 851 as amended, 59 Stat. 463 as amended, 68 Stat. 511–517 as amended, 72 Stat. 1785–1788 as amended, 76 Stat. 794 as amended, 82 Stat. 343–351, 90 Stat. 539–574 (21 U.S.C. 346, 346a, 348, 352, 353, 355, 356, 357, 360, 360b–360f, 360h–360j, 371(a), 376, and 381)) and the Public Health Service Act (secs. 215, 351, 354–360F, 58 Stat. 690, 702 as amended, 82 Stat. 1173–1186 as amended (42 U.S.C. 216, 262, 263b–263n)) and under authority delegated to him (21 CFR 5.1), the Commissioner amends Chapter I of 21 CFR as follows:

SUBCHAPTER A—GENERAL

PART 16—REGULATORY HEARING BEFORE THE FOOD AND DRUG ADMINISTRATION

1. Part 16 is amended in § 16.1 by redesignating paragraph (b)(30) as paragraph (c) and by adding new paragraph (b)(3), to read as follows:

§ 16.1 Scope.

*   *   *   *   *

(b) ***

(30) Section 58.204(b) of this chapter, relating to disqualifying a nonclinical laboratory testing facility.

(c) Any other provision in the regulations in this chapter under which a party who is adversely affected by regulatory action is entitled to an opportunity for a hearing, and no other procedural provisions in this part are by regulation applicable to such hearing.

———

2. Part 58 is added to read as follows:

PART 58—GOOD LABORATORY PRACTICE FOR NONCLINICAL LABORATORY STUDIES

Subpart A—General Provisions

Sec.
58.1   Scope.

Subparts H and I—[Reserved]

Subpart J—Records and Reports

Subpart K—Disqualification of Testing Facilities

AUTHORITY: Secs. 406, 408, 409, 502, 503, 505, 506, 507, 510, 512–516, 518–520, 701(a), 706, and 801, Pub. L. 717, 52 stat. 1049–1053 as amended, 1055, 1058 as amended, 55 Stat. 851 as amended, 59 Stat. 463, as amended, 68 Stat. 511–517 as amended, 72 Stat. 1785–1788 as amended, 76 Stat. 794 as amended, 82 Stat. 343–351, 90 Stat. 539–574 (21 U.S.C. 346, 346a, 438, 352, 353, 355, 356, 357, 360, 360b–360f, 360h–360j, 371(a), 376, and 381); secs. 215, 351, 354–360F, Publ. L. 410, 58 Stat. 690, 702 as amended, 82 Stat. 1173–1186 as amended (42 U.S.C. 216, 262, 263b–263n).

SUBPART A—GENERAL PROVISIONS

§ 58.1 Scope.

This part prescribes good laboratory practices for conducting nonclinical laboratory studies that support or are intended to support applications for research or marketing permits for products regulated by the Food and Drug Administration, including food and color additives, animal food additives, human and animal drugs, medical devices for human use, biological products, and electronic products. Compliance with this part is intended to assure the quality and integrity of the safety data filed pursuant to sections 406, 408, 409, 502, 503, 505, 506, 507, 510, 512–516, 518–520, 706, and 801 of the Federal Food, Drug, and Cosmetic Act and sections 351 and 354–360F of the Public Health Service Act.

§ 58.3 Definitions

As used in this part, the following terms shall have the meanings specified:

(a) "Act" means the Federal Food, Drug, and Cosmetic Act, as amended (secs. 201–902, 52 Stat. 1040 et seq., as amended (21 U.S.C. 321–392)).

(b) "Test article" means any food additive, color additive, drug, biological product, electronic product, medical device for human use, or any other article subject to regulation under the act or under sections 351 and 354–360F of the Public Health Service Act.

(c) "Control article" means any food additive, color additive, drug, biological product, electronic product, medical device for human use, or any other article other than a test article that is administered to the test system in the course of a nonclinical laboratory study for the purpose of establishing a basis for comparison with the test article.

(d) "Nonclinical laboratory study" means any in vivo or in vitro experiment in which a test article is studied prospectively in a test system under laboratory conditions to determine its safety. The term does not include studies utilizing human subjects or clinical studies or field trials in animals. The term does not include basic exploratory studies carried out to determine whether a test article has any potential utility or to determine physical or chemical characteristics of a test article.

(e) "Application for research or marketing permit" includes:

(1) A color additive petition, described in Part 71 of this chapter.

(2) A food additive petition, described in Parts 171 and 571 of this chapter.

(3) Data and information regarding a substance submitted as part of the procedures for establishing that a substance is generally recognized as safe for use, which use results or may reasonably be expected to result, directly or indirectly, in its becoming a component or otherwise affecting the characteristics of any food, described in §§ 170.35 and 570.35 of this chapter.

(4) Data and information regarding a food additive submitted as part of the procedures regarding food additives permitted to be used on an interim basis pending additional study, described in § 180.1 of this chapter.

(5) A "Notice of Claimed Investigational Exemption for a New Drug," described in Part 312 of this chapter.

(6) A "new drug application," described in Part 314 of this chapter.

(7) Data and information regarding an over-the-counter drug for human use, submitted as part of the procedures for classifying such drugs as generally recognized as safe and effective and not misbranded, described in Part 330 of this chapter.

(8) Data and information regarding a prescription drug for human use submitted as part of the procedures for classifying such drugs as generally

recognized as safe and effective and not misbranded, to be described in this chapter.

(9) Data and information regarding an antibiotic drug submitted as part of the procedures for issuing, amending, or repealing regulations for such drugs, described in Part 430 of this chapter.

(10) A "Notice of Claimed Investigational Exemption for a New Animal Drug," described in Part 511 of this chapter.

(11) A "new animal drug application," described in Part 514 of this chapter.

(12) Data and information regarding a drug for animal use submitted as part of the procedures for classifying such drugs as generally recognized as safe and effective and not misbranded, to be described in this chapter.

(13) An "application for a biological product license," described in Part 601 of this chapter.

(14) An "application for an investigational device exemption," described in Part 812 of this chapter.

(15) An "Application for Premarket Approval of a Medical Device," described in section 515 of the act.

(16) A "Product Development Protocol for a Medical Device," described in section 515 of the act.

(17) Data and information regarding a medical device submitted as part of the procedures for classifying such devices, described in section 513 of the act.

(18) Data and information regarding a medical device submitted as part of the procedures for establishing, amending, or repealing a performance standard for such devices, described in section 514 of the act.

(19) Data and information regarding an electronic product submitted as part of the procedures for obtaining an exemption from notification of a radiation safety defect or failure of compliance with a radiation safety performance standard, described in Subpart D of Part 1003 of this chapter.

(20) Data and information regarding an electronic product submitted as part of the procedures for establishing, amending, or repealing a standard for such product, described in section 358 of the Public Health Service Act.

(21) Data and information regarding an electronic product submitted as part of the procedures for obtaining a variance from any electronic product performance standard as described in § 1010.4 of this chapter.

(22) Data and information regarding an electronic product submitted as part of the procedures for granting, amending, or extending an exemption from any electronic product performance standard, as described in § 1010.5 of this chapter.

(f) "Sponsor" means:

(1) A person who initiates and supports, by provision of financial or other resources, a nonclinical laboratory study;

(2) A person who submits a nonclinical study to the Food and Drug Administration in support of an application for a research or marketing permit; or

(3) A testing facility, if it both initiates and actually conducts the study.

(g) "Testing facility" means a person who actually conducts a nonclinical laboratory study, i.e., actually uses the test article in a test system. "Testing facility" includes any establishment required to register under section 510 of the act that conducts nonclinical laboratory studies and any consulting laboratory described in section 704 of the act that conducts such studies. "Testing facility" encompasses only those operational units that are being or have been used to conduct nonclinical laboratory studies.

(h) "Person" includes an individual, partnership, corporation, association, scientific or academic establishment, government agency, or organizational unit thereof, and any other legal entity.

(i) "Test system" means any animal, plant, microorganism, or subparts thereof to which the test or control article is administered or added for study. "Test system" also includes appropriate groups or components of the system not treated with the test or control articles.

(j) "Specimen" means any material derived from a test system for examination or analysis.

(k) "Raw data" means any laboratory worksheets, records, memoranda, notes, or exact copies thereof, that are the result of original observations and activities of a nonclinical laboratory study and are necessary for the reconstruction and evaluation of the report of that study. In the event that exact transcripts of raw data have been prepared (e.g., tapes which have been transcribed verbatim, dated, and verified accurate by signature), the exact copy or exact transcript may be substituted for the original source as raw data. "Raw data" may include photographs, microfilm or microfiche copies, computer printouts, magnetic media, including dictated observations, and recorded data from automated instruments.

(1) "Quality assurance unit" means any person or organizational element, except the study director, designated by testing facility management to perform the duties relating to quality assurance of nonclinical laboratory studies.

(m) "Study director" means the individual responsible for the overall conduct of a nonclinical laboratory study.

(n) "Batch" means a specific quantity or lot of a test or control article that has been characterized according to § 58.105(a).

§ 58.10 Applicability to Studies Performed under Grants and Contracts

When a sponsor conducting a nonclinical laboratory study intended to be submitted to or reviewed by the Food and Drug Administration utilizes the services of a consulting laboratory, contractor, or grantee to perform an analysis or other service, it shall notify the consulting laboratory, contractor, or grantee that the service is part of a nonclinical laboratory study that must be conducted in compliance with the provisions of this part.

§ 58.15 Inspection of a Testing Facility

(a) A testing facility shall permit an authorized employee of the Food and Drug Administration, at reasonable times and in a reasonable manner, to inspect the facility and to inspect (and in the case of records also to copy) all records and specimens required to be maintained regarding studies within the scope of this part. The records inspection and copying requirements shall not apply to quality assurance unit records of findings and problems, or to actions recommended and taken.

(b) The Food and Drug Administration will not consider a nonclinical laboratory study in support of an application for a research or marketing permit if the testing facility refuses to permit inspection. The determination that a nonclinical laboratory study will not be considered in support of an application for a research or marketing permit does not, however, relieve the applicant for such a permit of any obligation under any applicable statute or regulation to submit the results of the study to the Food and Drug Administration.

SUBPART B—ORGANIZATION AND PERSONNEL

§ 58.29 Personnel

(a) Each individual engaged in the conduct of or responsible for the supervision of a nonclinical laboratory study shall have education, training, and experience, or combination thereof, to enable that individual to perform the assigned functions.

(b) Each testing facility shall maintain a current summary of training and experience and job description for each individual engaged in or supervising the conduct of a nonclinical laboratory study.

(c) There shall be a sufficient number of personnel for the timely and proper conduct of the study according to the protocol.

(d) Personnel shall take necessary personal sanitation and health precautions designed to avoid contamination of test and control articles and test systems.

(e) Personnel engaged in a nonclinical laboratory study shall wear clothing appropriate for the duties they perform. Such clothing shall be changed as often as necessary to prevent microbiological, radiological, or chemical contamination of test systems and test and control articles.

(f) Any individual found at any time to have an illness that may adversely affect the quality and integrity of the nonclinical laboratory study shall be excluded from direct contact with test systems, test and control articles and any other operation or function that may adversely affect the study until the condition is corrected. All personnel shall be instructed to report to their immediate supervisors any health or medical conditions that may reasonably be considered to have an adverse effect on a nonclinical laboratory study.

§ 58.31 Testing Facility Management

For each nonclinical laboratory study, testing facility management shall:

(a) Designate a study director as described in § 58.33, before the study is initiated.

(b) Replace the study director promptly if it becomes necessary to do so during the conduct of a study, and document and maintain such action as raw data.

(c) Assure that there is a quality assurance unit as described in § 58.35.

(d) Assure that test and control articles or mixtures have been appropriately tested for identity, strength, purity, stability, and uniformity, as applicable.

(e) Assure that personnel, resources, facilities, equipment, materials, and methodologies are available as scheduled.

(f) Assure that personnel clearly understand the functions they are to perform.

(g) Assure that any deviations from these regulations reported by the quality assurance unit are communicated to the study director and corrective actions are taken and documented.

§ 58.33 Study Director

For each nonclinical laboratory study, a scientist or other professional of appropriate education, training, and experience, or combination thereof, shall be identified as the study director. The study director has overall responsibility for the technical conduct of the study, as well as for the interpretation, analysis, documentation and reporting of results, and represents the single point of study control. The study director shall assure that:

(a) The protocol, including any change, is approved as provided by § 58.120 and is followed.

(b) All experimental data, including observations of unanticipated responses to the test system are accurately recorded and verified.

(c) Unforeseen circumstances that may affect the quality and integrity of the nonclinical laboratory study are noted when they occur, and corrective action is taken and documented.

(d) Test systems are as specified in the protocol.

(e) All applicable good laboratory practice regulations are followed.

(f) All raw data, documentation, protocols, specimens, and final reports are transferred to the archives during or at the close of the study.

§ 58.35   Quality Assurance Unit

(a) A testing facility shall have a quality assurance unit composed of one or more individuals who shall be responsible for monitoring each study to assure management that the facilities, equipment, personnel, methods, practices, records, and controls are in conformance with the regulations in this part. For any given study the quality assurance unit shall be entirely separate from and independent of the personnel engaged in the direction and conduct of that study.

(b) The quality assurance unit shall:

(1) Maintain a copy of a master schedule sheet of all nonclinical laboratory studies conducted at the testing facility indexed by test article and containing the test system, nature of study, date study was initiated, current status of each study, name of the sponsor, name of the study director, and status of the final report.

(2) Maintain copies of all protocols pertaining to all nonclinical laboratory studies for which the unit is responsible.

(3) Inspect each phase of a nonclinical laboratory study periodically and maintain written and properly signed records of each periodic inspection showing the date of the inspection, the study inspected, the phase or segment of the study inspected, the person performing the inspection, findings and problems, action recommended and taken to resolve existing problems, and any scheduled date for re-inspection. For studies lasting more than 6 months, inspections shall be conducted every 3 months. For studies lasting

less than 6 months, inspections shall be conducted at intervals adequate to assure the integrity of the study. Any significant problems which are likely to affect study integrity found during the course of an inspection shall be brought to the attention of the study director and management immediately.

(4) Periodically submit to management and the study director written status reports on each study, noting any problems and the corrective actions taken.

(5) Determine that no deviations from approved protocols or standard operating procedures were made without proper authorization and documentation.

(6) Review the final study report to assure that such report accurately describes the methods and standard operating procedures, and that the reported results accurately reflect the raw data of the nonclinical laboratory study.

(7) Prepare and sign a statement to be included with the final study report which shall specify the dates inspections were made and findings reported to management and to the study director.

(c) The responsibilities and procedures applicable to the quality assurance unit, the records maintained by the quality assurance unit, and the method of indexing such records shall be in writing and shall be maintained. These items including inspection dates, the study inspected, the phase or segment of the study inspected, and the name of the individual performing the inspection shall be made available for inspection to authorized employees of the Food and Drug Administration.

(d) A designated representative of the Food and Drug Administration shall have access to the written procedures established for the inspection and may request testing facility management to certify that inspections are being implemented, performed, documented, and followed-up in accordance with this paragraph.

(e) All records maintained by the quality assurance unit shall be kept in one location at the testing facility.

SUBPART C—FACILITIES

§ 58.41   General

Each testing facility shall be of suitable size, construction, and location to facilitate the proper conduct of nonclinical laboratory studies. It shall be designed so that there is a degree of separation that will prevent any function or activity from having an adverse effect on the study.

§ 58.43 Animal Care Facilities

(a) A testing facility shall have a sufficient number of animal rooms or areas, as needed, to assure proper: (1) Separation of species or test systems, (2) isolation of individual projects, (3) quarantine of animals, and (4) routine or specialized housing of animals.

(b) A testing facility shall have a number of animal rooms or areas separate from those described in paragraph (a) of this section to ensure isolation of studies being done with test systems or test and control articles known to be biohazardous, including volatile substances, aerosols, radio-active materials, and infectious agents.

(c) Separate areas shall be provided for the diagnosis, treatment, and control of laboratory animal diseases. These areas shall provide effective isolation for the housing of animals either known or suspected of being diseased, or of being carriers of disease, from other animals.

(d) When animals are housed, facilities shall exist for the collection and disposal of all animal waste and refuse or for safe sanitary storage of waste before removal from the testing facility. Disposal facilities shall be so provided and operated as to minimize vermin infestation, odors, disease hazards, and environmental contamination.

(e) Animal facilities shall be designed, constructed, and located so as to minimize disturbances that interfere with the study.

§ 58.45 Animal Supply Facilities

There shall be storage areas, as needed, for feed, bedding, supplies, and equipment. Storage areas for feed and bedding shall be separated from areas housing the test systems and shall be protected against infestation or contamination. Refrigeration shall be provided for perishable supplies or feed.

§ 58.47 Facilities for Handling Test and Control Articles

(a) As necessary to prevent contamination or mixups, there shall be separate areas for:

(1) Receipt and storage of the test and control articles.

(2) Mixing of the test and control articles with a carrier, e.g., feed.

(3) Storage of the test and control article mixtures.

(b) Storage areas for the test and/or control article and test and control mixtures shall be separate from areas housing the test systems and shall be adequate to preserve the identity, strength, purity, and stability of the articles and mixtures.

§ 58. 49 Laboratory Operation Areas

(a) Separate laboratory space shall be provided, as needed, for the performance of the routine procedures required by nonclinical laboratory studies, including specialized areas for performing activities such as aseptic surgery, intensive care, necropsy, histology, radiography, and handling of biohazardous materials.

(b) Separate space shall be provided for cleaning, sterilizing, and maintaining equipment and supplies used during the course of the study.

§ 58.51 Specimen and Data Storage Facilities

Space shall be provided for archives, limited to access by authorized personnel only, for the storage and retrieval of all raw data and specimens from completed studies.

§ 58.53 Administrative and Personnel Facilities

(a) There shall be space provided for the administration, supervision, and direction of the testing facility.

(b) Separate space shall be provided for locker, shower, toilet, and washing facilities, as needed.

SUBPART D—EQUIPMENT

§ 58.61 Equipment Design

Automatic, mechanical, or electronic equipment used in the generation, measurement, or assessment of data and equipment used for facility environmental control shall be of appropriate design and adequate capacity to function according to the protocol and shall be suitably located for operation, inspection, cleaning, and maintenance.

§ 58.63   Maintenance and Calibration of Equipment

(a) Equipment shall be adequately inspected, cleaned, and maintained. Equipment used for the generation, measurement, or assessment of data shall be adequately tested, calibrated and/or standardized.

(b) The written standard operating procedures required under § 58.81(b)(11) shall set forth in sufficient detail the methods, materials, and schedules to be used in the routine inspection, cleaning, maintenance, testing, calibration and/or standardization of equipment, and shall specify remedial action to be taken in the event of failure or malfunction of equipment. The written standard operating procedures shall designate the person responsible for the performance of each operation, and copies of the standard operating procedures shall be made available to laboratory personnel.

(c) Written records shall be maintained of all inspection, maintenance, testing, calibrating and/or standardizing operations. These records, containing the date of the operation, shall describe whether the maintenance operations were routine and followed the written standard operating procedures. Written records shall be kept of nonroutine repairs performed on equipment as a result of failure and malfunction. Such records shall document the nature of the defect, how and when the defect was discovered, and any remedial action taken in response to the defect.

SUBPART E—TESTING FACILITIES OPERATION

§ 58.81   Standard Operating Procedures

(a) A testing facility shall have standard operating procedures in writing setting forth nonclinical laboratory study methods that management is satisfied are adequate to insure the quality and integrity of the data generated in the course of a study. All deviations in a study from standard operating procedures shall be authorized by the study director and shall be documented in the raw data. Significant changes in established standard operating procedures shall be properly authorized in writing by management.

(b) Standard operating procedures shall be established for, but not limited to the following:

(1) Animal room preparation.

(2) Animal care.

(3) Receipt, identification, storage, handling, mixing, and method of sampling of the test and control articles.

(4) Test system observations.

(5) Laboratory tests.

(6) Handling of animals found moribund or dead during study.

(7) Necropsy of animals or postmortem examination of animals.

(8) Collection and identification of specimens.

(9) Histopathology.

(10) Data handling, storage, and retrieval.

(11) Maintenance and calibration of equipment.

(12) Transfer, proper placement, and identification of animals.

(c) Each laboratory area shall have immediately available laboratory manuals and standard operating procedures relative to the laboratory procedures being performed, e.g., toxicology, histology, clinical chemistry, hematology, teratology, necropsy. Published literature may be used as a supplement to standard operating procedures.

(d) A historical file of standard operating procedures, and all revisions thereof, including the dates of such revisions, shall be maintained.

§ 58.83 Reagents and Solutions

All reagents and solutions in the laboratory areas shall be labeled to indicate identity, titer or concentration, storage requirements, and expiration date. Deteriorated or outdated reagents and solutions shall not be used.

§ 58.90 Animal Care

(a) There shall be standard operating procedures for the housing, feeding, handling, and care of animals.

(b) All newly received animals from outside sources shall be placed in quarantine until their health status has been evaluated. This evaluation shall be in accordance with acceptable veterinary medical practice.

(c) At the initiation of a nonclinical laboratory study, animals shall be free of any disease or condition that might interfere with the purpose or conduct of the study. If, during the course of the study, the animals contract such a disease or condition, the diseased animals shall be isolated. If necessary, these animals may be treated for disease or signs of disease provided that such treatment does not interfere with the study. The diagnosis, authorizations of treatment, description of treatment and each date of treatment shall be documented and shall be retained.

(d) Warm-blooded animals, excluding suckling rodents, used in laboratory procedures that require manipulations and observations over an extended period of time or in studies that require the animals to be removed from and returned to their home cages for any reason (e.g., cage cleaning, treatment, etc.), shall receive appropriate identification (e.g., tattoo, toe clip, color code, ear tag, ear punch, etc.) All information needed to specifically identify each animal within an animal-housing unit shall appear on the outside of that unit.

(e) Animals of different species shall be housed in separate rooms when necessary. Animals of the same species, but used in different studies, should not ordinarily be housed in the same room when inadvertent exposure to control or test articles or animal mixup could affect the outcome of either study. If such mixed housing is necessary, adequate differentiation by space and identification shall be made.

(f) Animal cages, racks and accessory equipment shall be cleaned and sanitized at appropriate intervals.

(g) Feed and water used for the animals shall be analyzed periodically to ensure that contaminants known to be capable of interfering with the study and reasonably expected to be present in such feed or water are not present at levels above those specified in the protocol. Documentation of such analyses shall be maintained as raw data.

(h) Bedding used in animal cages or pens shall not interfere with the purpose or conduct of the study and shall be changed as often as necessary to keep the animals dry and clean.

(i) If any pest control materials are used, the use shall be documented. Cleaning and pest control materials that interfere with the study shall not be used.

SUBPART F—TEST AND CONTROL ARTICLES

§ 58.105   Test and Control Article Characterization

(a) The identity, strength, purity, and composition or other characteristics which will appropriately define the test or control article shall be determined for each batch and shall be documented before the initiation of the study. Methods of synthesis, fabrication, or derivation of the test and control articles shall be documented by the sponsor or the testing facility. In those cases where marketed products are used as control articles, such products will be characterized by their labeling.

(b) The stability of each test or control article shall be determined by the testing facility or by the sponsor before initiation or a nonclinical laboratory study. If the stability of the test and control articles cannot be determined before initiation of a study, standard operating procedures shall be established and followed to provide for periodic re-analysis of each batch.

(c) Each storage container for a test or control article shall be labeled by name, chemical abstract number or code number, batch number, expiration date, if any, and, where appropriate, storage conditions necessary to maintain the identity strength, purity, and composition of the test or control article. Storage containers shall be assigned to a particular test article for the duration of the study.

(d) For studies of more than 4 weeks' duration, reserve samples from each batch of test and control articles shall be retained for the period of time provided by § 58.195.

§ 58.107 Test and Control Article Handling

Procedures shall be established for a system for the handling of the test and control articles to ensure that:

(a) There is proper storage.

(b) Distribution is made in a manner designed to preclude the possibility of contamination, deterioriation, or damage.

(c) Proper identification is maintained throughout the distribution process.

(d) The receipt and distribution of each batch is documented. Such documentation shall include the date and quantity of each batch distributed or returned.

§ 58.113 Mixtures of Articles with Carriers

(a) For each test or control article that is mixed with a carrier, tests by appropriate analytical methods shall be conducted:

(1) To determine the uniformity of the mixture and to determine, periodically, the concentration of the test or control article in the mixture.

(2) To determine the stability of the test and control articles in the mixture. If the stability cannot be determined before initiation of the study, standard operating procedures shall be established and followed to provide for periodic re-analysis of the test and control articles in the mixture.

(b) For studies of more than 4 weeks' duration a reserve sample of each test or control carrier article mixture shall be taken and retained for the period of time provided by § 58.195.

(c) Where any of the components of the test or control article carrier mixture has an expiration date, that date shall be clearly shown on the container. If more than one component has an expiration date, the earliest date shall be shown.

SUBPART G—PROTOCOL FOR AND CONDUCT OF A NONCLINICAL LABORATORY STUDY

§ 58.120 Protocol

(a) Each study shall have an approved written protocol that clearly indicates the objectives and all methods for the conduct of the study. The protocol shall contain but shall not necessarily be limited to the following information:

(1) A descriptive title and statement of the purpose of the study.

(2) Identification of the test and control articles by name, chemical abstract number or code number.

(3) The name of the sponsor and the name and address of the testing facility at which the study is being conducted.

(4) The proposed starting and completion dates.

(5) Justification for selection of the test system.

(6) Where applicable, the number, body weight range, sex, source of supply, species, strain, substrain, and age of the test system.

(7) The procedure for identification of the test system.

(8) A description of the experimental design, including the methods for the control of bias.

(9) A description and/or identification of the diet used in the study as well as solvents, emulsifiers and/or other materials used to solubilize or suspend the test or control articles before mixing with the carrier. The description shall include specifications for acceptable levels of contaminants that are reasonably expected to be present in the dietary materials and are known to be capable of interfering with the purpose or conduct of the study if present at levels greater than established by the specifications.

(10) The route of administration and the reason for its choice.

(11) Each dosage level, expressed in milligrams per kilogram of body weight or other appropriate units, of the test or control article to be administered and the method and frequency of administration.

(12) Method by which the degree of absorption of the test and control articles by the test system will be determined if necessary to achieve the objectives of the study.

(13) The type and frequency of tests, analyses, and measurements to be made.

(14) The records to be maintained.

(15) The date of approval of the protocol by the sponsor and the signature of the study director.

(16) A statement of the proposed statistical methods to be used.

(b) All changes in or revisions of an approved protocol and the reasons therefor shall be documented, signed by the study director, dated, and maintained with the protocol.

§ 58.130 Conduct of a Nonclinical Laboratory Study

(a) The nonclinical laboratory study shall be conducted in accordance with the protocol.

(b) The test systems shall be monitored in conformity with the protocol.

(c) Specimens shall be identified by test system, study, nature, and date of collection. This information shall be located on the specimen container or shall accompany the specimen in a manner that precludes error in the recording and storage of data.

(d) Records of gross findings for a specimen from postmortem observations shall be available to a pathologist when examining that specimen histopathologically.

(e) All data generated during the conduct of a nonclinical laboratory study, except those that are generated as direct computer input, shall be recorded directly, promptly, and legibly in ink. All data entries shall be dated on the day of entry and signed or initialed by the person entering the data. Any change in entries shall be made so as not to obscure the original entry, shall indicate the reason for such change, and shall be dated and signed or identified at the time of the change. In computer driven data collection systems, the individual responsible for direct data input shall be identified at the time of data input. Any change in computer entries shall be made so as not to obscure the original entry, shall indicate the reason for change, and shall be dated and the responsible individual shall be identified.

SUBPARTS H-I—[Reserved]

SUBPART J—RECORDS AND REPORTS

§ 58.185 Reporting of Nonclinical Laboratory Study Results

(a) A final report shall be prepared for each nonclinical laboratory study and shall include, but not necessarily be limited to, the following:

(1) Name and address of the facility performing the study and the dates on which the study was initiated and completed.

(2) Objectives and procedures stated in the approved protocol, including any changes in the original protocol.

(3) Statistical methods employed for analyzing the data.

(4) The test and control articles identified by name, chemical abstracts number or code number, strength, purity, and composition or other appropriate characteristics.

(5) Stability of the test and control articles under the conditions of administration.

(6) A description of the methods used.

(7) A description of the test system used. Where applicable, the final report shall include the number of animals used, sex, body weight range, source of supply, species, strain and substrain, age, and procedure used for identification.

(8) A description of the dosage, dosage regimen, route of administration, and duration.

(9) A description of all circumstances that may have affected the quality or integrity of the data.

(10) The name of the study director, the names of other scientists or professionals, and the names of all supervisory personnel, involved in the study.

(11) A description of the transformations, calculations, or operations performed on the data, a summary and analysis of the data, and a statement of the conclusions drawn from the analysis.

(12) The signed and dated reports of each of the individual scientists or other professionals involved in the study.

(13) The locations where all specimens, raw data, and the final report are to be stored.

(14) The statement prepared and signed by the quality assurance unit as described in § 58.35(b)(7).

(b) The final report shall be signed by the study director.

(c) Corrections or additions to a final report shall be in the form of an amendment by the study director. The amendment shall clearly identify that part of the final report that is being added to or corrected and the reasons for the correction or addition, and shall be signed and dated by the person responsible.

§ 58.190 Storage and Retrieval of Records and Data

(a) All raw data, documentation, protocols, specimens, and final reports generated as a result of a nonclinical laboratory study shall be retained.

(b) There shall be archives for orderly storage and expedient retrieval of all raw data, documentation, protocols, specimens, and interim and final reports. Conditions of storage shall minimize deterioration of the documents or specimens in accordance with the requirements for the time period of their retention and the nature of the documents or specimens. A testing facility may contract with commercial archives to provide a repository for all material to be retained. Raw data and specimens may be retained elsewhere provided that the archives have specific reference to those other locations.

(c) An individual shall be identified as responsible for the archives.

(d) Only authorized personnel shall enter the archives.

(e) Material retained or referred to in the archives shall be indexed by test article, date of study, test system, and nature of study.

§ 58.195 Retention of Records

(a) Record retention requirements set forth in this section do not supersede the record retention requirements of any other regulations in this chapter.

(b) Except as provided in paragraph (c) of this section, documentation records, raw data and specimens pertaining to a nonclinical laboratory study and required to be made by this part shall be retained in the archive(s) for whichever of the following periods is shortest:

(1) A period of at least 2 years following the date on which an application for a research or marketing permit, in support of which the results of the nonclinical laboratory study were submitted, is approved by the Food and Drug Administration. This requirement does not apply to studies supporting notices of claimed investigational exemption for new drugs (IND's) or applications for investigational device exemptions (IDE's), records of which shall be governed by the provisions of paragraph (b)(2) of this section.

(2) A period of at least 5 years following the date on which the results of the nonclinical laboratory study are submitted to the Food and Drug Administration in support of an application for a research or marketing permit.

(3) Iin other situations (e.g., where the nonclinical laboratory study does not result in the submission of the study in support of an application for a research or marketing permit), a period of at least 2 years following the date on which the study is completed, terminated, or discontinued.

(c) Wet specimens, samples of test or control articles, samples of test or control article carrier mixtures and specially prepared material (e.g., histochemical, electron microscopic, blood mounts, teratological preparation, and uteri from dominant lethal mutagenesis tests), which are relatively fragile and differ markedly in stability and quality during storage, shall be retained only as long as the quality of the preparation affords evaluation. In no case shall retention be required for longer periods than those set forth in paragraphs (a) and (b) of this section.

(d) The master schedule sheet, copies of protocols, and records of quality assurance inspections, as required by § 58.35(c) shall be maintained by the quality assurance unit as an easily accessible system of records for the period of time specified in paragraphs (a) and (b) of this section.

(e) Summaries of training and experience and job descriptions required to be maintained by § 58.29(b) may be retained along with all other testing facility employment records for the length of time specified in paragraphs (a) and (b) of this section.

(f) Records and reports of the maintenance and calibration and inspection of equipment, as required by § 58.63(b) and (c), shall be retained for the length of time specified in paragraph (b) of this section.

(g) If a facility conducting nonclinical testing goes out of business, all raw data, documentation, and other material specified in this section shall be transferred to the archives of the sponsor of the study. The Food and Drug Administration shall be notified in writing of such a transfer.

SUBPART K—DISQUALIFICATION OF TESTING FACILITIES

§ 58.200 Purpose

(a) The purposes of disqualification are: (1) To permit the exclusion from consideration of completed studies that were conducted by a testing facility which has failed to comply with the requirements of the good laboratory practice regulations until it can be adequately demonstrated that such noncompliance did not occur during, or did not affect the validity or acceptability of data generated by, a particular study; and (2) to exclude from consideration all studies completed after the date of disqualification until the facility can satisfy the Commissioner that it will conduct studies in compliance with such regulations.

(b) The determination that a nonclinical laboratory study may not be considered in support of an application for a research or marketing permit does not, however, relieve the applicant for such a permit of any obligation under any other applicable regulation to submit the results of the study to the Food and Drug Administration.

§ 58.202   Grounds for Disqualification

The Commissioner may disqualify a testing facility upon finding all of the following:

(a) The testing facility failed to comply with one or more of the regulations set forth in this part (or any other regulations regarding such facilities in this chapter);

(b) The noncompliance adversely affected the validity of the nonclinical laboratory studies; and

(c) Other lesser regulatory actions (e.g., warnings or rejection of individual studies) have not been or will probably not be adequate to achieve compliance with the good laboratory practice regulations.

§ 58.204   Notice of and Opportunity for Hearing on Proposed Disqualification

(a) Whenever the Commissioner has information indicating that grounds exist under § 58.202 which in his opinion justify disqualification of a testing facility, he may issue to the testing facility a written notice proposing that the facility be disqualified.

(b) A hearing on the disqualification shall be conducted in accordance with the requirements for a regulatory hearing set forth in Part 16 of this chapter.

§ 58.206   Final Order on Disqualification

(a) If the Commissioner, after the regulatory hearing, or after the time for requesting a hearing expires without a request being made, upon an evaluation of the administrative record of the disqualification proceeding, makes the findings required in § 58.202, he shall issue a final order disqualifying the facility. Such order shall include a statement of the basis for that determination. Upon issuing a final order, the Commissioner shall notify (with a copy of the order) the testing facility of the action.

(b) If the Commissioner, after a regulatory hearing or after the time for requesting a hearing expires without a request being made, upon an evaluation of the administrative record of the disqualification proceeding, does not make the findings required in § 58.202, he shall issue a final order terminating the disqualification proceeding. Such order shall include a statement of the basis for that determination. Upon issuing a final order the Commissioner shall notify the testing facility and provide a copy of the order.

§ 58.210   Actions upon Disqualification

(a) Once a testing facility has been disqualified, each application for a research or marketing permit, whether approved or not, containing or relying upon any nonclinical laboratory study conducted by the disqualified testing facility may be examined to determine whether such study was or would be essential to a decision. If it is determined that a study was or would be essential, the Food and Drug Administration shall also determine whether the study is acceptable, notwithstanding the disqualification of the facility. Any study done by a testing facility before or after disqualification may be presumed to be unacceptable, and the person relying on the study may be required to establish that the study was not affected by the circumstances that led to the disqualification, e.g., by submitting validating information. If the study is then determined to be unacceptable, such data such be eliminated from consideration in support of the application; and such elimination may serve as new information justifying the termination or withdrawal of approval of the application.

(b) No nonclinical laboratory study begun by a testing facility after the date of the facility's disqualification shall be considered in support of any application for a research or marketing permit, unless the facility has been reinstated under § 58.219. The determination that a study may not be considered in support of an application for a research or marketing permit does not, however, relieve the applicant for such a permit of any obligation under any other applicable regulation to submit the results of the study to the Food and Drug Administration.

§ 58.213   Public Disclosure of Information Regarding Disqualification

(a) Upon issuance of a final order disqualifying a testing facility under § 58.206(a); the Commissioner may notify all or any interested persons. Such notice may be given at the discretion of the Commissioner whenever he believes that such disclosure would further the public interest or would promote compliance with the good laboratory practice regulations set forth in this part. Such notice, if given, shall include a copy of the final order issued under § 58.206(a) and shall state that the disqualification constitutes a determination by the Food and Drug Administration that nonclinical laboratory studies performed by the facility will not be considered by the Food and Drug Administration in support of any application for a research or marketing permit. If such notice is sent to another Federal Government agency, the Food and Drug Administration will recommend that the agency also consider whether or not it should accept nonclinical laboratory studies performed by the testing facility. If such notice is sent to any other person, it shall state that it is given because of the relationship between the testing facility and the person being notified and that the Food and Drug Administration is not advising or recommending that any action be taken by the person notified.

(b) A determination that a testing facility has been disqualified and the administrative record regarding such determination are disclosable to the public under Part 20 of this chapter.

§ 58.215 Alternative or Additional Actions to Disqualification

(a) Disqualification of a testing facility under this subpart is independent of, and neither in lieu of nor a precondition to, other proceedings or actions authorized by the act. The Food and Drug Administration may, at any time, institute against a testing facility and/or against the sponsor of a nonclinical laboratory study that has been submitted to the Food and Drug Administration any appropriate judicial proceedings (civil or criminal) and any other appropriate regulatory action, in addition to or in lieu of, and prior to, simultaneously with, or subsequent to, disqualification. The Food and Drug Administration may also refer the matter to another Federal, State, or local government law enforcement or regulatory agency for such action as that agency deems appropriate.

(b) The Food and Drug Administration may refuse to consider any particular nonclinical laboratory study in support of an application for a research or marketing permit, if it finds that the study was not conducted in accordance with the good laboratory practice regulations set forth in this part, without disqualifying the testing facility that conducted the study or undertaking other regulatory action.

§ 58.217 Suspension or Termination of a Testing Facility by a Sponsor

Termination of a testing facility by a sponsor is independent of, and neither in lieu of nor a precondition to, proceedings or actions authorized by this subpart. If a sponsor terminates or suspends a testing facility from further participation in a nonclinical laboratory study that is being conducted as part of any application for a research or marketing permit that has been submitted to any Bureau of the Food and Drug Administration (whether approved or not), it shall notify that Bureau in writing within 15 working days of the action; the notice shall include a statement of the reasons for such action. Suspension or termination of a testing facility by a sponsor does not relieve it or any obligation under any other applicable regulation to submit the results of the study to the Food and Drug Administration.

§ 58.219   Reinstatement of a Disqualified Testing Facility

A testing facility that has been disqualified may be reinstated as an acceptable source of nonclinical laboratory studies to be submitted to the Food and Drug Administration if the Commissioner determines, upon an evaluation of the submission of the testing facility, that the facility can adequately assure that it will conduct future nonclinical laboratory studies in compliance with the good laboratory practice regulations set forth in this part and, if any studies are currently being conducted, that the quality and integrity of such studies have not been seriously compromised. A disqualified testing facility that wishes to be so reinstated shall present in writing to the Commissioner reasons why it believes it should be reinstated and a detailed description of the corrective actions it has taken or intends to take to assure that the acts or omissions which led to its disqualification will not recur. The Commissioner may condition reinstatement upon the testing facility being found in compliance with the good laboratory practice regulations upon an inspection. If a testing facility is reinstated, the Commissioner shall so notify the testing facility and all organizations and persons who were notified, under § 58.213 of the disqualification of the testing facility. A determination that a testing facility has been reinstated is disclosable to the public under Part 20 of this chapter.

PART 71—COLOR ADDITIVE PETITIONS

3. Part 71 is amended:

a. § 71.1 by adding new paragraph (g), to read as follows:

§ 71.1   Petitions

*   *   *   *   *

   (g) If nonclinical laboratory studies are involved, petitions filed with the Commissioner under section 706(b) of the act shall include with respect to each nonclinical study contained in the petition, either a statement that the study was conducted in compliance with the good laboratory practice regulations set forth in Part 58 of this chapter, or, if the study was not conducted in compliance with such regulations, a statement that describes in detail all differences between the practices used in the study and those required in the regulations.

b. In § 71.6(b) by adding a new sentence at the end of the paragraph to read as follows:

§ 71.6 Extension of Time for Studying Petitions; Substantive Amendments; Withdrawal of Petitions without Prejudice

* * * * *

(b) * * *. If nonclinical laboratory studies are involved, additional information and data submitted in support of filed petitions shall include, with respect to each nonclinical laboratory study contained in the petition, either a statement that the study was conducted in compliance with the requirements set forth in Part 58 of this chapter, or, if the study was not conducted in compliance with such regulations, a statement that describes in detail all differences between the practices used in the study and those required in the regulations.

* * * * *

SUBCHAPTER B—FOOD FOR HUMAN CONSUMPTION

PART 170—FOOD ADDITIVES

4. Part 170 is amended:

a. In § 170.17 by adding new paragraph (c), to read as follows:

§ 170.17 Exemption for Investigational Use and Procedure for Obtaining Authorization to Market Edible Products from Experimental Animals

* * * * *

(c) If intended for nonclinical laboratory studies in food-producing animals, the study is conducted in compliance with the regulations set forth in Part 58 of this chapter.

* * * * *

b. In § 170.35 by adding new paragraph (c)(1)(vi) to read as follows:

§ 170.35   Affirmation of Generally Recognized as Safe (GRAS) Status

*   *   *   *   *

(c) * * *

(1) * * *

(vi) If nonclinical laboratory studies are involved, additional information and data submitted in support of filed petitions shall include, with respect to each nonclinical study, either a statement that the study was conducted in compliance with the requirements set forth in Part 58 of this chapter, or, if the study was not conducted in compliance, with such regulations, a statement that describes in detail all differences between the practices used in the study and those required in the regulations.

*   *   *   *   *

PART 171—FOOD ADDITIVE PETITIONS

5. Part 171 is amended:

a. In § 171.1 by adding new paragraph (k) to read as follows:

§ 171.1   Petitions

*   *   *   *   *

(k) If nonclinical laboratory studies are involved, petitions filed with the Commissioner under section 409(b) of the act shall include, with respect to each nonclinical study contained in the petition, either a statement that the study has been, or will be, conducted in compliance with the good laboratory practice regulations as set forth in Part 58 of this chapter, or, if any such study was not conducted in compliance with such regulations, a statement that describes in detail all differences between the practices used in conducting the study and the good laboratory practice regulations.

b. By revising § 171.6 to read as follows:

§ 171.6   Amendment of Petition

After a petition has been filed, the petitioner may submit additional information or data in support thereof. In such cases, if the Commissioner determines that the additional information or data amount to a substantive amendment, the petition as amended will be given a new filing date, and the time limitation will begin to run anew. Where the substantive amendment proposes a substantial change to any petition that may affect the quality of the human environment, the petitioner is required to submit an environmental analysis report pursuant to § 25.1 of this chapter. If nonclinical laboratory studies are involved, additional information and data submitted in support of filed petitions shall include, with respect to each nonclinical study, either a statement that the study was conducted in compliance with the requirements set forth in Part 58 of this chapter, or if the study was not conducted in compliance with such regulations, a statement that describes in detail all differences between the practices used in the study and those required in the regulations.

PART 180—FOOD ADDITIVES PERMITTED IN FOOD ON AN ITERIM BASIS OR IN CONTACT WITH FOOD PENDING ADDITIONAL STUDY

6. Part 180 is amended in § 180.1 by adding new paragraph (c)(4) to read as follows:

§ 180.1   General

* * * * *

(c) * * *

(4) If nonclinical laboratory studies are involved, studies filed with the Commissioner shall include, with respect to each study, either a statement that the study has been or will be conducted in compliance with the good laboratory practice regulations as set forth in Part 58 of this chapter, or, if any such study was not conducted in compliance with such regulations, a statement that describes in detail all differences between the practices used in conducting the study and the good laboratory practice regulations.

* * * * *

SUBCHAPTER D—DRUGS FOR HUMAN USE

PART 312—NEW DRUGS FOR INVESTIGATIONAL USE

7. In § 312.1 by adding new item 16 to Form FD-1571 in paragraph (a)(2) and by redesignating paragraph (d)(11) as (d)(12) and adding a new paragraph (d)(11), to read as follows:

§ 312.1   Conditions for Exemption of New Drugs for Investigational Use

(a) * * *

(2) * * *

Form FD-1571 * * *

16. A statement that all nonclinical laboratory studies have been, or will be, conducted in compliance with the good laboratory practice regulations set forth in Part 58 of this chapter, or, if such studies have not been conducted in compliance with such regulations, a statement that describes in detail all differences between the practices used in conducting the study and those required in the regulations.

*   *   *   *   *

(d) * * *

(11) All nonclinical laboratory studies were not conducted in compliance with the good laboratory practice regulations set forth in Part 58 of this chapter, or, of such studies were not conducted in compliance with such regulations, all differences between the practices used in conducting the study and the good laboratory practice regulations were not described in detail; or

*   *   *   *   *

PART 314—NEW DRUG APPLICATIONS

8. Part 314 is amended:

a. In § 314.1 by adding new item 16 to Form FD-365H in paragraph (c)(2), by redesignating paragraph (f)(7) as (f)(8) and by adding a new paragraph (f)(7) to read as follows:

§ 314.1 Applications

*   *   *   *   *

(c) * * *

(2) * * *

Form FD-356H-Rev. 1974 * * *

16. *Nonclinical laboratory studies.* With respect to each nonclinical laboratory study contained in the application, either a statement that the study was conducted in compliance with the good laboratory practice regulations set forth in Part 58 of this chapter, or, if the study was not conducted in compliance with such regulations, a statement that describes in detail all differences between the practices used in the study and those required in the regulations.

*   *   *   *   *

(f) * * *

(7) With respect to each nonclinical laboratory study contained in the application, either a statement that the study was conducted in compliance with good laboratory practice regulations set forth in Part 58 of this chapter, or, if the study was not conducted in compliance with such regulations, a statement that describes in detail all differences between the practices used in the study and those required in the regulations.

* * * * *

b. In § 314.8 by adding new paragraph (1) to read as follows:

§ 314.8 Supplemental Applications

* * * * *

(1) A supplemental application that contains nonclinical laboratory studies shall include, with respect to each nonclinical laboratory study, either a statement that the study was conducted in compliance with the requirements set forth in Part 58 of this chapter, or, if the study was not conducted in compliance with such regulations, a statement that describes in detail all differences between the practices used in the study and those required in the regulations.

c. In § 314.9 by adding paragraph (c) to read as follows:

§ 314.9 Insufficient Information in Application

* * * * *

(c) The information contained in an application shall be considered insufficient to determine whether a drug is safe and effective for use unless the application includes, with respect to each nonclinical laboratory study, either a statement that the study was conducted in compliance with the requirements set forth in Part 58 of this chapter, or, if the study was not conducted in compliance with such regulations, a statement that describes in detail all differences between the practices used in the study and those required in the regulations.

d. In § 314.12 by adding a new paragraph (c) to read as follows:

§ 314.12 Untrue Statements in Application

* * * * *

(c) All nonclinical laboratory studies contained in the application were not conducted in compliance with the good laboratory practice regulations as set forth in Part 58 of this chapter, or, if such studies were not conducted in compliance with such regulations, differences between the practices used in conducting the study and the good laboratory practice regulations were not described in detail.

e. In § 314.110 by adding new paragraph (a)(9) to read as follows:

§ 314.110   Reasons for Refusing to File Applications

(a) * * *

(9) The applicant fails to include in the application, with respect to each nonclinical laboratory study, either a statement that the study was conducted in compliance with the requirements set forth in Part 58 of this chapter, or, if the study was not conducted in compliance with such regulations, a statement that describes in detail all differences between the practices used in the study and those required in the regulations.

* * * * *

f. In § 314.111 by striking the period at the end of paragraph (a)(8), adding in lieu thereof a semicolon and the word "or" and adding new paragraph (a)(9) to read as follows:

§ 314.111   Refusal to Approve the Application

(a) * * *

(9) Any nonclinical laboratory study contained in the application was not conducted in compliance with the good laboratory practice regulations as set forth in Part 58 of this chapter, or, if such study was not conducted in compliance with such regulations, differences between the practices used in conducting the study and the good laboratory practice regulations were not described in detail.

* * * * *

g. In § 314.115 by adding new paragraph (c)(6) to read as follows:

§ 314.115   Withdrawal of Approval of an Application

* * * * *

(c) * * *

(6) That any nonclinical laboratory study contained in the application was not conducted in compliance with the good laboratory practice regulations as set forth in Part 58 of this chapter, or any differences between the practices used in conducting the study and those required in the regulations were not described in detail.

PART 330—OVER-THE-COUNTER (OTC) HUMAN DRUGS WHICH ARE GENERALLY RECOGNIZED AS SAFE AND NOT MIS-BRANDED

9. Part 330 is amended in § 330.10 by adding new paragraph (c) to read as follows:

§ 330.10 Procedures for Classifying OTC Drugs as Generally Recognized as Safe and Effective and Not Misbranded, and for Establishing Monographs

* * * * *

(c) Information and data submitted under this section shall include, with respect to each nonclinical laboratory study contained in the application, either a statement that the study was conducted in compliance with the good laboratory practice regulations set forth in Part 58 of this chapter, or, if the study was not conducted in compliance with such regulations, a statement that describes in detail all differences between the practices used in the study and those required in the regulations.

PART 430—ANTIBIOTIC DRUGS; GENERAL

10. In § 430.20 by adding new paragraph (e) to read as follows:

§ 430.20 Procedure for the Issuance, Amendment, or Repeal of Regulations

* * * * *

(e) No regulation providing for the certification of an antibiotic drug for human use shall be issued or amended unless each nonclinical laboratory study on which the issuance or amendment of the regulation is based was conducted in compliance with the good laboratory practice regulations as set forth in Part 58 of this chapter, or, if any such study has not been conducted in compliance with such regulations, differences between the practices used in conducting the study and the good laboratory practice regulations shall be described in detail.

PART 431—CERTIFICATION OF ANTIBIOTIC DRUGS

11. In § 431.17 by adding new paragraph (j) to read as follows:

§ 431.17   New Antibiotic and Antibiotic-Containing Products

*   *   *   *   *

(j) With respect to each nonclinical laboratory study contained in the application, either a statement that the study was conducted in compliance with the good laboratory practice regulations set forth in Part 58 of this chapter, or, if the study was not conducted in compliance with such regulations, a statement that describes in detail all differences between the practices used in the study and those required in the regulations.

———

SUBCHAPTER E—ANIMAL DRUGS, FEEDS, AND RELATED PRODUCTS

PART 511—NEW ANIMAL DRUGS FOR INVESTIGATIONAL USE

12. Part 511 is amended in § 511.1 by revising paragraph (b)(4)(ii), to read as follows:

§ 511.1   New Animal Drugs for Investigational Use Exempt from Section 512(a) of the Act

*   *   *   *   *

(b) * * *

(4) * * *

(ii) All labeling and other pertinent information to be supplied to the investigators. When such pertinent information includes nonclinical laboratory studies, the information shall include, with respect to each nonclinical study, either a statement that the study was conducted in compliance with the requirements set forth in Part 58 of this chapter, or, if the study was not conducted in compliance with such regulations, a statement that describes in detail all differences between the practices used in the study and those required in the regulations.

* * * * *

PART 514—NEW ANIMAL DRUG APPLICATIONS

13. Part 514 is amended:
   a. In § 514.1 by adding new paragraph (b)(12)(iii) to read as follows:

§ 514.1   Applications

* * * * *

   (b) * * *
   (12) * * *
   (iii) With respect to each nonclinical laboratory study contained in the application, either a statement that the study was conducted in compliance with the good laboratory practice regulations set forth in Part 58 of this chapter, or, if the study was not conducted in compliance with such regulations, a statement that describes in detail all differences between the practices used in the study and those required in the regulations.

* * * * *

   b. In § 514.8 by adding new paragraph (1) to read as follows:

§ 514.8   Supplemental New Animal Drug Applications

* * * * *

   (1) A supplemental application that contains nonclinical laboratory studies shall include, with respect to each nonclinical study, either a statement that the study was conducted in compliance with the requirements set forth in Part 58 of this chapter, or, if the study was not conducted in compliance with such regulations, a statement that describes in detail all differences between the practices used in the study and those required in the regulations.
   c. In § 514.15 by adding new paragraph (c) to read as follows:

§ 514.15   Untrue Statements in Applications

* * * * *

   (c) Any nonclinical laboratory study contained in the application was conducted in compliance with the good laboratory practice regulations as set forth in Part 58 of thic chapter, and differences between the practices used in the conduct of the study and those required in the regulations were not described in detail.

d. In § 514.110 by adding new paragraph (b)(8) to read as follows:

§ 514.110   Reasons for Refusing to File Applications

*   *   *   *   *

(b) * * *

(8) It fails to include, with respect to each nonclinical study contained in the application, either a statement that the study was conducted in compliance with the good laboratory practice regulations set forth in Part 58 of this chapter, or, if the study was not conducted in compliance with such regulations, a statement that describes in detail all differences between the practices used in the study and those required in the regulations.

*   *   *   *   *

e. In § 514.111 by adding new paragraph (a)(11) to read as follows:

§ 514.111   Refusal to Approve an Application

(a) * * *

(11) Any nonclinical laboratory study contained in the application was not conducted in compliance with the good laboratory practice regulations as set forth in Part 58 of this chapter, or any differences between the practices used in conducting the study and those required in the regulations were not described in detail.

*   *   *   *   *

f. In § 514.115 by adding new paragraph (b)(4) to read as follows:

§ 514.115   Withdrawal of Approval of Applications

*   *   *   *   *

(b) * * *

(4) That any nonclinical laboratory study contained in the application was not conducted in compliance with the good laboratory practice regulations as set forth in Part 58 of this chapter, and differences between the

practices used in conducting the study and the regulations were not described in detail.

* * * * *

PART 570—FOOD ADDITIVES

14. Part 570 is amended:
   a. In § 570.17 by adding new paragraph (c) to read as follows:

§ 570.17 Exemption for Investigational Use and Procedure for Obtaining Authorization to Market Edible Products from Experimental Animals

* * * * *

(c) If intended for nonclinical laboratory studies in food-producing animals, the study is conducted in compliance with the regulations set forth in Part 58 of this chapter.
   b. In § 570.35 by adding new paragraph (c)(1)(vi) to read as follows:

§ 570.35 Affirmation of Generally Recognized as Safe (GRAS) Status

* * * * *

(c) * * *
(1) * * *
(vi) If nonclinical laboratory studies are involved, additional information and data submitted in support of filed petitions shall include, with respect to each nonclinical study, either a statement that the study was conducted in compliance with the requirements set forth in Part 58 of this chapter, or, if the study was not conducted in compliance with such regulations, a statement that describes in detail all differences between the practices used in the study and those required in the regulations.

* * * * *

PART 571—FOOD ADDITIVE PETITIONS

15. Part 571 is amended:
   a. § 571.1 by adding paragraph (k) to read as follows:

§ 571.1   Petitions

* * * * *

(k) If nonclinical laboratory studies are involved, petitions filed with the Commissioner under section 409(b) of the act shall include, with respect to each study, either a statement that the study was conducted in compliance with the requirements set forth in Part 58 of this chapter, or, if the study was not conducted in compliance with such regulations, a statement that describes in detail all differences between the practices used in the study and those required in the regulations.

b. In § 571.6 by adding the following sentence to the end of the section to read as follows:

§ 571.6   Amendment of Petition

* * * If nonclinical laboratory studies are involved, additional information and data submitted in support of filed petitions shall include, with respect to each such study, either a statement that the study was conducted in compliance with the requirements set forth in Part 58 of this chapter, or, if the study was not conducted in compliance with such regulations, a statement that describes in detail all differences between the practices used in the study and those required in the regulations.

SUBCHAPTER F—BIOLOGICS

PART 601—LICENSING

16. Part 601 is amended:

a. In § 601.2 by revising paragraph (a) to read as follows:

§ 601.2   Applications for Establishment and Product Licenses; Procedures for Filing

(a) *General.* To obtain a license for any establishment or product, the manufacturer shall make application to the Director, Bureau of Biologics, on forms prescribed for such purposes, and in the case of an application for a product license, shall submit data derived from nonclinical laboratory and clinical studies which demonstrate that the manufactured product meets prescribed standards of safety, purity, and potency; with respect to each nonclinical laboratory study, either a statement that the study was conducted in compliance with the requirements set forth in Part 58 of this chapter, or, if the study was not conducted in compliance with such regulations, a statement that describes in detail all differences between the practices used in the study and those required in the regulations; a full description of manufacturing methods; data establishing stability of the product through the dating period; sample(s) representative of the product to be sold, bartered, or exchanged or offered, sent, carried or brought for sale, barter, or exchange; summaries of results of tests performed on the lot(s) represented by the submitted sample(s); and specimens of the labels, enclosures and containers proposed to be used for the product. An application for license shall not be considered as filed until all pertinent information and data shall have been received from the manufacturer by the Bureau of Biologics. In lieu of the procedures described in this paragraph, applications for radioactive biological products shall be handled as set forth in paragraph (b) of this section.

* * * * *

b. By revising § 601.30 to read as follows:

§ 601.30 Licenses Required; Products for Controlled Investigation Only

Any biological or trivalent organic arsenical manufactured in any foreign country and intended for sale, barter or exchange shall be refused entry by collectors of customs unless manufactured in an establishment holding an unsuspended and unrevoked establishment license and license for the product. Unlicensed products that are not imported for sale, barter or exchange and that are intended solely for purposes of controlled investigation are admissible only if the investigation is conducted in accordance with section 505 of the Federal Food, Drug and Cosmetic Act and the requirements set forth in Parts 58 and 312 of this chapter.

SUBCHAPTER J—RADIOLOGICAL HEALTH

PART 1003—NOTIFICATION OF DEFECTS OR FAILURE TO COMPLY

17. Part 1003 is amended in § 1003.31 by revising paragraph (b), to read as follows:

§ 1003.31 Granting the Exemption

* * * * *

(b) Such views and evidence shall be confined to matters relevant to whether the defect in the product or its failure to comply with an applicable

Federal standard is such as to create a significant risk of injury, including genetic injury, to any person and shall be presented in writing unless the Secretary determines that an oral presentation is desirable. Where such evidence includes nonclinical laboratory studies, the data submitted shall include, with respect to each nonclinical study, either a statement that each study was conducted in compliance with the requirements set forth in Part 58 of this chapter, or, if the study was not conducted in compliance with such regulations, a statement that describes in detail all differences between the practices used in the study and those required in the regulations.

*   *   *   *   *

PART 1010—PERFORMANCE STANDARDS FOR ELECTRONIC PROJECTS, GENERAL

13. Part 1010 is amended:
   a. In § 1010.4 by adding new paragraph (b)(1)(ix) to read as follows:

§ 1010.4 Variances

*   *   *   *   *

(b) * * *
(1) * * *
(ix) With respect to each nonclinical study contained in the application, either a statement that the study was conducted in compliance with the good laboratory practice regulations set forth in Part 58 of this chapter, or, if the study was not conducted in compliance with such regulations, a statement that describes in detail all differences between the practices used in the study and those required in the regulations.

*   *   *   *   *

   b. In § 1010.5 by revising paragraph (c)(12) to read as follows:

§ 1010.5 Exemptions for Products Intended for United States Government Use

*   *   *   *   *

(c) * * *
(12) Such other information required by regulation or by the Director, Bureau of Radiological Health, to evaluate and act on the application. Where such information includes nonclinical laboratory studies, the infor-

mation shall include, with respect to each nonclinical study, either a statement that each study was conducted in compliance with the requirements set forth in Part 58 of this chapter, or, if the study was not conducted in compliance with such regulations, a statement that describes in detail all differences between the practices used in the study and those required in the regulations.

*   *   *   *   *

Effective date. This rule is effective June 20, 1989.

(Secs. 406, 408, 409, 502, 503, 505, 506, 507, 510, 512–516, 518–520, 601, 701(a), 706, and 801, 52 Stat. 1049–1053 as amended, 1055, 1058 as amended, 55 Stat. 851 as amended, 59 Stat. 463 as amended, 68 Stat. 511–517 as amended, 72 Stat. 1785–1788 as amended, 76 Stat. 794 as amended, 82 Stat. 343–351, 90 Stat. 539–574 (21 U.S.C. 346, 346a, 348, 352, 353, 355, 356, 357, 360, 360b–360f, 360h–360i); secs. 215, 351, 354–360F, 58 Stat. 690, 702 as amended, 82 Stat. 1173–1186 as amended; 42 U.S.C. 216, 262, 263b–263n).

Dated: December 4, 1978.

DONALD KENNEDY,
Commissioner of
Food and Drugs.

Appendix F

PROPOSED REVISION OF THE CLINICAL LABORATORY REGULATIONS FOR MEDICARE, MEDICAID, AND CLINICAL LABORATORIES IMPROVEMENT ACT OF 1967 DEPARTMENT OF HEALTH AND HUMAN SERVICES: HEALTH CARE FINANCING ADMINISTRATION

Medicare, Medicaid and CLIA Programs; Revision of the Clinical Laboratory Regulations for the Medicare, Medicaid, and Clinical Laboratories Improvement Act of 1967 Programs

AGENCY: Health Care Financing Administration (HCFA), HHS.

42 CFR Parts 74, 405, 416, 440, 482, 483, 488, and 493

ACTION: Proposed rule.

SUMMARY: The proposed rule would revise regulations for laboratories
regulated under the Medicare, Medicaid and Clinical Laboratories
Improvement Act of 1967 (CLIA) programs. The revisions would recodify the
regulations for these programs into a new Part 493 in order to simplify
administration and unify the health and safety requirements for all programs
as much as possible. We propose to have a single set of regulations for the
three programs, with an additional subpart for the licensure procedure unique
to the CLIA program.
   We propose to revise the regulations to remove outdated, obsolete and
redundant requirements, make provision for new technologies, place increased
reliance on outcome measures of performance, and emphasize the
responsibilities and duties of personnel rather than the formal credentialing
requirements and detailed personnel standards in existing regulations. We
would provide for new uniform proficiency testing standards and add
requirements for additional specialties, such as cytogenetics.

DATE: To ensure consideration, comments must be received by November 3,
1988.

ADDRESSES: Address comments in writing to: Health Care Financing
Administration, U.S. Department of Health and Human Services, Attention:
HSQ-146-P, P.O. Box 26676, Baltimore, Maryland 21207.
   If you prefer, you may deliver your comments to one of the following
addresses:

Room 309-G, Hubert H Humphrey
   Building, 200 Independence Avenue,
   SW., Washington, DC, or
Room 132, East High Rise Building, 6325
   Security Boulevard, Baltimore,
   Maryland.
   Please address a copy of comments on information collection requirements
to:
Allison Herron, Office of Information and Regulatory Affairs, Room 3208 New
   Executive Office Building, Washington, DC 20503.
   In commenting, please refer to file code HSQ-146-P. Comments will be
available for public inspection as they are received, beginning approximately
three weeks after publication of this document, in Room 309-G of the
Department's offices at 200 Independence Avenue, SW., Washington, DC, on
Monday through Friday of each week from 8:30 a.m. to 5:00 p.m. (202-245-
7890).

FOR FURTHER INFORMATION CONTACT:
Pamela Renner, (301) 966-6818.

SUPPLEMENTARY INFORMATION:

I. BACKGROUND

Multiple Laboratory Activities

Under the Medicare program, we pay for diagnostic services furnished to beneficiaries by a variety of laboratories. These include a laboratory that is "hospital-based" (that is, it is located in or it is under the supervision of a hospital), located in a physician's office or is "independent" (not a hospital-based and not a rural health clinic, a group medical practice or a physician's office). By statute, the definition of a hospital contained in Section 1861(e) of the Social Security Act (the Act) extends Medicare participation to hospital laboratories. The paragraph following section 1861(s)(11) and sections 1861(s)(12) and (13) provide coverage for independent laboratory services.

Under the Clinical Laboratories Improvement Act of 1967 (CLIA), laboratories engaged in testing in interstate commerce must meet the requirements of Section 353 of the Public Health Service Act (42 U.S.C. 263a) in order to be licensed or remain licensed for testing in interstate commerce.

Medicaid, under the authority of Section 1902(a)(9)(C) of the Social Security Act, pays for services furnished only by laboratories that meet Medicare conditions for coverage. Because participation in the Medicaid program is governed by Medicare rules, henceforth when we refer to Medicare we are including Medicaid.

Various State laws govern licensure requirements for laboratories engaged in intrastate commerce. There are also Medicare-Medicaid requirements that laboratories must meet in terms of personnel qualifications and accuracy of test results. Under existing regulations, the laboratory requirements are integrated with other requirements applicable to the provider or supplier. Thus, conditions of participation for a hospital-based laboratory are found in the hospital conditions of participation in 42 CFR 482.27. Laboratories in skilled nursing facilities must meet the same conditions of participation as laboratories in hospitals. Conditions for coverage of independent laboratory services are found at 42 CFR 405.1310 to 405.1317. Except for laboratory services furnished by a physician, group medical practice, or rural health clinic or a few services furnished by an end-stage renal disease facility, which are not subject to our rules, laboratory services furnished by any entity other than a hospital or skilled nursing facility are governed by our rules concerning independent laboratories.

Regulations found at 42 CFR Part 74 implement Section 353 of CLIA.

A CLIA laboratory and any other entity identified as a laboratory under title XVIII of the Social Security Act that wishes to receive payment for its services from Medicare or Medicaid must meet Medicare's conditions of participation or conditions for coverage of services.

A laboratory that fails to meet the Medicare conditions for coverage for a given specialty is not "approved" for payment of services for that specialty. The loss of approval ("termination of approval") or failure to be approved initially results in no payment from Medicare or Medicaid for the services in the failed specialty. Failure to meet CLIA requirements for a category of services or a specific test results in the loss of licensure, or licensure is denied, for the category of services or that test. A laboratory may fail general conditions and fail to become approved for Medicare reimbursement for any specialty; similarly, a CLIA laboratory failing to meet general requirements would not be licensed for any tests.

Federal Oversight Activities

HCFA, under an interagency agreement and a memorandum of understanding (MOU) with the Public Health Service (PHS), has administrative responsibility for both the Medicare and CLIA programs. However, PHS (for the Food and Drug Administration (FDA)) has primary responsibility for the provision of technical advice on blood bank programs, including the revision of regulations concerning blood and blood products. An MOU in 1980 between HCFA and PHS (FDA) resulted in a reduction in the duplication of the inspection responsibilities in the area of blood banking and transfusion services and in a more coordinated follow-up on the transfusion related fatalities reported to FDA. Before the establishment of HCFA in 1977 and the signing of an interagency agreement, PHS, through the Centers for Disease Control (CDC), was responsible for the CLIA program. The most recent MOU between HCFA and CDC further delineated responsibilities by specifying that HCFA is responsible for developing regulations that relate to Medicare and CLIA and that CDC is responsible for assisting HCFA in obtaining technical and scientific expertise.

Over the past few years, we have been working with CDC and FDA in reviewing the various departmental regulations concerning laboratories to determine the appropriate modifications and revisions. This process has included meetings of HCFA, FDA and CDC with various private-sector organizations, concerned members of the laboratory industry and the public, State health officials, and other Federal agencies. Our review indicated the need for revising the regulations because of a lack of uniformity in the various laboratory program requirements, obsolete or redundant requirements, and the numerous changes in technology and data management systems that have occurred since the regulations were originally written.

The Department commissioned a study on clinical laboratories (Final Report on Assessment of Clinical Laboratory Regulations (April 1986)) that recommended that HHS review the existing regulations to determine how to improve the assurance of quality laboratory testing and to achieve as much program uniformity as possible. We determined that the best approach to achieve these goals was to revise both the CLIA and Medicare/Medicaid regulations concerning laboratories, since the various regulations are interrelated.

Legislation

Section 4064 of the Omnibus Budget Reconciliation Act of 1987 (OBRA'87). Pub. L. 100-203, enacted on December 22, 1987, amended the sentence following section 1861(s)(11) by requiring that physicians' offices that perform more than 5000 tests per year must meet conditions relating to the health and safety of individuals with respect for whom such tests are performed. This amendment applies to tests performed on or after January 1, 1990. We are not proposing regulations at this time to implement this legislation. However, we recognize that some of the standards in this proposed rule might also be applicable to tests run in physician office laboratories. Therefore, we ask commenters to inform us whether and on what basis they believe the standards we are proposing in this rule would be appropriate to apply to tests run in physician office laboratories.

Consolidation of Regulations

In this proposed rule we propose to consolidate all CLIA and Medicare-Medicaid laboratory requirements in a new 42 CFR Part 493. We propose to remove outdated and overly prescriptive requirements. We would require laboratories to comply with the health and safety standards of other Federal, State and local agencies and our decisions to approve or license laboratories would be affected by their compliance with these laws. We would also add provisions requiring facilities to develop and implement their own internal quality assurance programs, and we would also provide for increased reliance on outcome measures by using quality control and proficiency testing data in the assessment of laboratory performance.

Laboratory testing in both physicians' offices and rural health clinics that perform no tests on referral would not be subject to these revisions because both the Medicare and CLIA statutes (the sentence following § 1861(s)(11) of the Act and section 351(i) of the PHS Act, respectively) preclude the regulation at this time of physician office laboratories and of rural health clinics that perform tests only for their own patients. (As mentioned earlier, section 4064 of the Omnibus Budget Reconciliation Act of 1987 amended section 1861(s) of the Social Security Act to require the regulation of physician office laboratories performing more than 5000 tests a year effective January 1, 1990.)

II. PURPOSES OF THE PROPOSED REVISIONS

We propose to recodify and revise existing laboratory regulations to accomplish several goals.

• We intend, as stated above, to consolidate current laboratory requirements into one new part to make these requirements uniform—to the extent possible—from one provider or supplier to another and between CLIA and Medicare. Thus, a laboratory seeking CLIA licensure or approval for payment for services to Medicare or Medicaid beneficiaries would look in only one part of the regulations. Similarly, a laboratory operated by a skilled nursing facility would not be governed by regulations in two separate parts (one for hospitals and one for independent laboratories) as it is currently.

• Another major goal of the proposed regulation is to have, to the extent possible, the same requirements for both CLIA and Medicare. HCFA is responsible for implementing the requirements for both CLIA and Medicare; approximately 1300 of the 1500 CLIA laboratories also participate in Medicare. One agency imposing different program requirements on the same laboratory is confusing. As the major difference between CLIA and Medicare is that CLIA laboratories engage in interstate commerce, there is no compelling reason (other than where the statute requires) for having dissimilar requirements. These regulations would require that a laboratory that is denied approval for Medicare or Medicaid coverage also be denied licensure under CLIA.

To this end, we are proposing to revise requirements relating to personnel, proficiency testing, quality control, the applicability of the regulations, compliance with State and local laws, recordkeeping and inspection.

• We intend to revise our personnel standards so that personnel requirements are not focused primarily on qualifications but on the accurate performance of laboratory tests. Current hospital laboratory personnel requirements are not based on credentials for individuals other than the laboratory director, but current independent laboratory requirements specify credentials for laboratory director, technical and general supervisors, technologists, cytotechnologists, technicians and technician trainees. We would combine hospital-based and independent laboratory requirements and establish consistent requirements for all laboratories.

● We would impose a new quality assurance program on all laboratories. This provision would require a laboratory to be responsible for the quality of its services and provide the laboratory with the flexibility to evaluate the competency of technical staff.

● We also intend to update current internal quality control requirements for each specialty and subspecialty, taking into consideration current and future technological advances. We propose to emphasize the importance of quality control and we would make failure of quality control in a specialty or subspeciality result in the loss of approval or licensure in that specialty or subspecialty.

● We intend to revise the current Medicare and CLIA proficiency testing requirements considerably. We would require every laboratory to enroll and participate successfully in an approved proficiency testing program for each specialty and subspecialty for which there is an approved program and for which the laboratory seeks or has Medicare approval or CLIA licensure. The proficiency testing programs would have to meet our requirements, including grading criteria, in order to be considered an approved program for purposes of our regulations.

● We propose to update the licensure requirements applicable only to CLIA laboratories by eliminating overly-prescriptive requirements, such as those involved with annual license renewal.

● We propose to require Medicare laboratories to comply with Federal laws concerning health and safety and CLIA laboratories to comply with Federal, State and local laws concerning health and safety. Currently, a laboratory out of compliance with a State or local law may lose its Medicare approval but retain CLIA licensure, an inconsistency that this regulation would eliminate.

III. PROPOSED RULE

General Approach

We propose to revise the standards for all laboratories participating in Medicare or Medicaid or licensed under CLIA to provide as much uniformity as possible within these programs. (For these proposed regulations we would not consider a physician or group of physicians that performs tests only on his, her, or their own patients as a laboratory; however, the OBRA's change concerning physicians performing 5000 tests does not distinguish between tests done on referral and those that are not.) There are certain limitations on the extent to which the regulations can be unified because of differences in the Medicare and CLIA statutes.

The new Part 493 would have ten subparts dealing with general provisions, administration, proficiency testing, proficiency testing programs, patient test management, quality control, personnel, quality assurance, inspection, and requirements unique to CLIA laboratories and CLIA licensure procedures. The regulations affecting other facilities that have requirements for laboratory services would be modified by cross-referring them to the new regulations.

Under the Medicare program, conditions of participation or for coverage are the requirements that an entity, such as a laboratory, must meet in order to participate and have tests paid for by Medicare or Medicaid. Each condition is usually comprised of one or more standards, which enumerate activities, outcomes, or other requirements that, upon evaluation by HCFA or a State survey agency under contract with HCFA, serve as the basis for determining that a particular condition has been satisfied. If the laboratory fails to comply with any condition for coverage, we initiate an adverse action to terminate the laboratory's participation in Medicare or revoke the laboratory's licensure under CLIA. The adverse action may be taken by terminating a laboratory's participation or licensure in a specialty or subspecialty if the deficiencies are limited to particular categories of testing or the laboratory's approval for all services may be terminated if the deficiencies are pervasive, affecting the overall services offered by the laboratory.

Unless otherwise identified in our regulations, standards refer to the condition immediately preceding them. Depending on the complexity of the subject matter a standard may be designated as a regulation section or subsection. A subordinate designation does not imply that some standards are more important than others.

SUBPART A—GENERAL PROVISIONS

I. Section 493.1, Basis and Scope

In a new § 493.1, Basis and scope, we would indicate the sections of the laws that apply to the new part—sections 1861(e) and (j), the sentence following section 1861(s)(11), sections 1861(s)(12) and (13) and 1902 of the Social Security Act and section 353 of the Public Health Service Act. We intend for this part to apply to all independent and hospital-based laboratories, intermediate care facilities for the mentally retarded, skilled nursing facilities and ambulatory surgical centers that perform laboratory services, rural health clinics that perform tests on referral, and physicians' offices that perform any tests on referral.

2. Section 493.2, Definitions

It is our practice to define terms whose meanings may not be clear from their context or where we apply an interpretation that may not be commonly used. This section would contain definitions that are applicable to both Medicare and CLIA. We would eliminate requirements of present regulations that we feel are unnecessarily prescriptive; hence, many definitions found in current §§ 405.1310 and 74.2 are no longer necessary.

• Independent laboratory—
Currently, "independent laboratory" is defined in § 405.1310(a). We propose to delete the definition of "independent laboratory" from the definition section because it placed emphasis on location and ownership, conditions no longer considered relevant under this proposal. We would remove the exception in the present definition for laboratories maintained by physicians that accept no more than 100 specimens on referral in any category during any calender year because our experience has not shown that the definition is effective in assuring that physician office laboratories limit testing of specimens received on referral. Rather, we will continue to enforce rigorously our requirement that Medicare-approved laboratories refer tests only to approved laboratories. We would maintain the exception for physician office and rural health clinic laboratories that perform no tests on referral. Under this proposed rule these rural health clinics and physician office laboratories that perform any tests on referral would be required to meet these conditions for coverage of services of laboratories.

In addition, on January 1, 1990, the legislative provision requiring that physicians performing 5,000 or more tests per year must meet health and safety conditions will go into effect.

• Clinical Laboratory—
Currently, under § 405.1310 a clinical laboratory is a facility for the examination of the microbiological, serological, chemical, hematological, radiobioassay, cytological, immunohematological, pathological, or other examination of materials derived from the human body for purpose of diagnosis, prevention or treatment of disease or assessment of a medical condition. Section 74.2(a) makes Part 74 applicable to laboratories engaged in the laboratory examination of, or other laboratory procedures relating to, specimens solicited or accepted in interstate commerce for the purpose

of providing information for the diagnosis, prevention or treatment of any disease or impairment, or the assessment of the health, of human beings.

Our purposed definition of a clinical laboratory would remove the word "clinical". It does not add anything to the definition and may create confusion, since we also pay for the services of independent laboratories, hospital laboratories, etc., which perform anatomic services, which are generally not considered part of clinical laboratory services. In addition, our definition would differ from that currently in § 405.1310(b) and § 74.2(a) since we would include in the examination of material derived from the human body screening tests and any test where a measurement is made on an analyte (test) or where a procedure is performed.

We would make this modification because, in the case of CLIA laboratories, there have been questions concerning the applicability of CLIA to certain screening procedures, such as drug testing or HIV (AIDS) testing. Some people believe that these tests are not subject to regulation because they would indicate a drug level or presence of HIV virus and are not a diagnosis or an assessment of health and thus subject to regulation. However, these tests are currently reimbursed under Medicare and Medicaid and are considered test procedures for health assessment. Therefore, in moving the definition in § 74.2(a) and § 405.1310(b) to new § 493.2 we would expand it to provide uniformity between Medicare and CLIA and to clarify the type of procedures covered by the regulations.

We would not consider as laboratories those facilities that only collect specimens and do not perform testing. We also would not consider mailing services as laboratories if they just send out reports, the testing laboratory assumes responsibility for the report, and the name and address of the testing laboratory is on the report.

• Proficiency testing—

We would not include a definition of proficiency testing because the two proposed subparts, C and D, include extensive discussion of proficiency testing. We propose to provide criteria for State and private-sector programs to meet in order to be approved for use in the Medicare and CLIA programs. Laboratories would have to enroll in these approved programs to meet DHHS' requirements for successful participation in a proficiency testing program for both CLIA and for Medicare.

• Personnel requirements—

Terms currently contained in paragraphs (d), (e), (g), (h), (i) and (k) in § 405.1310 and in § 74.1(c) would not appear in our proposed rule. Therefore, these definitions are no longer required.

• Radiobioassay and histocompatibility—

We would not include the current definitions for radiobioassay and histocompatibility in § 405.1310(l) and (m) and for radiobioassay in § 74.1(i) since these terms are clearly explained elsewhere in the proposed regulations.

● Authorized person—

We would add a definition of "authorized person" as the person authorized to order and receive tests. Under Medicare that person is a physician as defined in section 1861(r) of the Act. We would permit other individuals, including patients, to be "authorized persons" when State law and Medicaid allows. This would reduce the conflict between Federal and State law over who can order and receive tests and would defer to the States for Medicaid purposes as well as to the State where Federal funding under Medicare or Medicaid is not involved.

● Consultation—

We would not include two definitions currently contained in § 405.1310. "Consultation" and "Secretary" do not appear in proposed new Part 493. (We use, instead of "Secretary," "HHS," which is defined in Part 400.)

● Miscellaneous—

+ "Referee laboratory", which is a laboratory used for comparison of proficiency test results, would be defined as it is currently in § 74.1(j).

+ Reference laboratory", currently defined in Part 74.1(k), would be deleted because the proposed proficiency testing program does not include the term "reference laboratory" in defining grading criteria.

+ We would explain what "CLIA" stands for.

+ We would define "accredited laboratory" based on current § 74.1(c), including the current definition of "approved accreditation body" that is now in § 74.1(e)

+ Accredited institution," "director," "health insurance program," and "physician" now defined in Part 74 would be excluded from our proposed rule as duplicative or unnecessary. "Specimen" would be defined as it appears in current § 405.1310 and § 74.1(b).

+ We would define "challenge" and "target value," which are related to the proficiency testing requirements.

SUBPART B—ADMINISTRATION

Section 493.11 Condition—Compliance with Federal, State and Local Laws

Present Medicare requirements in § 405.1311 relating to compliance with State and local laws do not include compliance with Federal laws related to health and safety nor do current requirements under CLIA. We do not believe that a laboratory should be certified if it is not in compliance with other Federal regulations concerning patient health and safety. New § 493.11 would include this requirement.

The requirement to comply with Federal, State and local laws would also apply to CLIA laboratories. This requirement is more stringent that CLIA requirements currently in § 74.30, which requires compliance with State laws for personnel standards only, by cross-referring to the Medicare standards. The inclusion of these provisions for CLIA laboratories would eliminate the requirement for a State survey agency to inspect the laboratory for CLIA licensure purposes when the laboratory cannot operate in the State because of noncompliance with State laws. In the case of laboratories participating in both the Medicare and CLIA programs, which are the majority of interstate laboratories, unifying the requirements to assure compliance with State and local laws would eliminate the possibility of laboratories being acceptable under CLIA and unacceptable in the Medicare program for the same test procedures.

We would require compliance with the existing regulatory programs covered by State and local laws for fire safety and with State, Federal and local laws on environment and health-related matters.

Proposed § 493.11 would not include the requirements in present § 405.1316(c) concerning who may draw specimens from a patient, as this issue is governed by State law if such requirements exist.

SUBPART C—PARTICIPATION IN PROFICIENCY TESTING

Present regulations at 42 CFR 405.1314(a) and Part 74, Subpart E require laboratories to participate in proficiency testing (PT) programs. We would greatly expand these requirements. A new Subpart C would contain the general requirements a laboratory must meet for PT and would elevate current requirements for enrollment and successful participation to the condition level.

The new PT requirements would emphasize the increased importance attached to obtaining a passing score on samples with known contents as an outcome measure of the quality of testing in laboratories. CDC, under an Interagency Agreement with HCFA, established a Task Force that developed proposed PT requirements in response to our request for a PT program that could be applied to both Medicare and CLIA laboratories.

Although specific requirements for a PT program are included for CLIA laboratories in 42 CFR Part 74, minimally acceptable requirements for Medicare-approved laboratories have never been described by HHS in terms of program content, challenge, frequency of PT shipments or testing events and grading criteria for individual analytes and specialties and subspecialties. As a consequence, State and private organizations have offered PT programs that have been approved for Medicare purposes. These programs vary considerably in degree of challenge and grading criteria from each other and from the existing CLIA PT requirements.

Our proposed regulatory revisions would focus on assessing the quality of laboratory tests that are commonly performed or have results critical to the patient's health (e.g., an incorrect result has a moderate to high risk of death), or both. In the future we hope to include additional tests in areas such as histopathology, cytogenetics and drug abuse testing. We also propose to exclude from the proposed PT program those test areas for which no performance problems are evident, such as routine urinalysis, and to continue to reevaluate the necessity for PT for other test areas in the future.

The regulations would include a PT grading system by specialty and subspecialty. The current CLIA requirements in 42 CFR Part 74 include a composite score for a few specialty areas but do not include a scoring system for most areas that would provide for a determination to be made of the overall performance of the laboratory in each specialty and subspecialty.

The proposed PT requirements would apply to selected tests, named in the regulations. It is neither possible nor currently feasible to evaluate every test a laboratory can perform by PT, since many tests are not available in PT programs presently offered. The tests and analytes we selected for challenge are representative of the laboratory's ability to perform in each speciality and subspecialty.

These proposed regulations would make adverse actions based on unsuccessful PT performances consistent for laboratories approved under Medicare and those licensed under CLIA.

The Medicare program currently requires successful participation in a PT program in each specialty and subspecialty. We do not currently provide States with criteria for determining successful participation for specialties and subspecialties. Hence, under Medicare, States may establish individual protocols for evaluating PT performance and these evaluations vary from State to State.

Under CLIA, HCFA may take an adverse action against a laboratory by removing licensure for individual tests. However, current CLIA PT regulations do not provide for grading PT performance by specialty or subspecialty except for bacteriology, mycology and parasitology, which are not graded by test but are graded by overall subspecialties.

Therefore, with the exception of the subspecialties of bacteriology, mycology and parasitology, adverse actions proceed test by test as unsatisfactory performance occurs with no provisions for removing licensure by specialty or subspecialty even though the laboratory may be performing related methodologies in the same specialty or subspecialty. Moreover, a laboratory could fail all PT testing in a particular specialty or subspecialty and lose its license for those tests but would still be allowed to perform related testing in the specialty or subspecialty because PT testing is not offered for all tests categorized in a specialty or subspecialty.

In the Medicare program, we approve and disapprove independent laboratories by specialties and subspecialties rather than by individual tests. The Medicare carrier is notified of the laboratory's approval by specialty and subspecialty, not by test(s) performed. It is not feasible under Medicare to terminate approval for individual tests since thousands of tests are involved. Monitoring laboratories' continuous changes in approval status by individual tests is not effective since test offerings change frequently and updates to the States, regions and carriers would always lag behind changes in laboratory services.

We have decided that the development of a specialty and subspecialty grading system is the most reasonable and appropriate mechanism to monitor quality of testing. Since our intent is to make the Medicare and CLIA programs as consistent as possible and to provide an overall assurance of the quality, we propose to require identical mechanisms of assessing PT performance for CLIA and Medicare laboratories. We propose a unified assessment to be made by the Medicare State survey agency or HHS in determining the licensure and Medicare approval status of a laboratory by specialty and subspecialty. The State survey agencies would receive a single set of PT data and would consistently determine compliance for the two programs at the same time. A laboratory failing the PT requirements and the enhanced PT requirements would lose Medicare approval for the speciality or subspecialty. Similarly, a CLIA licensed laboratory would be notified of any PT failure and, if a hearing is not requested, the license would be revoked for the specialty or subspecialty.

Currently, a laboratory may enroll in one or more PT programs to satisfy Medicare requirements and a different program to satisfy requirements under CLIA. The proposed modification of the regulation would require laboratories to specify the PT program it is using to meet the Medicare or CLIA requirements (or both). It would not preclude a State or private-sector accreditation program from having different standards for its own program purposes but it would require that laboratories be enrolled and successfully participate in a PT program that meets the Federal criteria.

We would introduce requirements for an enhanced PT program in § 493.25. Condition: Enhanced proficiency testing. If a laboratory were to perform unsuccessfully in PT, it would be able to request enrollment in an enhanced PT program within 15 days of notification of unsuccessful performance in order to prevent immediate termination for that failed specialty or subspecialty. While the laboratory is participating in the enhanced PT program, Medicare approval and CLIA licensure, as applicable, for the tests performed under the specialty or subspecialty would continue. If the laboratory also fails the enhanced PT, we would initiate termination of Medicare approval or revocation of CLIA licensure, as applicable, for the failed speciality or subspecialty. By instituting the enhanced PT program, we would not be penalizing a laboratory for one set of aberrant results. The laboratory duing this period of participation in enhanced PT would be able to undertake any necessary corrective action while also demonstrating sustained successful performance in the failed specialty or subspecialty.

In the new section on enhanced PT (§ 493.25), we propose that laboratories that fail PT must successfully participate in an enhanced PT program for three consecutive shipments, testing events or combination thereof in the failed specialty or subspecialty in order to prevent immediate termination of approval or licensure for the specialty or subspecialty. The enhanced PT program would consist of more specimens per shipment or testing event than what is required in the routine PT and may be at a frequency greater than quarterly. Enhanced PT may consist of challenges to cover the overall specialty or subspecialty or it may consist of individual analyte challenges, depending upon the laboratory's PT failure. If laboratory performance falls below the acceptable level after one, two or three shipments or testing events of enhanced PT, approval of the applicable specialty or subspecialty would be terminated, licensure revocation proceedings instituted, or both. If a laboratory's Medicare participation were terminated or its CLIA license were revoked as a result of a PT failure, the laboratory's participation or licensure for the applicable specialty or subspecialty could not be reinstated until the laboratory demonstrated sustained successful performance in three consecutive shipments or testing events (or combination) of enhanced PT. The proposed acceptable levels for enhanced PT (80 percent of results within limits for categories of testing; and for individual analysis: 5 of 6 results for one shipment, 10 of 12 for two shipments and 15 of 18 for three shipments) are based on statistical calculations and intended to detect substandard performance without falsely identifying competent laboratories. For these calculations, a competent laboratory is assumed to have less than a one percent chance of exceeding the acceptable limits established for each analyte. We ask for comments regarding other alternative enhanced PT schemes, for example, the desirability of requiring satisfactory scores on five of six results for each of two shipments, rather than requiring satisfactory scores on five of six results for each of three shipments. We seek comments on the ability of laboratories, including physician office laboratories, to meet these criteria, and the adequacy of these criteria for protecting health and safety.

We are not considering enhanced PT for cytology at this time but are instead proposing different remedial actions a laboratory failing PT in cytology would have to take to remain approved or licensed. See "Remedial actions for cytology" below.

The revised regulations would establish an evaluation system based on successful participation for each testing event or shipment of PT at approximately quarterly intervals. Under current Medicare practices we disapprove a facility if three out of four shipment scores are unsatisfactory. Currently, if a laboratory has three unsatisfactory shipments out of four, it takes over a year to initiate a termination. Under the proposed revision we would decrease this time interval to as short as two months if the laboratory does not enroll in enhanced PT. (The PT program has one month to notify us of PT results; we notify the laboratory of its loss of approval within one month.)

The proposal for revision of the PT requirements also would include a provision for action against a laboratory when there is unsatisfactory performance for the same single analyte in any PT shipment or testing event or when there is unsatisfactory performance for one of two challenges for the same analyte in each of two consecutive shipments or testing events (or combination). We would terminate approval of the entire specialty or subspecialty in which the failed analyte is categorized unless the laboratory requests enrollment in the enhanced PT program, since under the Medicare program there is no mechanism for taking action to deny payment for a single test.

Since the tests that we have selected for participation in the PT program are critical to patient health and are representative of the universe of tests frequently performed in the specialty or subspecialty, poor performance in even one test, if uncorrected, is indicative of more widespread problems in testing, management and proper quality assurance procedures. If a laboratory performs unsatisfactorily on the individual analytes selected for PT and the same analytes in the enhanced PT program, we would take termination action against the laboratory within 15 days of notification by the PT program to protect the health and safety of the individuals being tested.

We would also institute an overall PT evaluation by specialty in chemistry, immunology and microbiology. Our objective is to assess overall performance to the extent possible. Thus, poor performance in one subspecialty would result in the laboratory failing the specialty as well and the laboratory would be unable to obtain approval for any of the other subspecialties in the same category. The laboratory would therefore have an additional incentive to participate successfully in PT.

Laboratory Requirements

• General

In new § 493.21, Condition: Enrollment and testing of samples, we would have two standards: (1) If a PT program has been approved under new Subpart D for a specialty or subspecialty for which a laboratory seeks or has approval or licensure, a laboratory must enroll in an approved PT program for each specialty and subspecialty for which it seeks approval (Medicare or Medicaid) or licensure (CLIA); and (2) the laboratory must test or examine the PT samples in the laboratory's routine manner.

This section would require the laboratory to notify HHS of the PT program it has chosen; it would be able to designate no more than one PT program per specialty (in specialties without subspecialties) or subspecialty for the purposes of meeting the PT enrollment requirements. A laboratory could change its selection of PT programs after four quarterly shipments but would have to notify us. The laboratory would have to agree to allow the PT program to release any data to us that we need to evaluate the laboratory's performance.

Section 493.21 would contain a standard concerning how the laboratory is to run the PT samples it receives from the PT program. The laboratory would have to run the samples with its regular work and by personnel who ordinarily perform the laboratory's testing; it could not run tests in replicate unless it usually does so; and it could not send the samples to another laboratory for analysis.

In § 493.22, Condition: Successful participation, a laboratory that does not successfully participate in PT for a given specialty and subspecialty for which it seeks approval for Medicare or Medicaid participation or licensure under CLIA may request enrollment in an enhanced PT program to prevent immediate termination of approval or instition of license revocation proceedings for the failure specialty or subspecialty. If we institute an enhanced PT program, good laboratories would not be penalized for a single PT failure since the laboratory would be afforded the opportunity to demonstrate sustained successful PT performance through enrollment in the enhanced PT program. The enhanced PT program would identify marginal and poorly performing laboratories (those that cannot demonstrate sustained successful performance) for expeditious adverse action.

In § 493.24, Reinstatement after failure to participate successfully, we propose that a laboratory failing PT in any specialty or subspecialty must demonstrate successful performance for three consecutive shipments or testing events (or combination) of enhanced PT in that failed specialty or subspecialty before it may discontinue enhanced PT, or, if it has been terminated, before it can be reinstated in the specialty or subspecialty.

When a laboratory performs poorly over an extended period of time, it should not be allowed back into the program until it has demonstrated its ability to perform successfully for a sufficient length of time to assure that it has not made just a temporary improvement in its testing performance. Therefore, we determined that approval for the failed specialty or subspecialty would be terminated for a period of no less than six months. Reinstatement of approval in the specialty or subspecialty that had been the subject of an adverse action would not be considered until the laboratory has demonstrated successful performance over three consecutive enhanced PT shipments or testing events (or combination). We are extending the successful performance period for reinstatement from two shipments to three shipments or testing events to assure that the laboratory demonstrates sustained improvement.

● Proficiency Testing by Specialty and Subspecialty

+ General

Sections 493.31 through 493.63 would contain the criteria for acceptable performance a laboratory would have to meet to participate successfully in a PT program for each specialty and subspecialty. The specialties and subspecialties named in these proposed regulations would be microbiology (bacteriology, mycobacteriology, mycology, and parasitology), diagnostic immunology (syphilis serology and general immunology), chemistry (routine chemistry, endocrinology, and toxicology), hematology, pathology (including cytology for gynecologic examinations) and immunohematology.

As stated earlier, PT would consist of four annual shipments from the PT program for each speciality and subspecialty. In addition, we are considering conducting onsite PT, through the State survey agencies. (All cytology PT might be conducted onsite; see proposal below.)

Whenever possible, we determined a composite performance score for specialties and subspecialties of quantitative testing after evaluating historical accreditation program PT performance data from 1986. We consider a grade of 80% for an overall specialty or subspecialty of testing to be a reasonable, achievable level of performance. In immunohematology in which even one error may have serious and immediate consequences, we would require a performance level of 100% for subspecialties of testing.

The 80% composite score for specialties and subspecialties of service together with the evaluation of analyte performance, which focuses on identifying analyte testing problems, would provide the basis for assessing laboratory performance to permit detection of laboratories whose results are consistently marginal or worse, while not penalizing a laboratory that rarely produces marginal results. The unique features of including the enhanced PT program, although not currently offered by PT programs, would provide an improved measure for separating a rare single laboratory error from a true laboratory performance problem ensuring that poor performers are recognized as such and that good performers retain Medicare approval or CLIA licensure (or both).

After we have implemented the PT program and collected and examined PT performance data, we anticipate that additional refinements in the PT program grading criteria would be necessary.

Whenever changes are necessary, we would expedite the rulemaking process to ensure the most rapid implementation in order to have dynamic requirements to respond timely to new testing procedures and methodologies as well as refinements in performance and evaluation criteria.

We would publish a notice in the Federal Register before revising PT program requirements.

+ Cytology

While we recognize especially the importance of developing a national consistent standard for a PT program for cytology, we also acknowledge the complexity and uniqueness of this area of laboratory testing. Cytology PT differs from most other PT programs because (1) cytology testing remains a manual process that is totally dependent upon the judgment of the person viewing the specimen; (2) a cytology PT program tests the accuracy of each individual engaged in the examination of cytologic preparations more than the overall performance of the laboratory; and (3) there is a substantial degree of difficulty in obtaining cytology specimens for PT.

Currently, most PT programs do not test cytology services. As a result, we do not have as much information as in other areas from which to assess cytology PT. We are therefore proposing three options (onsite testing, mailed shipments of specimens, and a combination) for interested parties to comment upon. All options include ranges for accuracy rates, number of challenges per testing event and number of testing events per year. We would appreciate comments concerning which accuracy rate in each range commenters prefer (and why) as well as which PT option is preferred and why.

We are proposing at this time to require cytological PT only for gynecologic examinations; thus we are proposing standards at this time for a national PT program in cytology for Pap smears only. We are proposing that Cytology: Gynecologic examinations be a subspecialty under the specialty of pathology. Other tests, such as fine needle aspiration, breast fluid, etc., which are non-gynecologic cytology tests, constitute a small portion of cytology tests, and all cases are ultimately reviewed by qualified pathologists. Therefore, we are not proposing standards for a PT program for non-gynecological cytology testing at this time.

We considered numerous approaches to cytology PT, some of which were found to be impractical as the basis for a national standard for PT in cytology. For example, we considered "blind" PT as one approach. In blind PT, samples are submitted through cooperative clients (physicians) to the laboratory. These slides are theoretically undetected as PT test cases because they resemble routine specimens received by the laboratory. We decided this approach would not be practical to use on a national basis because of the difficulty in obtaining enough specimens as well as the cooperation of physicians to distribute PT samples to the laboratories and assuring that these specimens are, in actuality, not easily identified as PT challenges. While we agree that a true "blind" PT program is a valuable methodology, we could not implement such a complex program on a national basis because of the logistics. Therefore, we propose to include the "blind" submittal of known specimens as an option for the laboratory to use as part of its own internal quality assurance program to evaluate cytotechnologist performance and to document the quality of laboratory performance (See Subpart H—Quality Assurance).

We also considered a retrospective PT program approach and likewise found it to be impractical to implement on a large scale. This system involves the selection of previously screened slides for screening by either exchanging slides with another laboratory as a "round robin" scheme, rescreening one laboratory's cytology cases by a reference laboratory with recognized expertise in cytology, or rescreening by a panel of cytology experts in a State. The slides selected would have to be representative of the laboratory's workload and would have to include some abnormal specimens. Since managing a retrospective program would be highly labor-intensive and complicated to conduct and maintain, and criteria do not exist for selecting the laboratories which would have final responsibility for determining correctness of diagnosis, we have also decided to propose including the retrospective PT approach as an option for the laboratory to use as part of its own quality assurance program. Retrospective PT may be used to evaluate individual cytotechnologist performance and overall laboratory performance and as one of several approaches for remedial action if any cytotechnologists fail the national standards for the cytology PT program ultimately selected.

– Options

We would like to develop a program that will be feasible and practical and that will serve as a strong measure of performance evaluation in cytology testing. We are seeking public comment on three possible approaches to cytology PT.

1. The first approach to cytology PT standards about which we seek comment involves an "onsite" methodology. This system involves taking actual PT slides to the laboratory or testing the individuals engaged in examination of cytologic preparations at an offsite testing center and directly observing the individuals as they review the slides. This system, to be conducted by the State survey agency, would allow us to test all individuals in the laboratory who are involved in the examination of slides and has the advantage of permitting immediate feedback if a performance problem is detected.

Disadvantages are: slides are prestained and may differ from the laboratory's usual staining quality; positives may sometimes be present in greater frequency than would ordinarily be found in the laboratory's regular screening workload; and a set time limit for review per slide, imposed to expedite the testing process, may be discomforting to some examinees and may not reflect the conditions under which cytotechnologists routinely examine slides. In addition, though the onsite PT methodology is practical as a standard for a national PT program it is expensive to operate, especially in terms of the personnel needed to transport the slides and conduct the PT program.

We are proposing that satisfactory performance for each individual would be based on a 80 to 100 percent correct response on each PT survey and over the span of three testing events. Successful performance for the laboratory, which includes all individuals engaged in slide examination, would also be based on 80 to 100 percent correct responses on each testing event. A correct response would be an 80 percent or greater agreement of participants. We are interested in receiving comments regarding instances when there is not agreement of a minimum 80% of the response or when a graded response may be close to the consensus but not exact.

2. The second approach to cytology PT standards about which we seek comment involves mailed specimens. This system involves mailing sets of slides or photographs or interactive video discs that represent a spectrum of possible cytologic findings. Each individual in the laboratory who is engaged in slide examination analyzes the set of samples and reports his or her results in a standardized format.

Problems with this approach include slide breakage in transit, poor reproducibiity of colors in photographs and variations in slide staining. However, video discs may offer solutions to these problems. Overall, this option would tend to reflect the best rather than the usual performance by individual and laboratory.

We are proposing the same scoring system for this option as for the "onsite" option, but hope to receive comments on more stringent grading criteria.

3. The third approach is simply a combination of onsite and mailed cytology PT. We are seeking detailed comments on how we could set standards for a cytology PT program that would contain features of both onsite and mailed PT.

+ Remedial Actions for Cytology (Gynecologic Preparations Only)

The principal purpose of any PT program is to identify areas of performance that need correction or improvement and to ensure that good performance is maintained over time. Because correct cytology performance involves the skill and judgment of individuals viewing slide preparations (as opposed to a machine making correct analyses of fluids, for example) and because the results are especially critical to diagnosis and treatment, we are proposing different remedial actions that would be taken when individuals fail the cytology PT program.

We are proposing remedial actions applicable to individuals who do not demonstrate satisfactory performance and penalties applicable to laboratories that fail to maintain overall successful performance in their cytology testing programs. The remedial actions would be the same for any of the cytology PT options: mailed specimens, onsite testing and a combination of the two.

We are inviting comments on this proposal, especially in terms of ways we can improve the process of remedial actions when individuals fail the PT program.

1. Remedial Actions Concerning Individuals

The first time an individual fails any part of a cytology PT survey, the laboratory would have to provide the individual with immediate remedial training and education in the area of the failure and a review in those areas passed. If the individual's score is 50 percent or less in each of two testing events, we would require a more stringent form of remedial action up to prohibiting the individual from reporting negative slides until the individual has been retrained and demonstrates necessary accuracy by scoring 100 percent on two consecutive PT testing events. If either two or more or ten percent or more of the individuals in a laboratory, whichever number is greater, fail any PT testing event, all individuals engaged in the examination of slides would have to undergo additional training and education in addition to that required in Subpart G (personnel requirements) *and* the laboratory would have to participate in a retrospective PT program until the laboratory achieves an overall score of 95 percent or more correct responses over three subsequent consecutive PT testing events. The 95 percent score for the laboratory represents the composite score of all cytotechnologists.

2. Fiscal Penalties for Laboratories

We are proposing that if the laboratory fails to take required remedial actions (as described above) when an individual fails the PT program or if either two or more or ten percent or more of the individuals in a laboratory, whichever number is greater, fail two or more of PT testing events, we would terminate the laboratory's Medicare approval for the subspecialty of gynecologic examinations, revoke its CLIA licensure, or both as applicable.

SUBPART D—PROFICIENCY TESTING PROGRAMS

A new Subpart D would contain the requirements a PT program would have to meet before a laboratory could use it to meet the PT requirements of Subpart C. Subpart D would indicate for each specialty and subspecialty (a) program content and frequency of challenge; (b) the number of challenges per quarter; and (c) how to evaluate analytes or test performance.

Before 1966 CDC operated a PT program for CLIA licensed laboratories. The revised regulations would not include a federally-operated PT program. We would depend on private-sector and State-operated programs to provide the PT. We would evaluate the programs against the standards established in these regulations in Subpart D.

Ideally, the criteria for acceptable laboratory performance should be based on clinical usefulness of test results; however, there are no generally agreed upon medical usefulness criteria. In the absence of medical criteria for acceptable performance, we would define acceptable performance using fixed criteria, when appropriate, and statistical limits elsewhere. This approach corresponds to current laboratory performance evaluations based on the state of the practice. We would use a three standard deviation limit to identify laboratories requiring regulatory intervention; this is similar to laboratory internal quality control systems that currently use a three standard deviation limit to identify serious problems requiring remedial action. We are interested, however, in receiving comments on the marginal effectiveness of using two standard deviations for any analyte.

Basically, programs wishing to qualify as a PT program under these regulations would have to offer a minimum of at least two challenges per quarter for each test or analyte for the subspecialties included in the specialties of diagnostic immunology, chemistry, hematology, and immunohematology and six challenges per quarter for the specialty of microbiology, which includes the subspecialties of bacteriology, mycology, mycobacteriology and parasitology. For the enhanced PT program, we would require six challenges per shipment for each test or analyte for the subspecialties, include in the specialties of diagnostic immunology, chemistry, hematology and immunohematology and twelve challenges per shipment for the specialty of microbiology. Where appropriate, the regulations would describe the different types of services a laboratory could perform for that specialty or subspecialty. The laboratory would be expected to perform tests on PT samples to the same extent that testing is performed on patient samples.

Subpart D would describe criteria for acceptable performance. The criteria for grading was developed through an evaluation of the current criteria in use by States and private sector programs and an evaluation of data CDC had for the performance characteristics of laboratories.

A PT program would evaluate a laboratory in a manner that reflects the scope and level of services the laboratory offers as opposed to the current CLIA PT requirements that specify that laboratories enroll only in programs of fixed size, regardless of their scope of services.

After the PT program has been in operation for two years we would consider revisions to the program based on the performance of laboratories. We would solicit comments from all concerned groups regarding the need to modify the PT program requirements. Changes in the PT program might be made to incorporate new analytes, tests, or organisms of clinical significance, to delete obsolete or well-performed tests, or to improve the evaluation scheme based on new data describing actual distributions of test scores, and the relationship of test errors to physician practices and patient outcomes. When we have decided to include new challenges or evaluation criteria in future PT, we would expect these changes to be provided by approved PT programs within two years of our approval and announcement. We would review the standards for PT on a regular basis and make such changes as are necessary and provide notice of these changes to all affected.

The requirements for program content and number of challenges per quarter would be implemented through an expedited rulemaking process to enable us to drop or add tests timely to reflect current technologies.

- Cytology (Gynecologic Examinations)

As noted above in the discussion concerning Subpart C, we are proposing three options for cytology. For all options, we are proposing one to four PT testing events per year, with five to 12 slide preparations per individual per testing event. We are proposing that the type of challenges include "normals," infectious agents, benign reactive processes, pre-malignant processes, and malignant processes.

We would require the program to provide previously "referenced" slides: "positive" slides would have been confirmed by tissue biopsy and "negative" slides would have been confirmed by 95 percent consensus agreement.

SUBPART E—PATIENT TEST MANAGEMENT

We would establish a new subpart E and a new condition—§ 493.201, Condition—Patient test management. This condition would provide a uniform set of requirements for all laboratories (CLIA and Medicare) for specimen submission and requisition and would more clearly define the actual records that must be kept and why they are required.

The proposed requirement would be based on current Medicare requirements dealing with clinical laboratory management (§§ 405.1316(d), (e), (f), and (g)), quality control (405.1317(a)(7)), and CLIA laboratory requirements dealing with reports and records (§§ 74.53 and 74.54).

The existing requirements would be modified to allow for electronic requisition of laboratory tests to keep pace with modern technology and the advances that allow for computer systems to interact directly. We expect these computer systems to be provided with security systems and "keys" to assure that only authorized persons can order tests. In addition, we would add to the specimen requisition requirement for cytology examinations, in § 493.201(b)(5), the provision that pertinent clinical information necessary for accurate diagnosis of cytology specimens must be provided to the laboratory, including for Pap smear testing, an indication of whether the patient is at risk for developing cervical cancer or its precursors.

The existing requirement for retention of reports in pathology would be increased from two years to ten years because a two-year time period is insufficient to assure adequate patient tracking for cancer screening, diagnosis and follow up.

The proposed standard on specimen records would indicate that the critical requirement is for the laboratory to have a system that ensures identification of the specimen being treated through all stages of testing. The laboratory would not be required to maintain the name of patients as long as there is some mechanism in place to assure specimen identification and enable results to be reported to the person ordering or requiring the test. The regulations would require patient confidentially to be maintained but still provide for the correct identification and reporting of the specimen results. However, the Medicare carriers and intermediaries would continue to be able to require names or other identifiers in assuring payment of a claim.

The proposed rule would remove the restrictive standards under Medicare that only persons authorized under the Medicare program to request or receive results could obtain such results even if they were not seeking Medicare payment. Several State laws permit individuals to order and obtain their own test results and this conflicts with the Medicare regulations. The proposed modification would allow flexibility for those tests performed on non-Medicare beneficiaries, provided State laws have no restrictions on who can order tests.

The laboratory would have to determine or verify normal ranges used for reporting patient test results through validation studies required in § 493.235.

The new section also would require the laboratory to make available to clients, information on factors that may affect the interpretation of test results (if they are known), including interferences, detection limits, sensitivities, specificity, accuracy, precision and validity of these test measurements. In addition, laboratories would be required to notify clients whenever changes occur in testing methodology that affects test results or interpretation of test results. This would enable the individual requesting and receiving test results to evaluate the laboratory's quality and any limitations on the information they receive. This is the type of information that any laboratory should know in order to perform and report tests and is obtained through its quality control and quality assurance program and, in many instances, is available from the manufacturers of the test systems.

We would add a requirement on test referral (standard (e)) to indicate that each laboratory performing tests either directly or on referral must have its name and other identifier on the report to the individual requesting or receiving test results so that the individual receiving the report will know which laboratory actually performed the test. The current regulations are not clear on whether all testing facilities must be indicated on the report so that it is possible that multiple referrals do not include the name of all testing facilities on the report.

Under the proposed requirements the laboratory must maintain a legally reproduced copy, rather than an exact duplicate as is required by current § 405.1316(g), to make these regulations consistent with other Medicare recordkeeping requirements and to allow for the use of new technologies in the storage and transmittal of data.

SUBPART F—QUALITY CONTROL

Existing quality control requirements are in § 405.1317 and Part 74, Subpart C. We would revise and move them to Part 493 and form separate conditions. New § 493.221, Condition; General quality control, would be applicable to all the specialties and subspecialties. New § 493.241 would specify that failure to meet the condition unique to a particular specialty or subspecialty would result in the loss of Medicare approval or CLIA licensure (or both) of that specialty or subspecialty.

The revision of these regulations would reflect changes in technology as well as clarify the specific requirements for each standard. The clarifications would reflect the current Medicare guidelines and would more explicitly inform the laboratories of their responsibilities under the regulations.

1. General Quality Control (§ 493.221)

The general quality control regulations in §§ 405.1317(a) and 74.20 would be combined into one uniform condition as § 493.221. We have made the general quality control a condition to indicate its importance. We would divide the requirements in current §§ 405.1317(a)(1) and 74.20(a) into several standards in new sections in order to define more clearly what the requirements are and to separate the various requirements in the current regulatory factor into several distinct and related categories so that each is equal in importance.

New § 493.221 would elaborate on what such items as "adequacy of equipment" and "test systems" consist of beyond descriptions in current § 405.1317(a).

We would add a requirement that the laboratory specify the procedure the staff is to follow in case quality control results or patterns do not follow the expected patterns established by the laboratory. There would also be procedures for reporting patient results when test method limitations are exceeded. These are critical elements in the performance and reporting of test results and are necessary to assure that accurate and reliable results are obtained and reported. Since these factors are essential, the laboratory staff should be aware of these procedures and the quality assurance program (described in § 493.451) would assure that these procedures are in place and followed.

We would also require the laboratory to verify the validity of its procedures. This requirement is contained in the current regulations in § 405.1317(a)(1) but we would spell out in detail what constitutes the validation of each test method.

We would also add a requirement that the laboratory have a mechanism in place to verify the accuracy and reliability of data management and reporting systems to assure that the data is accurately analyzed, processed and reported. This is a critical requirement since no matter how accurate and precise a method is, data not analyzed properly or reported correctly could result in incorrect patient diagnosis or treatment. We also would revise the regulations to indicate the importance of detecting errors in test results and reporting and promptly correcting these errors since the detection of the errors is a critical element in assuring accurate and reliable test results.

We would add a requirement under the general quality control condition for the frequency of running quality control materials. The frequencies are currently indicated in § 405.1317(b). We would place the requirement in the general section, rather than with each specialty, to provide flexibility and eliminate the arbitrary classification of procedures to match the placement of certain quality control requirements in a specialty area. This revision would also reflect the changes in laboratory technology. We would also add provisions to these regulations to allow for lesser frequencies as changes in technology lead to methodological improvements.

Our new requirements on equipment maintenance and function checks would indicate that the laboratory must define its own program based on the manufacturer's instructions. This is a revision from current requirements at § 405.1317(a). The laboratory would have to demonstrate that its procedures produce accurate and reliable test results. We have decided not to specify detailed requirements for these procedures in the regulations; we would allow flexibility and the utilization of the manufacturer's protocols. We are seeking comment on the appropriateness of relying on manufacturer's protocols. We are seeking comment on the appropriateness of relying on manufacturers' protocols. The new requirement would also provide for technological change. We would not specify performance characteristics but would specify that the laboratory must determine its own performance characteristics based on validation studies and must adhere to these established performance characteristics. We would require the laboratory to make the performance characteristics available upon request to individuals ordering and receiving test results. The quality assurance subpart, Subpart H, would require the laboratory to adhere to its quality assurance program and established protocols.

We would define validation of methods and remedial actions and specifically indicate what is required. The requirements would match our current guidelines and would better inform the laboratories of their responsibilities.

Whenever possible we would place similar requirements in the general section that apply to more than one specialty area. For example, we would combine the references on recalibration from the sections on hematology, chemistry, and radiobioassay, into a general section. In addition, because of the advent of certain new technolgies, we would no longer require daily instrument verifications separate from quality control checks. Rather, we would specify the basis and frequency for performing instrumental checks, which correspond to our current guidelines in this area.

We would also define the timeframes in which control samples must be tested with patient specimens to assure accurate results. The current requirement for including controls with each test run of patient specimens is clarified in these proposed requirements. We would include alternatives to the use of two standards or two controls since these materials are not always available.

2. Specific Quality Control Requirements (§ 493.243—§ 493.315)

We would transfer the contents of current § 405.1317(b). Standard: Quality control system methodologies, to this new subpart and create for each specialty a new condition and for each subspecialty new standard. A laboratory must meet the conditions corresponding to the specialties and the standards corresponding to the subspecialties for which it wishes to be approved or licensed. These conditions and standards would appear in §§ 493.243 through 493.315. Any laboratory found out of compliance with a condition for a specialty or standard for a specialty would not become or remain approved or licensed for that specialty or subspecialty.

In § 493.243 we would include revised microbiology requirements to indicate that the frequency of performing controls has been changed to reflect standards for current technology and state of the art developed by HCFA and CDC working with the National Committee for Clinical Laboratory Standards (NCCLS). The NCCLS is a national voluntary standards organization consisting of representatives including manufacturer, laboratories and Federal and State regulatory agencies that develop and publish guidelines and standards for laboratories. The NCCLS standards for microbiology media quality control and susceptibility testing quality control have been implemented through the HCFA State Operations Manual for Survey and Certification and have been in place for approximately two years with no alteration in the quality of testing observed. The regulations would be modified to reflect these changes. The laboratory community has estimated that the implementation of these changes has saved several million dollars per year in quality control costs with no adverse effects on microbiology testing.

HHS is continuing to work with NCCLS and other groups to develop beneficial and cost effective standards. The standards would be modified as data become available and can be assessed to justify additional changes.

We would also revise (in § 493.255) the serology (diagnostic immunology) requirements to reflect the fact that the CDC no longer publishes a reference manual on tests for syphilis serology. We would require the laboratories to follow the manufacturers' instructions. We intend to revise this section at a later date if the CDC revises its manual. We would also add the hepatitis testing and human immunodeficiency virus test requirements for facilities performing this testing on blood and blood products used for transfusions. We would consolidate the requirements for serologic testing of the blood in this section since they relate to this area and we would not require reference laboratories performing this testing for blood banks to obtain additional certification in immunohematology. The requirements would coincide with the PT categories being developed for these regulations.

We would revise the chemistry regulations (in § 493.261) by adding requirements for three subspecialty areas: routine chemistry, endocrinology and toxicology. All three areas, at this time, would have to meet the applicable general quality control requirements and, for routine chemistry, laboratories performing blood gases, urinalysis, or both, would have to meet the subspecialty requirements.

The revised hematology section (§ 493.267) would reflect the fact that most of the requirements have been moved to the general quality control section. We would not retain current provisions for allowing an exemption from running specimens in duplicate for coagulation tests such as prothrombin time since there is no scientific evidence at this time available to justify retaining the current provisions. When criteria are available they would be published.

The requirements for cytology would be revised to assure that the laboratory has a quality control program to detect errors and assure accurate laboratory diagnosis. We would specify that gynecologic preparations must be stained using the Papanicolaou stain because it is the stain of choice for demonstrating abnormal cells. We would impose requirements for staining procedures to protect slide preparations from cross-contamination from other specimens.

We are seeking comments on whether we should establish workload requirements for individuals examining cytology slides. Because of the complexity and difficulty of establishing regulations in this area, we have decided not to include a specific provision in the proposed regulation. However, depending upon the comments, we may do so in the final regulation. Our deliberations on establishing workload requirements resulted in consideration of several options. We seek specific comments on each option as well as the effectiveness and workability of any limitation of this kind.

One option we are considering is setting a limit on the number of slides that may be reviewed by each cytotechnologist in a day. Specifically, we could require that each full-time cytotechnologist who reviews only gynecologic slides may review no more than from 80 to 100 slides in each 8 hour workday of each twenty-four hour period. For part-time cytotechnologists, we would require the number of slides that may be reviewed to be prorated using the same 80 to 100 slide limit per eight hour workday.

If the full-time cytotechnologist performs duties other than slide examination, we would require the 80 to 100 slide limit to be prorated to correlate with the actual time spent in slide examinations using the formula:

$$\frac{\text{Hours worked on gynecologic slides}}{\text{Total hours worked}} \times (80 \text{ to } 100 \text{ slide limit})$$

With respect to the 80 to 100 slide limit, we are identifying what we believe to be a range of acceptable workloads and are interested in comments about the precise number that we should select, or if we should retain a range, leaving the choice of the specific number of slides for each cytotechnologist to review up to the technical supervisor.

Under this option we would require that each full-time cytotechnologist who reviews only nongynecologic slides may review no more than 30 slides in each 8 hour workday or each twenty-four hour period. If the cytotechnologist is part-time or performs duties other than slide examination, we would require the 30 slide limit to be prorated to correlate with the actual time spent in slide examination using the formula:

$$\frac{\text{Hours worked on nongynecologic slides}}{\text{Total hours worked}} \times 30$$

We would also require that the number of slides of both gynecologic and nongynecologic preparations a full-time cytotechnologist may examine be prorated by multiplying the ratio of hours worked on a given type of slide to the total number hours worked by the selected value in the 80 to 100 range for gynecological slides and 30 for nongynecological slides. (See above paragraphs for formulas.)

A second option we are considering is setting an annual volume limit based on reasonable staffing patterns for cytology laboratories. For example, we could set a limit of 12,000 gynecology cases with 2 slides per case as the maximum for each full-time cytotechnologist who reviews only gynecologic slides and a limit of 3,000 cases for each full-time cytotechnologist who reviews only nongynecologic slides. Calculations of annual volume for cytotechnologists who are not engaged in examining slides on a full-time basis would be prorated as previously described in the first option.

An alternative that we also are considering to specifying a Federal workload limitation for cytotechnologists is placing with the technical supervisor responsibility for determining the number of slides that can be reviewed competently and accurately by each full-time cytotechnologist in an eight hour day or for part-time cytotechnologists in a lesser time period. The technical supervisor would document the competency of each cytotechnologist and would base the individual workload of each cytotechnologist on assessments of the accuracy of diagnosis for each cytotechnologist at a particular workload rate. We seek comments on this alternative proposal particularly on whether it is feasible for a technical supervisor to determine workload limits based on accuracy assessments.

Finally, we are interested in systematic studies that demonstrate a link between quality and the number of slides examined per day, and whether there is an optimum relationship. We also seek comments on whether such a requirement would necessitate record systems and workload supervision so complex that it would interfere with laboratory performance; whether changes in technology (e.g., computer assisted imaging) would soon make any such limitation counter productive; and whether proficiency testing would render such a limitation unnecessary.

We would revise our current requirements for rescreening of gynecologic or Pap smears interpreted to be negative from our current requirement of a ten percent random sample to either a ten percent rescreen of all negative cases screened by each cytotechnologist or a rescreen of all cases from women who are at risk for developing cervical cancer or its precursors. We would strengthen our requirements for rescreening in response to many comments that our current rescreening requirement is ineffective in detecting false-negative cases and to improve the quality of slide review. We would specify that the laboratory must complete the rescreening before it issues final reports in order to detect and correct any false-negative results in a timely manner. Also, the laboratory would not have to report Pap smear examinations immediately except in cases of viral infections in pregnant patients, dysplasia or abnormal results.

The proposed standards for correlating abnormal Pap smear findings with tissue biopsy reports and for retrospectively reviewing previous cytology results whenever abnormal results are identified are the best approaches to quality control in cytology. However, many laboratories do not have access to tissue biopsy reports and also may not have slides from previous cytology cases; therefore, these requirements, although preferred, may have limited applicability.

We are proposing that laboratories calculate annual data reflecting volume of cases processed by specimen type, number of cases by diagnosis, error rates and number of unsatisfactory specimens in order to develop a statistical approach to evaluate laboratory performance. Evaluating cytotechnologist case diagnosis with overall laboratory performance by diagnosis is an approach to determining outlier performance and focusing case reviews and interaction between cytotechnologists and technical supervisors to improve slide examination performance.

We would specify information that must be on the laboratory report to assure that the individual ordering the cytology examination has all of the facts needed to interpret the results reported. The importance of cytology testing and the significance of the examination results makes it critical for laboratories to provide all specimen findings including follow up recommendations, if indicated.

We are increasing our requirements for retention of slides and reports to assure that laboratories correlate previous diagnosis with current findings.

The requirements for histocompatibility testing (§ 493.277) are the same as those that already apply to Medicare laboratories but would now also apply to CLIA laboratories. We would update technical requirements now in Medicare regulations and include the explicit requirements for which HLA antigens are to be identified. We would not retain certain frequency checks for the components of the serum trays found in current regulations because we now believe the requirements are overly burdensome and are no longer necessary to assure quality. We would also make the requirements more explicit with regard to what is required under each section of the regulation. These changes would reflect the current guidelines for the practice of histocompatibility.

We would also consolidate all the requirements for histocompatibility testing into one section. Previously, the requirements were divided between a section on end-stage renal disease and the independent laboratory general quality control requirements. We would move these to one location in order to clarify the exact nature of the requirements and to make the requirements more readily accessible to those seeking requirements for histocompatibility testing. We would also make these requirements applicable to other areas currently mentioned in the regulations, such as liver transplantation and bone marrow transplantation as well as other transplantation areas. We therefore propose to move material from § 405.2171 and include it in this section.

The new section would also provide uniform standards for Medicare and CLIA laboratories for histocompatibility testing. This requirement would impose no additional burden on laboratories currently in both programs but would provide consistency in the regulations and requirements since CLIA histocompatibility testing laboratories are currently classified in serology and Medicare laboratories are classified in a separate category for histocompatibility.

In new § 493.281 we propose to revise the immunohematology requirements now in § 74.25 to make them consistent with the Medicare requirements. The CLIA requirements were not revised at the time Medicare adopted the FDA standards in 1981. We propose to adopt these same standards for CLIA laboratories. We would also cross-refer this standard to all of 21 CFR Part 640 to provide total consistency between Medicare and FDA regulations and to assure that any changes in the FDA regulations are reflected in the Medicare regulations. We also propose to add a requirement that laboratories collecting, processing and transfusing blood and blood products meet the requirements in 21 CFR Part 606 to make the Medicare regulations consistent with the FDA regulations on this subject. The need to refer to Part 606 was inadvertently overlooked during the promulgation of the Medicare regulations in 1981. The revision would clarify the precise requirements that are expected of the facilities.

We would also add explicit requirements for cytogenetics testing because of the importance of this area in testing for genetic defects and the fact that the existing general quality control requirements do not adequately address this area. In addition, laboratories seeking payment for these services have been classified in the anatomic pathology or serology areas under a less than precise methodology and the anatomic pathology and serology requirements are not applicable to cytogenetics testing. The revisions of the regulations would be based on the standards of the private sector and New York State, which are the only existing models for quality control standards.

We propose to add a condition to quality control on blood banking and transfusion services (§ 493.301) to consolidate all the requirements in the regulations relating to the collection, processing, transfusion and storage of blood and blood products in one section of the regulations. Currently there are requirements relating to blood collection and storage in both the quality control sections of the independent laboratory conditions for coverage and in the hospital conditions of participation. The provisions contained in § 405.1317(b)(4)(ii) are more appropriate to a section on collection and processing than a section on the quality control of the testing performed in blood banks and transfusion services.

The regulations in § 405.1317(b)(4)(ii) were adopted as part of the revision of the requirements under the MOU with FDA. However, the establishment of a new condition on blood banking and transfusion services and the consolidation of these requirements would simplify and clarify the regulations; those subject to the provisions of the regulations would be better able to determine the specific requirements applicable to them.

We have reviewed the causes of fatal transfusion reactions since 1976, based on the files of the FDA and HCFA, and have concluded that the facilities' own internal quality assurance programs are an important segment in preventing transfusion reactions and deaths. Therefore, we would modify the existing regulations that were adopted from 42 CFR 482.27(d) to include the importance of a quality assurance review. This would also be emphasized in the general quality assurance standards (Subpart H). It is our intent for the condition on blood banking and transfusion service to apply to all facilities were these services are offered. Therefore, we would not retain in § 493.393 the hospital-specific language contained in § 482.27(d) and would move the remainder to Subpart F and cross-refer all other applicable regulations to this subpart. Facilities not offering these types of services would not have to comply with these requirements.

We also propose to cross-refer the condition on bloodbanking and transfusion services to all of 21 CFR Part 640 rather than just certain sections, as does § 405.1317(b)(4)(ii), since the broader cross-reference would (1) give greater assurance of quality for the bloodbanking activities, (2) reduce the necessity for revising the regulations whenever FDA adds requirements into these sections and (3) reduce duplicative requirements, policies and procedures between HCFA and FDA.

We would also include the reference to 21 CFR Part 606 in this new condition since it contains the specific FDA recordkeeping requirements for this area. This would reduce the uncertainty of what records must be kept, consolidate recordkeeping requirements for collection, processing, and transfusion into the section where they are applicable, and make the survey of the facility and the citation of specific deficiencies easier for the survey agency.

SUBPART G—PERSONNEL STANDARDS

The current Medicare independent laboratory regulations (§§ 405.1312, 405.1313, 405.1314(b) and 405.1315), and CLIA personnel standards (§§ 74.30 through 74.31) contain detailed education and experience requirements for individuals at the director, technical supervisor, general supervisor, technologist and technician level (CLIA does not have a technician level requirement). The Medicare conditions of participation for hospitals (42 CFR Part 482) have specific requirements only for the laboratory director, who has responsibility for determining the qualifications of the supervisory personnel and the individuals performing the tests at the bench. These latter regulations provide the director with the maximum flexibility in the selection and utilization of personnel.

We believe it important to retain technical supervision qualifications for some specialty and subspecialty testing areas currently contained in the regulations. Therefore, qualifications for individuals providing technical supervision of tests in the areas of histopathology, including skin pathology, and oral pathology currently in § 482.27(a)(3)(ii)-(iv) would be moved to § 493.403, as would be the qualifications for individuals supervising transfusion and blood banking services now found at § 482.27(d)(1). In addition, qualifications for individuals providing technical supervision in cytology currently in § 405.1314(b)(9) would be repeated at new § 493.403, as are the qualifications for individuals providing technical supervision in histocompatibility currently in § 405.1314(b)(13). In order that new § 493.403 contain all qualifications necessary for the range of tests performed, we would also add new requirements for those who supervise cytogenetics testing.

We note that these requirements are new for hospital-based laboratories. Therefore, we solicit comments on all of the requirements for technical supervision, with specific reference to the need and justification for them, and whether these provisions would, if they were to be applied to physician office or small hospital laboratories, prevent the performance of tests that these laboratories are in fact effectively conducting.

We intend to adopt the same personnel requirements in the new Part 493 for all laboratories that participate in Medicare or Medicaid or are licensed under CLIA. The proposed rule contains the provision required by section 9339(d) of the Omnibus Budget Reconciliation Act of 1986 to accept for Medicare purposes individuals that meet State licensure requirements for laboratory directors. This provision specifies that if a State provides licensing or other standards with respect to the operation of laboratories (including those in hospitals) in the State and establishes qualifications under which an individual may direct a laboratory, title XVIII of the Act may not be construed as authorizing the Secretary of HHS to require other qualifications; this provision was effective January 1, 1987. We are developing a proposed revision to the independent laboratory requirements for Medicare approval, CLIA licensure requirements and hospital-based laboratory rules to incorporate this provision in a separate document.

We would also include a provision to enable individuals who qualify as laboratory directors under current regulations to continue to qualify as such.

We intend to continue to allow for the recognition of private sector certification programs for director level personnel as an alternative mechanism for qualifications as is currently contained in §§ 74.30 and 405.1313. This provision reduces the need for the program to evaluate the credentials of individuals who have already been evaluated by a private sector organization approved by HHS and provides recognition for many of the programs in existence.

We would revise the current personnel requirements for the following reasons:

  • It is necessary to emphasize the responsibility of the director for assuring the quality of the services of the laboratory and to allow the director the maximum flexibility to choose the personnel required to achieve this goal.

● Changes in technology make it difficult to develop detailed specific standards and revise them as needed to cover the wide variety of instruments, methodology and test systems currently performed and to be performed in the future in laboratories.

● It is more reliable to depend on outcome measures such as quality control, proficiency testing and quality assurance programs, rather than detailed personnel standards, as a mechanism to assure the quality of testing.

Although it is generally believed that degreed individuals are better prepared to assume technical responsibilities to assure quality, there is limited evidence available to correlate the degree level of an individual with the quality of the test results produced.

There have been several studies on the relationship between personnel standards and quality of testing, including one commissioned by the Office of the Assistant Secretary for Planning and Evaluation of the Office of the Secretary of the Department of Health and Human Services, but there are no definitive data to show the nature of the relationship of specific standards to quality of testing. The studies have been applied to limited areas. Although evidence exists of some improvements in performance as a function of credentialing, these studies are limited in scope, and they are not all based on the same assessment techniques. They also do not indicate that inaccurate or medically unacceptable results were produced by any particular type of level of individual.

Our decision to set standards for the director, supervisor and cytotechnologist is based on the model for the current hospital standards with additional requirements for supervisor and cytotechnologist. The hospital laboratory personnel requirements have not resulted in any known adverse effects on patient health and safety in these facilities over the past twenty years. We have heard from some parties that such a model is more relevant in the hospital setting because of the direct oversight by the medical staff. However, the majority of users of laboratory services are physicians and, if the physician is the key to the quality assurance in hospitals, there is no reason to assume that these same physicians will be less concerned when testing is performed in independent laboratories. Supervision of day-to-day testing is necessary to assure accurate and reliable test results. Since the director may not always be present when testing is performed, we would add requirements for laboratory supervision. Moreover, the current cytotechnologist requirements would be maintained because they are essential in assuring the quality of cytology test performance and reporting. We would also require each cytotechnologist to record the number and types of slides screened each day in order to determine compliance with the workload requirements in § 493.271(b).

Proposed rules applicable to the personnel levels below the director and supervisor would provide for maximum flexibility for the director in choosing the laboratory staff, except for cytology where specified personnel qualifications would be required to assure the quality of cytology results. The individuals employed in laboratories would still have to meet State standards, if any exist. This would place responsibility with the States to set specific criteria for personnel to meet local needs. The director would have to ensure that the personnel have the necessary training, experience, continuing education, receive continuous evaluation and monitoring of performance levels and meet any State licensure requirements. The proposed personnel requirements would allow the flexibility to utilize the various private-sector credentialing programs, State licensure programs and private-sector examinations as a guide in selecting individuals for employment purposes.

In proposed § 493.405 we specify laboratory director responsibilities and emphasize the duties required of the laboratory director. The laboratory director would have the overall responsibility for the quality of testing performed by the laboratory and would be responsible for establishing and maintaining a quality assurance program and establishing performance characteristics for the test systems employed by the laboratory. The director would also be responsible for providing evidence that the laboratory can maintain these performance levels: the director would assess factors such as staff performance, quality control results, proficiency testing results, validation of test procedures and methodologies, and assure that the laboratory corrects all problems before reporting test results. In addition, if errors are detected after results are reported, the director would be responsible for providing the necessary corrected information to the individual requesting or utilizing the test results.

SUBPART H—QUALITY ASSURANCE

In new § 493.451 we would add a quality assurance condition for both the Medicare and CLIA laboratories to require the laboratories to establish and follow protocols that assess the effectiveness of their operations. The new condition would require the laboratory to establish procedures for monitoring the quality of its testing and staff performance and to assure that the laboratory's performance is within established acceptable criteria. This section would add an additional level of quality control and place the burden on the laboratory to accept responsibility for monitoring its own performance as an adjunct to the checks placed on the facility by the regulatory agency. The laboratory would utilize its quality control and PT data and regular staff performance evaluations to monitor and assure the quality of testing and reporting.

The responsibility for establishing and implementing a quality assurance program would be placed on the laboratory director; it would serve as an additional outcome measurement of quality and would assure accurate and reliable test performance and reporting.

The director would also have the responsibility for having a program in place to monitor and control various health and safety hazards from a variety of biological, chemical, environmental and radiological materials or factors, which may affect testing as well as patient and worker safety. This would obviate the need for specifying detailed Federal regulations under Medicare and CLIA such as those now contained in 42 CFR 405.1316, which may not cover all possible contingencies. The new requirement would place responsibility on the laboratory director for setting up and implementing an appropriate program, which would include assuring the compliance with the existing Federal, State, and local laws. This would reduce the need for duplicative standards and allow the facility to develop programs to meet its own needs. This is not meant to imply a lessening of concern for employee health and safety, including protection from various hazards, but the provision would implement a more efficient mechanism for monitoring laboratory safety and employee health.

As part of the quality assurance initiative, we would also encourage laboratories to enroll in PT programs for analytes other than those included in the current grading scheme. For the subspecialty of cytology, laboratories would be able to insert "blind samples" into their workload or exchange slides with another laboratory for retrospective screening and comparison.

In addition to soliciting comments on all the proposed revisions we are requesting comments on alternate mechanisms of quality assurance that can be used in a Federal regulatory program. We anticipate that these regulations will provide the flexibility to take into account future changes but we are reviewing and intend to continue to review the need for further changes and improvements in the regulatory system. We solicit specific suggestions for changes in quality assurance requirements and data to support these changes.

SUBPART I—INSPECTION

We propose to add a condition on inspection of the laboratories, § 493.501, which specifies the requirements a laboratory must meet for inspections and record retention and availability. Provisions concerning inspection are currently in several different sections of the Medicare and CLIA regulations (42 CFR 405.1317(a)(6), 42 CFR 405.1909, and Part 74, Subpart G) and some of them are obsolete.

Under this proposal, we would require the laboratory to demonstrate satisfactory performance on quality control and PT before inspection or approval.

The proposed § 493.501 would allow us to inspect a laboratory during any hours of operation or business, would state that HHS has the right of access to all records required to make a determination of a facility's status and would require that the laboratory make these records available to us for a reasonable period of time during the course of the inspection.

The regulations would extend our authority to require the laboratory to test specimens or undertake quality control actions on patient materials in the laboratory during the inspection to allow us to determine the competency of the personnel and ability of the laboratory to perform tests. In addition to the Subpart C requirement for participation in a mailed PT program, we are considering the development of a methodology for evaluating laboratory performance through onsite PT to enhance the survey process. In 1990, we plan to select a limited number of States in which a random sampling of laboratories would be chosen for the State survey agencies to conduct unannounced onsite PT surveys to evaluate the feasibility of this type of PT.

In addition, under the proposed requirements, DHHS would be able to reinspect the laboratories at such frequencies as are necessary to determine compliance or continued compliance with the regulations. We propose to indicate that denial of access could result in revocation or denial of licensure, termination, or denial of initial approval under Medicare.

Under this proposal the laboratory would also be required to notify us of changes in ownership, direction, location or services so that we can determine the status of the laboratory and its ability to provide reliable and accurate test results. These provisions would not add new requirements but would serve to clarify and unify the existing Medicare and CLIA requirements.

The new § 493.501 would not include a number of the requirements currently in § 405.1909. What constitutes a laboratory test (in § 405.1909(a)) would not be retained since it refers to a section of our regulations deleted earlier. The provisions relating to specialties and subspecialties for licensure and approval that we would retain would be relocated in Subpart F, Quality Control. The content of § 405.1909(b) would be deleted since the date to which it refers has expired. The content of § 405.1909(c), the requirements for successful participation in proficiency testing, would be revised and moved to Subpart C, Participation in Proficiency Testing and Subpart D, Proficiency Testing Program.

In addition, the specific requirement in § 405.1909(c) for onsite PT would be revised in Subpart D, Proficiency Testing Programs, to allow us flexibility in determining the mechanism for compliance with PT requirements after a failure and the time period in which we would have to revisit the laboratory. Specifying a specific three-month interval for onsite PT performance or a six-month interval for evaluation of mailed PT samples does not allow sufficient flexibility. The proposed requirements in Subpart C would clarify the conditions for successful and unsuccessful performance and this subpart would specify that HHS may perform whatever follow-up and inspections are required to determine a laboratory's compliance with the standards.

SUBPART J—CLIA-ONLY REQUIREMENTS

In Subpart J we would place requirements applicable only to laboratories engaged in interstate commerce. For example, the CLIA requirements concerning recognition of accreditation programs are different from those for Medicare laboratories due to differences in the two statutes; CLIA allows HHS more flexibility on the type and scope of information that can be requested.

We also propose to make several modifications in the licensure procedures for laboratories under CLIA. We would revise the regulations to indicate that (1) licenses will be issued or revoked by specialty and subspecialty rather than by individual test procedures in order to achieve uniformity between programs, (2) we are placing increased reliance on overall outcome measures by restructuring of the regulations, and (3) only certain tests are subject to PT because not all tests are currently included in PT programs and some tests, such as chemical screening in urinalysis, may provide such consistent results that continued PT monitoring is not useful.

We would also revise the exemption applicable to certain physician office laboratories that examine specimens on referral so that we would grant exemptions in cases only in which the *total* number of tests performed annually is 100 or fewer rather than granting exemptions for each specialty or subspecialty in which 100 tests or fewer are performed. This would simplify the accounting system for determining if a laboratory requires licensure. We cannot eliminate the test limit since CLIA requires us to provide a low test volume exemption mechanism. We propose to specify that HHS may determine that certain categories or types of tests pose a hazard to public health and no exemption will be granted in these cases. This addition to the regulation would be consistent with the statute and its intent. It will make the Medicare and CLIA regulations more congruent on this matter since the Medicare statute does not have a provision for a low volume exemption.

We are also proposing to add a provision (§ 493.704) that would allow us to issue a notice that a license can continue in effect for another year, rather than reissuing the formal license every year. This would meet the intent of the statute and still permit annual renewal without inordinate paperwork when no changes in licensure status have occurred.

III. OTHER REVISIONS AFFECTING LABORATORIES

• We propose to revise § 405.1909, Special requirements applicable to independent laboratories, to delete the current paragraphs (b) through (d), as the comparable content would be in Part 493, and to add a definition of "independent laboratory." The term "independent laboratory" is still necessary for payment purposes (i.e., Medicare's supplementary medical insurance program (Part B), rather than hospital insurance program (Part A), pays for independent laboratory services); § 405.1909 discusses independent laboratory services and supplementary medical insurance. We would define an independent laboratory as a facility meeting the requirements of Part 493 that is maintained for the purpose of performing diagnostic laboratory tests. It would be a facility that is not controlled, managed or supervised by a hospital, a hospital's medical staff, or the attending or consulting physician's office. (A physician's office performing tests on referral for the attending or consulting physician would be considered an independent laboratory.)

We would also revise the last sentence in paragraph (a), which currently indicates that diagnostic tests performed by an attending or consulting physician are physician's services rather than clinical laboratory services. Section 1833 (h)(1)(A) and (h)(5)(A) of the Social Security Act, as amended on July 18, 1984, requires clinical laboratory services furnished by any person or entity (other than a provider of services for its inpatients) to be paid for based on a fee schedule. The current rule implies that laboratory services are "physicians' services", which are not subject to a fee schedule. Because we pay for these laboratory services based on a fee schedule, we propose to revise the last sentence so that it no longer calls laboratory services "physician services."

• We would revise the definitions found in current 42 CFR 405.2102, to clarify that histocompatibility testing determines compatibility between a potential organ donor and recipient and not between a donor organ and a recipient. This would be a minor technical amendment to clarify the intent of the regulations.

• We would also revise current § 405.2171(d), Condition: Minimal service requirements for a renal transplantation center, to cross-refer it to the new unified regulations for clinical laboratories, including histocompatibility testing for renal transplantation centers. This would simplify the requirements by placing all related requirements in one section of the regulation. However, we would leave the requirement concerning 24-hours availability of services (§ 405.2171(d)(1)) as it would not apply to other laboratories.

• We would revise current 42 CFR 416.49, Condition for coverage—Laboratory and radiological services, to require laboratories in ambulatory surgical centers to comply with the conditions of coverage of laboratory services in new Part 493 if they perform laboratory testing or perform transfusion of blood and blood products.

• Our revisions to § 482.27 (the condition of participation concerning hospital-based laboratories) would only delete requirements that would be in the new part. We intend to retain the requirements currently in § 482.27(a) (1) and (2) as they would continue to apply only to hospital-based laboratories.

IV. REGULATORY IMPACT ANALYSIS

A. Introduction

Executive Order 12291 (E. O. 12291) requires us to prepare and publish an initial regulatory impact analysis for any proposed regulation that meets one of the E. O. criteria for a "major rule"; that is, that would be likely to result in: an annual effect on the economy of $100 million or more; a major increase in costs or prices for consumers, individual industries, Federal, State, or local government agencies, or geographic regions; or, significant adverse effects on competition, employment, investment, productivity, innovation, or on the ability of United States-based enterprises to compete with foreign-based enterprises in domestic or export markets. In addition, we generally prepare an initial regulatory flexibility analysis that is consistent with the Regulatory Flexibility Act (RFA) (5 U.S.C. 601 through 612), unless the Secretary certifies that a proposed regulation would not have a significant economic impact on a substantial number of small entities. For purposes of the RFA, we treat all hospital-based and independent laboratories as small entities. For purposes of this regulation, physician laboratories that perform any tests on referral from other physicians also are small entities. Individuals and states are not included in the definition of a small entity.

In addition, Section 1102(b) of the Act requires the Secretary to prepare a regulatory impact analysis for any proposed rule that may have a significant impact on the operations of a substantial number of small rural hospitals. Such an analysis must conform to the provisions of section 603 of the RFA. For purposes of Section 1102(b) of the Act, we define a small rural hospital as a hospital with fewer than 50 beds located outside a metropolitan statistical area.

We do not believe that the provisions of this regulation constitute a major rule. However, because we expect that this regulation could have a significant impact on some laboratories, may affect some personnel employed by laboratories, and may have an effect on some States regarding State requirements, licensure and certification of laboratories, we have performed the following analysis voluntarily.

B. Anticipated Effects

1. Affected Entities

There are approximately 12,000 Federally regulated laboratories located in hospitals and independent settings. These facilities range from large medical centers and corporate-operated independent laboratories to small, independent laboratories and physician's office laboratories.

Although not small entities, we expect States to be affected by some additional administrative burden because they may have to adapt or establish a methodology for assessment of PT requirements and ensure compliance with these requirements. States may have to make more recommendations for termination of Medicare approval if the PT standards are not met.

We expect entities providing PT programs to be affected because of program changes that may be necessary in order to meet the criteria for an approved PT program and additional documentation required to perform in the Federal program. It is estimated that there may be more than 40,000 physician office laboratories that perform more than 5,000 tests annually and that will be subject to Federal standards under current law beginning in 1990. We have not attempted to estimate the effects of the standards in this proposal on physician office laboratories that do not perform referral tests. However, the possibility exists that certain of these standards may also be found to be applicable to physician office testing. Therefore, we are soliciting comments on the effect on physician office testing of the proposed proficiency testing, quality control, and personnel requirements were these requirements to be applied to physician office laboratories. If there is substantial evidence that any of the standards contained in this proposed rule would so affect the availability of physician office testing that patient care would be adversely affected, that evidence would be appropriate in the context of this rulemaking.

2. Costs/Savings

We expect our standards to be achievable by the majority of laboratories although some may have to incur costs to achieve the required compliance with PT standards. Depending upon the actual costs in upgrading a specific lab to meet PT standards, the charges of that laboratory for its services may rise to offset the costs of improvements. However, we believe that in an area with sufficient competition, including that from physicians' office laboratories, the charges for services will remain stable. Therefore, since charge increases are unlikely, we assume that laboratories will seek to minimize cost increases through increased efficiencies. The regulation provides increased flexibility over existing regulations to permit different approaches to achieving efficiencies.

This proposed regulation would expand Medicare coverage from physician office laboratories receiving 100 or more referrals to those laboratories receiving any referrals. This change may create a slight increase in the number of laboratories requiring Medicare approval.

We expect that some physician office laboratories would seek Medicare approval and thus comply with Medicare requirements to be reimbursed for tests. We also expect that those physician office laboratories that may incur a substantial increase in costs to upgrade to elect to stop doing tests on referral.

There may also be additional increases in the purchases of automated laboratory equipment and computers when laboratory managers make decisions regarding the methodology of achieving the PT standards.

3. Proficiency testing effects

This regulation establishes consistent PT requirements based on data from professional organizations that operate PT programs. We would require specific minimum PT passing scores for each specialty and subspecialty. Currently, passing levels for PT are set by each State and the levels vary. We expect the PT standards to enable us to identify and take consistent action against Medicare and CLIA laboratories whose PT performances are below the range of acceptability achieved by the vast majority of laboratories. A laboratory's poor PT performance would result in a denial of Medicare or Medicaid payment for a failed specialty or subspecialty of testing or in loss of CLIA licensure.

Even though we lack definitive data, we do not expect this regulation to affect most laboratories adversely because most laboratories already have in place PT mechanisms, both for their own benefit and because Medicare requires it. However, laboratories in States will less rigorous proficiency testing program standards than we would require would be the most likely to be adversely affected. We also expect the number of tests for which payment is denied by the Medicare and Medicaid programs for payment purposes to increase for those laboratories that do not improve to our designated performance level.

Some laboratories may incur greater costs to achieve the required PT standards than others because of the costs that would be incurred to improve their quality control activities and quality assurance programs.

If a laboratory's PT performance is determined to be unsuccessful, payment would not be made for tests in the failed specialty or subspecialty. We expect the number of laboratories not receiving payment to increase. However, we also expect the quality of laboratory services to improve as a result of implementation of these standards.

At present, Medicare and CLIA laboratories are not required to participate in PT programs for cytology, because no such program has been established. This regulation would establish a nationwide cytology proficiency testing program. We are proposing certain limits and specifications concerning the developing cytology program. This may mean additional expenses for some laboratories that have to participate in a cytology PT program for the first time or incur additional costs in order to perform successfully in the cytology PT program.

4. Quality Control

This proposed regulation includes an update of quality control requirements to account for changes in technology and instrumentation that have occurred in the laboratory field since 1971. Laboratories would be required to implement their own quality assurance programs that they would be expected to follow. As a result of this regulation, laboratory directors should have more flexibility in meeting the requirements and more opportunities to assess laboratory performance and staff competency.

We expect this regulation to improve laboratory testing in terms of the quality of the end result or outcome while removing many of the process requirements that current regulations specify to achieve that outcome. The laboratory would have more discretion over what type of internal controls, methodology, and equipment (such as computerized rather than manual equipment) are necessary to ensure that the required quality control standards are met.

5. Personnel Standards

Our proposed regulations would place less emphasis on formal credentialing for laboratory personnel and focus on the responsibilities of the laboratory personnel. As a result, the administrative burden on the States to document and maintain more detailed personnel records to meet Federal requirements would be reduced. These regulations would require laboratory directors to be responsible for ensuring that the laboratory staff is competent to perform tests. We would specify education and experience requirements for the laboratory director, technical supervision, laboratory supervisor and cytotechnologists. The qualifications of the remainder of the staff are to be determined by the laboratory director and individual State requirements.

6. Conclusion

We believe that these changes would result in clearer, more uniformly applied criteria for determining acceptability of laboratory performance. We expect some laboratories to incur costs to upgrade their performance. We expect these costs to be somewhat offset by savings from removal of detailed process requirements. Overall benefits, in terms of consistent laboratory requirements and improved quality would increase benefits to patients and would more than offset the costs of upgrading and improvements.

We conclude, based on the analysis above, that the proposed rule is not a major rule under Executive Order 12291. Although some laboratories would be adversely effected, we believe that benefits to society will outweigh the adverse effects. We expect most laboratories not to incur substantial costs to comply with our conditions. The Secretary certifies that this proposed regulation would not have a significant impact on the operations of a substantial number of small rural hospitals.

V. PAPERWORK REDUCTION ACT

Sections 493.21, 493.93, 493.99, 493.101, 493.103, 493.105, 493.201, 493.225, 493.231, 493.233, 493.235, 493.237, 493.240, 493.241, 493.271, 493.273, 493.277, 493.279, 493.315, 493.417, 493.451, 493.501, 493.701, 493.704, 493.708, and 493.710 of this proposed rule contain information collection requirements that are subject to the Office of Management and Budget (OMB) approval under the Paperwork Reduction Act of 1980 (44 U.S.C. 3504, et seq.)

Organizations and individuals desiring to submit comments on the information collection requirements should direct them to the agency official whose name appears in the **ADDRESS** section of the preamble.

Section 405.1317 of the current regulations, as recodified into §§ 493.229, 493.251 and 493.253 of the proposed rule, also contains information collection requirements that are subject to the OMB review. These requirements were approved by that Office in August of 1987 in accordance with the Paperwork Reduction Act under OMB approval number 0938–0368.

VI. RESPONSE TO COMMENTS

Because of the large number of items of correspondence we normally receive on rules requesting public comment, we are not able to acknowledge or respond to them individually. A summary of comments and our responses will be included in our final rule.

VII. LIST OF SUBJECTS

42 CFR Part 74

Administrative practice and procedure, health, laboratories, reporting and recordkeeping requirements.

42 CFR Part 405

Administrative practice and procedure, Health facilities, Health professions, Kidney diseases, Laboratories, Medicare, Nursing homes, Reporting and recordkeeping requirements, Rural areas, X-rays.

42 CFR Part 416

Health facilities, Health professions, Medicare, Reporting and recordkeeping requirements.

42 CFR Part 440

Grant programs-health, Medicaid.

42 CFR Part 482

Hospitals, Medicaid, Medicare, Reporting and recordkeeping requirements.

42 CFR Part 488

Health facilities, Survey and certification, Forms and guidelines.

42 CFR Part 493

Laboratories, Medicare, Medicaid, Health facilities, Reporting and recordkeeping requirements.

Title 42 of the Code of Federal Regulations would be amended, as set forth below:

1. Chapter I is amended by removing Part 74 and reserving it as follows:

PART 74—[RESERVED]

II. Chapter IV is amended as set forth below:

PART 405—[AMENDED]

A. Part 405 is amended as follows:

1. Subpart E is amended as follows:

a. The authority citation continues to read as follows:

Authority: Secs. 1102, 1.814(b), 1832, 1833(a), 1842 (b) and (h), 1861 (b) and (v), 1862(a)(14), 1866(a), 1871, 1881, 1886, 1887, and 1889 of the Social Security Act as amended (42 U.S.C. 1302, 1395f(b), 1395k, 1395l(a), 1395u (b) and (h), 1395x (b) and (v), 1395y(a)(14), 1395cc(a), 1395hh, 1395rr, 1395ww, 1395xx, and 1395zz).

b. Paragraph (c) of pst 405.556 is revised to read as follows:

§ 405.556 Conditions for Payment of Charges: Physician Laboratory Charges

*   *   *   *   *

(c) *Independent laboratory services furnished to hospital inpatients.* Laboratory services furnished to a hospital inpatient by an independent laboratory (as defined in § 405.1909(b) will be reimbursed on a reasonable charge basis under this subpart only if they are physician laboratory services as described in paragraph (a) of this section. Payment for nonphysician services furnished to a hospital inpatient by an independent laboratory will be made by the intermediary to the hospital in accordance with Part 413 of this chapter.

2. Subpart K is amended as follows:

a. The authority citation continues to read as follows:

Authority: Secs. 1102, 1814, 1832, 1833, 1861, 1863, 1865, 1866, 1871 of the Social Security Act; 42 U.S.C. 1302; 1395f, 1395k, 1395l, 1395x, 1395z, 1395bb, 1395cc, 1395hh.

b. Section 405.1128 is revised to read as follows:

§ 405.1128 Condition of Participation—Laboratory and Radiologic Services

The skilled nursing facility has provision for promptly obtaining required laboratory, X-ray, and other diagnostic services.

(a) *Standard: Provision for services.*

(1) If the skilled nursing facility furnishes its own x-ray services, it must meet the applicable conditions established for certification of hospitals in § 482.26 of this chapter. If the facility does not provide x-ray services, it makes arrangements to obtain these services from a physician's office, a participating hospital or skilled nursing facility, or a portable x-ray supplier.

(2) If the skilled nursing facility furnishes its own laboratory services, it must meet the applicable conditions established for certification of hospitals and for approval of laboratories found in §§ 482.27 and Part 493 of this chapter respectively. If the facility does not provide laboratory services, it makes arrangements to obtain these services from a participating hospital or skilled nursing facility, or a laboratory meeting the requirements of Part 493 of this chapter.

(3) All x-ray and laboratory services are provided only on the orders of the attending physician, who is notified promptly of the findings. The facility assists the patient, if necessary, in arranging for transportation to and from the source of service. Signed and dated reports of a clinical laboratory, X-ray, and other diagnostic services are filed with the patient's medical record.

(b) *Standard: Blood and blood products.* Blood handling and storage facilities are safe, adequate, and properly supervised. If the facility provides for maintaining and transfusing blood and blood products, it meets the conditions established in §§ 493.301 through 493.315 of this chapter. If the facility does not provide its own facilities but does provide transfusion services alone, it meets at least the requirements of §§ 493.305, 493.307, 493.309 and 493.315 of this chapter.

3–4. Subpart M (§§ 405.1310–405.1317) is removed and reserved and the table of contents is amended to reflect this change.

SUBPART M [§§ 405.1310–405.1317]—[RESERVED]

5. Subpart U is amended as follows:

a. The authority citation continues to read as follows:

Authority: Secs. 1102, 1861, 1862(a), 1871, 1874, and 1881 of the Social Security Act (42 U.S.C. 1302, 1395x, 1395y(a), 1395hh, 1395kk, and 1395rr).

b. The definition of histocompatibility testing in § 405.2102 is revised to read as follows:

§ 405.2102 Definitions

Histocompatibility testing. Laboratory test procedures which determine compatibility between a potential organ donor and a potential organ transplant recipient.

*   *   *   *   *

c. Paragraph (d) of § 405.2171 is revised to read as follows:

§ 405.2171 Condition: Minimal Services Requirements for a Renal Transplantation Center

*   *   *   *   *

(d) *Standard: laboratory services.* (1) The Renal Transplantation Center makes available, directly or under arrangements, laboratory services to meet the needs of ESRD patients. Laboratory services are performed in a laboratory facility approved in accordance with Part 493 of this chapter to participate in the Medicare program and, for histocompatibility testing purposes, also meets §§ 493.221 through 493.239, 493.257, 493.259, 493.277, and 493.281 of this chapter and, when services are furnished in the subspecialty of histopathology, § 493.273 of this chapter.

(2) Laboratory services for cross-matching of recipient serum and donor lymphocytes for preformed antibodies by an acceptable technique are available on a 24-hour emergency basis.

B. Part 416 is amended as follows:

PART 416—[AMENDED]

1. The authority citation continues to read as follows:

Authority: Secs. 1102, 1832(a)(2), 1833, 1863 and 1864 of the Social Security Act (42 U.S.C. 1302, 1395k(a)(2), 1395l, 1395z and 1395aa).

2. Section 416.49 is revised to read as follows:

§ 416.49 Condition for Coverage—Laboratory, and Radiologic Services

The ASC must have procedures for obtaining routine and emergency laboratory and radiologic services, from Medicare approved facilities, to meet the needs of patients. The laboratory offering the services must have been approved in accordance with Part 493 of this chapter, except for urinalyses, hemoglobins and hematocrits performed within a few days before, or on, the day of the surgery.

C. Part 440 is amended as follows:

PART 440—[AMENDED]

1. The authority citation continues to read as follows:

Authority: 42 U.S.C. 1302.

2. Section 440.30 is amended by republishing the introductory text and revising paragraph (a) to read as follows:

§ 440.30 Other Laboratory and X-ray Services

"Other laboratory and X-ray services" means professional and technical laboratory and radiological services—

(a) Ordered and provided by or under the direction of a physician or other licensed practitioner of the healing arts within the scope of his practice as defined by State law or ordered and billed by a physician but provided by an independent laboratory as defined in § 405.1909(b) of this chapter.

* * * * *

D. Part 482 is amended as follows:

PART 482—[AMENDED]

1. The authority citation continues to read as follows:

Authority: Secs. 1102, 1814(a)(6), 1861(e), (f), (k), (r), (v)(1)(G), and (z), 1864, 1871, 1883, 1886, 1902(a)(30), and 1905(a) of the Social Security Act (42 U.S.C. 1302, 1395f(a)(6), 1395x (e), (f), (k), (r), (v)(1)(G), and (z), 1395aa, 1395hh, 1395tt, 1395ww, 1396a(a)(30), and 1396d(a)).

Section 482.27 is revised as follows:

§ 482.27 Condition of Participation: Laboratory Services

(a) The hospital must maintain, or have available, adequate laboratory services to meet the needs of its patients. The hospital must ensure that all laboratory services provided to its patients are performed in a facility approved by Medicare in accordance with Part 493 of this chapter.

(b) *Standard: Adequacy of laboratory services.* The hospital must have laboratory services available, either directly or through a contractual agreement with a Medicare approved hospital or independent laboratory, that meet the needs of the patients and the medical staff.

(1) Emergency laboratory services must be available 24 hours a day.

(2) A written description of services provided must be available to the medical staff.

(3) The laboratory must make provision for proper receipt and reporting of tissue specimens.

(4) The medical staff and a pathologist must determine which tissue specimens require a macroscopic (gross) examination and which require both macroscopic and microscopic examinations.

E. Part 483 is amended as follows:

PART 483—[AMENDED]

1. The authority citation continues to read as follows:

Authority: Secs. 1102, 1905 (c) and (d) of the Social Security Act (42 U.S.C. 1302, 1396d (c) and (d)).

2. Section 483.460(n) is revised to read as follows:

§ 483.460 Conditions of Participation: Health Care Services

(n) *Standard: Laboratory services.* (1) For purposes of this section, "laboratory" means an entity for the microbiological, serological, chemical, hematological, radiobioassay, cytological, immunohematological, pathological or other examination of materials derived from the human body, for the purpose of providing information for the diagnosis, prevention, or treatment of any disease or assessment of a medical condition.

(2) If a facility chooses to provide laboratory services, the laboratory must—

(i) Meet the management requirements specified in Part 493 of this chapter; and

(ii) Provide personnel to direct and conduct the laboratory services.

(A) The laboratory director must be technically qualified to supervise the laboratory personnel and test performance and must meet licensing or other qualification standards established by the State with respect to directors of clinical laboratories. For those States that do not have licensure or qualification requirements pertaining to directors of clinical laboratories, the director must be either—

(1) A pathologist or other doctor of medicine or osteopathy with training and experience in clinical laboratory services; or

(2) A laboratory specialist with a doctoral degree in physical, chemical or biological sciences, and training and experience in clinical laboratory services.

(B) The laboratory director must provide adequate technical supervision of the laboratory services and assure that tests, examinations and procedures are properly performed, recorded and reported.

(C) The laboratory director must ensure that the staff—

(1) Has appropriate education, experience, and training to perform and report laboratory tests promptly and proficiently;

(2) Is sufficient in number for the scope and complexity of the services provided; and

(3) Receives in-service training appropriate to the type and complexity of the laboratory services offered.

(D) The laboratory technologists must be technically competent to perform test procedures and report test results promptly and proficiently.

(3) The laboratory must meet the proficiency testing requirements specified in Part 493 of this chapter.

(4) The laboratory must meet the quality control requirements specified in Part 493 of this chapter.

(5) If the laboratory chooses to refer specimens for testing to another laboratory, the referral laboratory must be approved by the Medicare program either as a hospital or an independent laboratory.

F. Part 488 is amended as follows:

PART 488—[AMENDED]

1. The authority citation continues to read as follows:

Authority: Sec. 1102, 1814, 1861, 1865, 1866, 1871, 1880, 1881, and 1883 of the Social Security Act (42 U.S.C. 1302, 1395f, 1395x, 1395bb, 1395cc, 1395hh, 1395qq, 1395rr, and 1395tt).

2. Section 488.52 is revised to read as follows:

§ 488.52 Special Requirements Applicable to Independent Laboratories

(a) The services of a qualified independent laboratory for which reimbursement may be made under the supplementary medical insurance program relate only to diagnostic tests performed in an independent laboratory. Diagnostic laboratory tests for purposes of section 1861(s) (13) and

(14) of the Act and for purposes of this Subpart S shall include only those clinical and anatomical pathology diagnostic tests and procedures listed in § 493.2 of this chapter under "laboratory". Such diagnostic tests performed by out-of-hospital physicians whose primary practice is directly attending patients and/or consultation (i.e., furnishing an attending physician with an opinion about a patient's condition or diagnosis), even though conducted partly through diagnostic procedures, are not considered services of an independent laboratory except when they are done on referral.

(b) For purposes of this section, an independent laboratory is a facility meeting the requirements of Part 493 of this chapter and maintained for the purpose of performing diagnostic laboratory tests. An independent laboratory is not a facility that is controlled, managed or supervised by a hospital as defined by section 1861(e) of the Act, a hospital's organized medical staff, or the attending or consulting physician's office.

G. A new Part 493 is added as follows:

PART 493—LABORATORY REQUIREMENTS

Subpart A—General Provisions

Sec.
493.1   Basis and scope.
493.2   Definitions.

Subpart B—Administration

493.11   Condition: Compliance with Federal, State and local laws.

Subpart C—Participation in Proficiency Testing

493.21   Condition: Enrollment and testing of samples.
493.22   Condition: Successful participation.
493.23   Condition: Successful participation before initial approval or licensure.
493.24   Reinstatement after failure to participate successfully.
493.25   Condition: Enhanced proficiency testing.

Proficiency Testing by Specialty and Subspecialty

493.31   Condition: Microbiology.
493.33   Standard: Bacteriology.
493.35   Standard: Mycobacteriology.
493.37   Standard: Mycology.
493.39   Standard: Parasitology.
493.43   Condition: Diagnostic immunology.
493.45   Standard: Syphilis serology.

Subpart G—Personnel

Authority: Secs. 1102, 1861(e), the sentence following 1861(a)(11), 1861(s)(12) and 1861(s)(13) of the Social Security Act and sec. 353 of the Public Health Service Act (42 U.S.C. 263a, 1302, the sentence following sec. 1395 x(s)(11), and secs. 1395x(s)(12) and (13).)

SUBPART A—GENERAL PROVISIONS
§ 493.1   Basis and Scope

This part set forth the conditions that laboratories must meet in order for their tests to be approved for coverage under the Medicare and Medicaid programs and in order for laboratories to be licensed to perform testing on specimens received in interstate commerce. It implements sections 1861(e) and (j), the sentence following section 1861(s)(11), sections 1861(s)(12) and (13), and 1902 of the Social Security Act, and section 353 of the Public

Health Service Act. This part applies to: laboratories located in physicians' offices (including group medical practices) that perform any tests on referred specimens; hospitals meeting at least the requirements specified in section 1861(e) of the Act to qualify for emergency hospital services under section 1814 of the Act; skilled nursing facilities; intermediate care facilities for the mentally retarded; rural health clinics that perform tests on referral; ambulatory surgical centers except as provided in § 416.49 of this chapter; end-stage renal disease facilities except as provided in § 405.2163 of this chapter; and independent laboratories, as defined in § 405.1909 of this chapter. It does not apply to laboratories operated by a rural health clinic or physician's office exclusively for its own patients.

§ 493.2   Definitions

As used in this part—

"Accredited laboratory" means a laboratory (including a laboratory in a hospital) accredited by, with respect to hospitals, the Joint Commission on the Accreditation of Healthcare Organizations or the American Osteopathic Association and, with respect to interstate licensed laboratories, the Commission on Laboratory Accreditation of the College of American Pathologists, or any other national accreditation organization that has been approved by HHS as provided in section 353 of the Public Health Service Act.

"Authorized person" means a person authorized under section 1861(r) of the Act to order and to receive test results. With respect to tests performed on individuals not receiving or seeking Medicare reimbursement, an authorized person is an individual not excluded under State law or by Medicaid.

"Challenge" means, for quantitative tests, an assessment of the amount of substance or analyte present in a sample. For qualitative tests, a challenge means the determination of the presence or the absence of an analyte, organism, or substance in s sample.

"CLIA" means the Clinical Laboratories Improvement Act of 1967 (Section 353 of the Public Health Service Act).

"Laboratory" means a facility for the microbiological, serological, chemical, hematological, cytological, histological, pathological, immunohematological, radiobioassay, cytogenetical, toxicological, histocompatibility or other examination of materials derived from the human body for the purpose of providing information for the diagnosis, prevention, or treatment of any disease or impairment, or the assessment of the health, of human beings. These examinations also include screening procedures to determine the presence or absence of various substances and organisms in the body.

Facilities only collecting specimens or only serving as a mailing service and not performing testing are not considered laboratories.

"Referee laboratory" means a laboratory that has had a record of accurate performance for at least one year in a specific test specialty or subspecialty and has been designated by HHS as a referee laboratory for that specialty or subspecialty.

"Sample" means the material contained in a vial, a slide, or other unit that contains material to be tested by proficiency testing program participants. When possible, samples are of human origin.

"Target value" means either the mean of all responses after removal of outliers (those responses greater than 3 standard deviations from the original mean) or the mean established by groups of 20 or more participants that use the same methodology (to be used only when the method bias results in a skewed distribution of responses). If the method group is less than 20 participants, "target value" means the overall mean after outlier removal (as defined above) unless acceptable scientific reasons are available to indicate that such an evaluation is not appropriate.

SUBPART B—ADMINISTRATION

§ 493.11   Condition—Compliance with Federal, State and Local Laws

The laboratory must be in compliance with all applicable Federal, State and local laws.

(a) *Standard; Federal laws.* The laboratory must be in compliance with applicable Federal laws related to the health and safety of individuals whose specimens are submitted to it for testing.

(b) *Standard; State licensure.* The laboratory must be (1) licensed if State or applicable local law requires licensure; or (2) approved as meeting standards for licensing established by the agency of the State or locality responsible for licensing laboratories.

(c) *Standard; licensed staff.* All personnel, including those individuals who collect specimens, must be licensed or meet other applicable standards that are required by State and local laws.

(d) *Standard; fire safety.* The laboratory must comply with State and local laws related to fire safety.

(e) *Standard; environment and health.* The laboratory must comply with Federal, State and local laws relating to the storage, handling and disposal of chemical, biological and radioactive materials.

SUBPART C—PARTICIPATION OF PROFICIENCY TESTING

§ 493.21 Condition: Enrollment and Testing of Samples

A laboratory must enroll in a proficiency testing program that meets the criteria in Subpart D of this part and is approved by HHS. The laboratory must enroll in such a program for each of the specialties and subspecialties for which it seeks or have approval for Medicare or Medicaid participation or for licensure under CLIA. The laboratory must test the samples in a routine manner.

(a) Standard; Enrollment. The laboratory must notify HHS of the approved program or programs in which it chooses to participate to meet proficiency testing requirements of this subpart. The laboratory must—

(1) Designate the program to be used for each specialty and subspecialty to determine compliance with this subpart if the laboratory participates in more than one proficiency testing program approved by HHS;

(2) For each specialty and subspecialty, participate in one approved proficiency testing program for four quarters before designating a different program and notify HHS before any change in designation; and

(3) Authorize the proficiency testing program to release to HHS all data required by HHS to determine the laboratory's compliance with this subpart.

(b) Standard; Testing of proficiency tesing samples. The laboratory must examine or test, as applicable, the proficiency testing samples it receives from the proficiency testing program in the same manner as it tests patient specimens.

(1) The samples must be examined or tested with the laboratory's regular patient workload by personnel who routinely perform the testing in the laboratory, using the laboratory's routine methods.

(2) The laboratory may not test the samples in duplicate unless it routinely tests patient samples in duplicate.

(3) The laboratory may not send the samples or portions of samples to another laboratory for analysis.

(4) The laboratory must document the handling, processing, examination, testing and reporting of results for all proficiency testing samples and the records must be maintained for a minimum of two years from the date of the proficiency testing shipment or testing event.

§ 493.22   Condition: Successful Participation

(a) Each laboratory must successfully participate in a proficiency testing program approved by HHS as described in Subpart D of this part for each specialty and subspecialty in which the laboratory seeks Medicare approval or licensure under CLIA.

(b) If the laboratory fails to participate successfully in proficiency testing for a given specialty or subspecialty, the laboratory's Medicare approval or licensure under CLIA, or both, will be terminated for the specialty or subspecialty unless the laboratory enrolls within 15 days of notification of unsuccessful specialty or subspecialty performance and successfully participates in an approved enhanced proficiency testing program as described in § 493.25.

(c) If the laboratory fails to perform satisfactorily for the challenges on a given analyte or test procedure, the laboratory performance for the specialty or subspecialty in which the analyte is categorized is considered unsuccessful. The laboratory's Medicare-approval or licensure under CLIA, or both, for the specialty or subspecialty will be terminated unless the laboratory enrolls within 15 days of notification of unsatisfactory analyte performance and satisfactorily performs in an approved enhanced proficiency testing program as described in § 493.25 for the failed analyte.

§ 493.23   Condition: Successful Participation before Initial Approval or Licensure

Laboratories must successfully participate for three consecutive proficiency testing shipments for each specialty and subspecialty before initial Medicare or Medicaid approval or CLIA, licensure of the specialty or subspecialty.

§ 493.24   Reinstatement after Failure to Participate Successfully

(a) If a laboratory fails to participate successfully in one or more specialties or subspecialties, and does not request enrollment in the enhanced proficiency testing program described in § 493.25 of this subpart or voluntarily withdraws its participation from Medicare or Medicaid or its licensure under CLIA (or any applicable combination of the three) for the failed specialty or subspecialty, the laboratory's participation or licensure for the applicable specialty or subspecialty will be terminated. The laboratory must then demonstrate sustained successful performance on three consecutive enhanced proficiency testing shipments or onsite testing events (or combination thereof) in a period of no less than six months before HHS will consider it for reinstatement in the specialty or subspecialty.

(b) If the laboratory enrolls in and fails to participate successfully in the enhanced proficiency testing program for the failed specialty or subspecialty, the laboratory's Medicare approval or licensure under CLIA, or both, for the applicable specialty or subspecialty will be terminated. The laboratory must then demonstrate sustained successful performance in three consecutive enhanced proficiency testing shipments or testing events (or combination thereof) before HHS will consider it for reinstatement in that specialty or subspecialty.

(c) The termination period for Medicare participation or period for revocation of licensure under CLIA for the failed specialty or subspecialty is for a period of not less than six months from the date of termination or revocation.

§ 493.25 Condition: Enhanced Proficiency Testing

Each laboratory that fails proficiency testing in a specialty, subspecialty (except cytology) or analyte must successfully participate in an enhanced proficiency testing program for three consecutive shipments or onsite testing events (or combination). (These shipments or testing events may be provided quarterly or more frequently.) Enhanced proficiency testing consists of six specimens per shipment or testing event for the subspecialties included within the specialties of diagnostic immunology, chemistry, hematology, and immunohematology and consists of twelve specimens per shipment for bacteriology, mycobacteriology, mycology and parasitology. (Because cytology has its own remedial program, there is no enhanced proficiency testing for it. See § 493.61 of this subpart.) The enhanced proficiency testing shipment or testing event consists of challenges for the overall specialty or subspecialty when the failure in proficiency testing is for a specialty or subspecialty or consists of analyte challenges when individual analyte(s) are failed in proficiency testing. The samples must be tested as in § 493.21(b).

(a) To participate successfully in enhanced proficiency testing, the laboratory—

(1) Must perform perform successfully on each shipment or testing event of enhanced proficiency testing by achieving an overall shipment score of at least 80% on each of the three shipments or testing events when unsuccessful proficiency testing performance is in a specialty or subspecialty; and

(2) Must perform satisfactorily on each shipment or testing event of

enhanced proficiency testing by achieving correct responses on five of six specimens for the first shipment or testing event; ten of 12 specimens for the first and second shipments or testing events (or combination) and 15 of 18 specimens for all three shipments or testing events (or combination) of the failed analyte when unsatisfactory proficiency testing performance is for any analyte.

(b) Failure to participate in a shipment or testing event, except when services are not offered for patient testing, results in a score of 0 for the shipment or testing event.

(c) Failure to return enhanced proficiency testing results to the proficiency testing program in the timeframes specified by the program results in a score of 0 for the survey.

(d) Evaluation of a laboratory's analyte or test performance and the criteria for acceptable analyte performance in the enhanced proficiency testing program for each specialty, subspecialty and analyte is the same as that described in Subpart D for the proficiency testing program.

(e) Following three shipments or testing events (or combination) of enhanced proficiency testing in which the laboratory performs successfully for the failed specialty or subspecialty or satisfactorily for each failed analyte, the laboratory returns to proficiency testing as specified in § 493.21.

(f) If the laboratory performs unsuccessfully on any of the three enhanced proficiency testing shipments or testing events for the failed specialty or subspecialty or performs unsatisfactorily on any of the three enhanced proficiency testing shipments or testing events for a failed analyte that results in unsuccessful performance for the specialty or subspecialty in which the analyte is categorized, termination in the applicable specialty or subspecialty will be effective 15 days after notification of unsuccessful or unsatisfactory performance in the enhanced proficiency testing program.

Proficiency Testing by Specialty and Subspecialty

§ 493.31   Condition: Microbiology

The specialty of microbiology includes, for purposes of proficiency testing, the subspecialties of bacteriology, mycobacteriology, mycology and parasitology.

(a) To participate successfully in microbiology the laboratory must attain an average score of 80% for a given testing event or shipment and may not have a score in any subspecialty of less than 80% for any testing event or shipment. The average score is the sum of shipment or testing event scores for bacteriology, mycobacteriology, mycology and parasitology divided by the total number of the subspecialties for which the laboratory seeks, or has Medicare or Medicaid approval of CLIA licensure.

(b) Failure to participate in a shipment or testing event, except when services are not offered for patient testing, results in a score of 0 for the shipment or testing event.

(c) Failure to return proficiency testing results to the proficiency testing service within the timeframes specified results in a score of 0 for the shipment.

§ 493.33 Standard: Bacteriology

(a) To participate successfully in a bacteriology proficiency testing program, the laboratory must attain a score of 80% acceptable responses for a given testing event or shipment.

(b) Failure to participate in a shipment or testing event, except when services are not offered for patient testing, results in a score of 0 for the shipment or testing event.

(c) Failure to return proficiency testing results to the proficiency testing program within the timeframes specified by the program results in score of 0 for the shipment.

§ 493.35 Standard; Mycobacteriology

(a) To participate successfully in a mycobacteriology proficiency testing program, the laboratory must attain a score of 80% acceptable responses for a given testing event or shipment.

(b) Failure to participate in a shipment or testing event, except when services are not offered for patient testing results in a score of 0 for the shipment or testing event.

(c) Failure to return proficiency testing results to the proficiency testing program within the timeframes specified by the program results in a score of 0 for the shipment.

§ 493.37 Standard; Mycology

(a) To participate successfully in a mycology proficiency testing program, the laboratory must attain a score of 80% acceptable responses for a given testing event or shipment.

(b) Failure to participate in a shipment or testing event, except when services are not offered for patient testing, results in a score of 0 for the shipment or testing event.

(c) Failure to return proficiency testing results to the proficiency testing program within the timeframes specified by the program results in a score of 0 for the shipment.

§ 493.39   Standard; Parasitology

(a) To participate successfully in a parasitology proficiency testing program, the laboratory must attain a score of 80% acceptable responses for a given testing event or shipment.

(b) Failure to participate in a shipment or testing event, except when services are not offered for patient testing, results in a score of 0 for the shipment or testing event.

(c) Failure to return proficiency testing results to the proficiency testing program within the timeframes specified by the program results in a score of 0 for the shipment.

§ 493.43   Condition: Diagnostic Immunology

The specialty of diagnostic immunology includes for the purposes of proficiency testing the subspecialties of syphilis serology and general immunology.

(a) To participate successfully in diagnostic immunology the laboratory must attain an average score of 80% for a given testing event or shipment and may not have a score in any subspecialty of less than 80% for a given testing event or shipment. The average score is the sum of shipment or testing event scores for syphilis serology and general immunology divided by the total number of subspecialties for which the laboratory seeks, or has, Medicare or Medicaid approval or CLIA licensure.

(b) Failure to participate in a shipment or testing event, except when services are not offered for patient testing, results in a score of 0 for the shipment or testing event.

(c) Failure to return proficiency testing results to the proficiency testing program within the time frames specified results in a score of 0 for the shipment by the program.

§ 493.45   Standard; Syphilis Serology

(a) To participate successfully in a syphilis serology proficiency testing program, the laboratory—

(1) For any testing event or shipment, may not have a test or analyte for which all results have been unacceptable; and

(2) Must attain a score of 80% acceptable responses in a given testing event or shipment.

(b) Failure to participate in a shipment or testing event, except when services are not offered for patient testing, results in a score of 0 for the shipment or testing event.

(c) Failure to return proficiency testing results to the proficiency testing program within the timeframes specified by the program results in a score of 0 for the shipment.

§ 493.47 Standard; General Immunology

(a) To participate successfully in a general immunology proficiency testing program, the laboratory—

(1) For any testing event or shipment, may not have a test or analyte for which all results have been unacceptable;

(2) For the same analyte in two consecutive testing events, or shipments, may not have an unsatisfactory result on one of the two challenges; and

(3) Must attain a score of 80% of acceptable responses for a given event or shipment.

(b) Failure to participate in a shipment or testing event, except when services are not offered for patient testing, results in a score of 0 for the shipment or testing event.

(c) Failure to return proficiency testing results to the proficiency testing program in the timeframes specified by the program results in a score of 0 for the shipment.

§ 493.49 Condition: Chemistry

The specialty of chemistry includes for the purposes of proficiency testing the subspecialties of routine chemistry endocrinology and toxicology.

(a) To participate successfully in chemistry the laboratory must attain an average score of 80% for a given testing event or shipment and may not have a score in any subspecialty of less than 80% for a testing event or shipment. The average score is the sum of the shipment scores of routine chemistry, endocrinology and toxicology divided by the total number of the subspecialties for which the laboratory seeks, or has, Medicare or Medicaid approval or CLIA licensure.

(b) Failure to participate in a shipment or testing event, except when services are not offered for patient testing, results in a score of 0 for the shipment or testing event.

(c) Failure to return proficiency testing results to the proficiency testing program within the time frames specified by the program results in a score of 0 for the shipment.

§ 493.51   Standard; Routine Chemistry

(a) To participate successfully in a routine chemistry proficiency testing program, the laboratory—

(1) For any testing event or shipment, may not have a test or analyte for which all results have been unacceptable;

(2) For the same analyte in two consecutive testing events or shipments, may not have an unsatisfactory result on one of the two challenges; and

(3) Must attain a score of 80% acceptable responses for a given testing event or shipment.

(b) Failure to participate in a shipment or testing event, except when services are not offered for patient testing, results in a score of 0 for the shipment or testing event.

(c) Failure to return proficiency testing results to the proficiency testing program in the timeframes specified by the program results in a score of 0 for the shipment.

§ 493.53   Standard; Endocrinology

(a) To participate successfully in an endocrinology proficiency testing program, the laboratory:

(1) For any testing event or shipment, may not have a test or analyte for which all results have been unacceptable;

(2) For the same analyte in two consecutive testing events or shipments, may not have an unsatisfactory result on one of the two challenges; and

(3) Must attain a score of 80% acceptable responses for a given testing event or shipment.

(b) Failure to participate in a shipment or testing event, except when services are not offered for patient testing, results in a score of 0 for the shipment or testing event.

(c) Failure to return proficiency testing results to the proficiency testing program in the timeframes specified by the program results in a score of 0 for the shipment.

§ 493.55 Standard; Toxicology

(a) To participate successfully in a toxicology proficiency testing program, the laboratory—

(1) For any testing event or shipment, may not have a test or analyte for which all results have been unacceptable;

(2) For the same analyte in two consecutive testing events or shipment, may not have an unsatisfactory result on one of the two challenges; and

(3) Must attain a score of 80% acceptable responses for a given testing event or shipment.

(b) Failure to participate in a shipment or testing event, except when services are not offered for patient testing, results in a score of 0 for the shipment or testing event.

(c) Failure to return proficiency testing results to the proficiency testing program in the timeframes specified by the program results in a score of 0 for the shipment.

§ 493.57 Condition: Hematology

(a) To participate successfully in a proficiency testing program for the specialty of hematology, the laboratory—

(1) For any testing event or shipment, may not have a test or analyte for which all results have been unacceptable;

(2) For the same analyte in two consecutive testing events or shipments, may not have an unsatisfactory result on one of the two challenges; and

(3) Must attain an average score of 80% for a given testing event or shipment.

(b) Failure to participate in a shipment or testing event, except when services are not offered for patient testing, results in a score of 0 for the shipment or testing event.

(c) Failure to return proficiency testing results to the proficiency testing program in the timeframes specified by the program results in a score of 0 for the shipment.

§ 493.59 Condition; Pathology

The specialty of pathology includes, for purposes of proficiency testing, the subspecialty of cytology limited to gynecologic examinations.

§ 493.61   Standard; Cytology: Gynecologic Examinations

(Option 1. Onsite proficiency testing)

To participate successfully in a cytology proficiency testing program for gynecologic examinations (Pap smears), the laboratory must meet the requirements of paragraphs (a) through (d) of this section.

(a) The laboratory must require each individual engaged in the examination of cytologic preparations to be tested directly either onsite at the laboratory or at an offsite location that is chosen by HHS.

(b) The laboratory may have no individual who achieves grades of less than (80–100)[1] percent for any testing event and (80–100) percent overall for three consecutive testing events.

(c) The first time an individual fails any part of a proficiency testing event, the laboratory must provide him or her with immediate remedial training and education in the area of the failure and must review the areas passed. If the individual fails 50 percent or more of two testing events, the laboratory must implement a more stringent form of remedial action, up to prohibiting the individual from reporting negative slides, until the individual has been retrained and scores 100 percent on two consecutive proficiency testing events.

(d) If either two or more or ten percent or more of the individuals in a laboratory, whichever number is greater, fail any proficiency testing event, all individuals engaged in the examination of cytologic preparations must undergo additional training and education in addition to that required in Subpart G of this part and the laboratory must participate in a retrospective proficiency testing program until the laboratory achieves 95 percent correct responses over three subsequent consecutive proficiency testing events.

(e) If the laboratory fails to take required remedial actions as described in paragraphs (c) and (d) of this section when individuals are found to be failing the proficiency testing program, HHS will terminate the laboratory's Medicare approval for gynecologic cytology testing or revoke its licensure under CLIA, or both if applicable. If either two or more or 10 percent or more of the individuals in a laboratory, whichever number is greater, fail two or more consecutive testing events, HHS will terminate the laboratory's Medicare approval or revoke the laboratory's CLIA license, or both if applicable, for gynecologic cytology.

(Option 2. Proficiency testing by mail)

---

[1]We are identifying what we believe to be a range of acceptable scores and are interested in comments about the precise value that we should select.

To participate successfully in a cytology proficiency testing program for Pap smears, the laboratory must meet the requirements of paragraphs (a) through (d) of this section.

(a) The laboratory may test each individual engaged in the examination of cytologic preparations using prepared materials received from the proficiency testing program.

(b) The laboratory may have no individual who achieves grades of less than (80–100) percent for any shipment and (80–100) percent overall for three consecutive shipments.

(c) The first time an individual fails any part of a proficiency testing shipment, the laboratory must provide the individual with immediate remedial training and education in the area of the failure and review the areas passed. If the individual fails 50 percent or more of two shipments, the laboratory must implement a more stringent form of remedial action up to prohibiting the individual from reporting negative slides until the individual has been retrained and scores 100 percent on two consecutive proficiency testing shipments.

(d) If either two or more or ten percent or more of the individuals in a laboratory, whichever number is greater, fail any shipment, all individuals engaged in the examination of cytologic preparations must undergo additional training and education in addition to that required in Subpart G of this part and the laboratory must participate in a retrospective proficiency testing program until the laboratory achieves 95 percent correct responses over three subsequent shipments.

(e) If the laboratory fails to take required remedial actions as described in paragraphs (c) and (d) of this section when individuals fail the proficiency testing program, HHS will terminate the laboratory's Medicare approval for gynecologic cytology testing or revoke the laboratory's CLIA license (or both if applicable). If either two or more or ten percent or more of the individuals in a laboratory, whichever number is greater, fail two or more consecutive shipments, HHS will terminate the laboratory's Medicare approval or revoke the laboratory's CLIA license (or both if applicable) for gynecologic cytology.

(Option 3. Combination of Options 1 and 2)

(Reserved—Will be determined by comments to NPRM)

§ 493.63   Condition: Immunohematology

(a) For the specialty of immunohematology, there is no overall specialty score for each shipment or testing event, since the proficiency testing in immunohematology is defined by performance in subspecialty services. To participate successfully in an immunohematology proficiency testing program, the laboratory must attain a score of 100% acceptable responses for the subspecialties of grouping, typing, antibody detection and crossmatch compatibility for incompatible crossmatches. More than one false incompatible result (for a compatible sample) in any shipment or testing event will result in failure of the shipment. For the subspecialty of antibody identification, a score of 80% is required for a given testing event or shipment.

(b) Failure to participate in a shipment or testing event, except when services are not offered for patient testing, results in a score of 0 for the shipment or testing event.

(c) Failure to return proficiency testing results to the proficiency testing program in the timeframes specified by the program results in a score of 0 for the shipment.

SUBPART D—PROFICIENCY TESTING PROGRAMS

§ 493.91 Approval of Proficiency Testing Programs

In order for a proficiency testing program to receive HHS approval, the program must, for each specialty and subspecialty for which it provides testing—

(a) Assure the quality of test samples, appropriately evaluate the testing results, and identify performance problems in a timely manner; and

(b) Demonstrate to HHS that it has—

(1) The technical ability required to prepare and distribute samples, using rigorous quality control to assure that samples mimic actual patient specimens when possible and that samples are homogeneous and will be stable within the time frame for analysis by proficiency testing participants;

(2) A scientifically defensible process for determining the correct answer for each challenge offered by the program;

(3) A program of sufficient annual challenge and frequency to establish that a laboratory has met minimum performance requirements; and

(4) The resources needed to provide, statewide or nationwide, reports to regulatory agencies on individual laboratory performance on shipments or testing events, cumulative reports about laboratory performance, and reports of specific laboratory failures using grading criteria acceptable to HHS on a timely basis; and

(c) Meet the specific criteria for proficiency testing programs listed by specialty and subspecialty of services contained in §§ 493.91–943.153 for initial approval and thereafter provide HHS, on an annual basis, with a description of program content and grading criteria.

§ 493.93 Administrative Responsibilities

The proficiency testing program must—

(a) Issue reports in a format approved by HHS on each laboratory's performances for the individual Medicare, Medicaid or CLIA-licensed specialty or subspecialty of service within 30 days from the date by which the laboratory must report proficiency testing results to the proficiency testing program. Copies of these laboratory reports must be sent to the State survey agency at the same time reports are sent to the laboratory.

(b) Furnish to HHS cumulative reports on an individual laboratory's performance and aggregate data on Medicare approved and CLIA-licensed laboratories;

(c) Provide HHS with additional information and data upon request; and

(d) Maintain records of Medicare-approved and CLIA-licensed laboratories' performance for a period of five years or such time as may be necessary for any legal proceedings.

§ 493.95 Disapproved Proficiency Testing Programs

If a proficiency testing program fails to meet the criteria contained in §§ 493.97–493.153 for approval of the proficiency testing program, HHS will notify the program and all laboratories that are Medicare-approved or CLIA-licensed of the non-approval and the reasons for non-approval.

§ 493.96 Process for Updating Proficiency Testing Program

HHS reviews the requirements for proficiency testing on a regular basis and considers revisions to the program based on the performance of laboratories. It will change requirements after soliciting comments from all concerned groups regarding the need to modify the criteria for an approved proficiency testing program. Changes in the program may be made to incorporate new analytes, tests, or organisms of clinical significance, to delete obsolete or well-performed tests, or to improve the valuation scheme. When HHS decides to include new challenges or evaluation criteria in future proficiency testing, it will notify all proficiency testing programs of the necessary changes in proficiency testing and require these changes to be provided by approved proficiency testing programs within two years of the notice of change.

Proficiency Testing Programs by Specialty and Subspecialty

§ 493.97 Microbiology

The subspecialties under the specialty of microbiology for which a program may offer proficiency testing are bacteriology, mycobacteriology, mycology, and parasitology. Specific criteria for these subspecialties are found at 493.99 through 493.153.

§ 493.99 Bacteriology

(a) *Types of laboratories.* In bacteriology, for proficiency testing purposes, there are three types of laboratories:

(1) Those that interpret only Gram stains, use direct antigen techniques to detect an organism, perform primary inoculation, or perform any combination of these;

(2) Those that—

(i) May use direct antigen techniques to detect an organism or isolate aerobic and anaerobic bacteria from mixed bacterial populations; and

(ii) Perform limited identification, perform antimicrobial susceptibility tests on selected microorganisms isolated, or both; and

(3) Those that—

(i) Are able to identify aerobic and anaerobic bacteria from mixed bacterial populations to both genus and species in most instances and perform antimicrobial susceptibility tests on the microorganisms isolated; or

(ii) May use direct antigen techniques to detect an organism.

(b) *Program content and frequency of challenge.* To be approved for proficiency testing for bacteriology, the annual program must provide a minimum of six samples per quarter. The samples may be provided to the laboratory through mailed shipments or, at HHS' option, may be provided to HHS for on-site testing. An annual program must include bacterial species that are representative of the six major groups of bacteria: anaerobes, Enterobacteriaceae, Gram-positive bacilli, Gram-positive cocci, Gram-negative cocci, and miscellaneous Gram-negative bacteria.

(1) An approved program must, during each calendar year, furnish HHS with a description of samples that it plans to include in its annual program. Some of the bacterial species representative of the six major groups must be varied from year to year. At least 50% of the samples must be mixtures of the principal organism and appropriate normal flora. The program must include other important emerging pathogens (as determined by HHS) and

either organisms commonly occurring in patient specimens or opportunistic pathogens. The program must include two types of samples and each must meet the 50 percent mixed culture criterion:

(i) Samples that require laboratories to report only organisms that the testing laboratory considers to be a significant pathogen that is clearly responsible for a described illness. The program determines the reportable isolates, including antimicrobial susceptibility for the isolate.

(ii) Samples that require laboratories to report all organisms present. Samples must contain multiple organisms frequently found in specimens such as urine, blood, abscesses, and aspirates where multiple isolates are clearly significant or where specimens are derived from immunocompromised patients. The program determines the reportable isolates.

(2) An approved program may vary over time. For example, the types of organisms that might be included in an approved program over time are—

Anaerobes
Bacteroides fragilis group
Clostridium perfringens
Peptostreptococcus anaerobius
Enterobacteriaceae
Klebsiella pneumoniae
Salmonella typhimurium
Serratia marcescens
Shigella sonnei
Yersinia enterocolitica
Gram-positive bacilli
Listeria monocytogenes
Corynebacterium Species CDC group JK
Gram-positive cocci
Staphylococcus aureus
Streptococcus Group A
Streptococcus Group B
Streptococcus Group D (S. *bovis* and *enterococcus*)
Streptococcus pneumoniae
Gram-negative cocci
Branhamella catarrhalis
Neisseria gonorrhoeae
Neisseria meningitidis
Miscellaneous Gram-negative bacteria
Campylobacter jejuni
Haemophilis influenza, Type B

(3) For antimicrobial susceptibility testing, the program must provide

at least one sample per quarter that includes Gram-positive or Gram-negative strains that have a predictable pattern of sensitivity or resistance to the common antimicrobial agents.

(c) *Evaluation of a laboratory performance.* HHS approves only those programs that assess the accuracy of a laboratory's responses in accordance with paragraphs (c) (1) through (5) of this section.

(1) To determine the accuracy of a laboratory's response, the program must compare the laboratory's response for each sample with the response which reflects agreement of at least 80% of ten or more referee laboratories or all participating laboratories agree. Sample scores must be averaged to determine the score for the shipment or testing event.

(2) For samples described in paragraph (b)(1)(i) of this section, failure to report the specific pathogen or reporting an additional organism as a pathogen must receive equal penalty. For samples described in paragraph (b)(1)(ii) of this section, misidentification or failure to report an isolate receives equal penalty. The total number of correct responses divided by the number of organisms present plus the number of incorrect organisms reported is multiplied by 100 to establish a score for each sample in each shipment or testing event. For example, if a sample contained one principal organism and the laboratory reported it correctly but reported the presence of an additional organism, which was not considered reportable, the sample grade would be $1/(1 + 1) \times 100 = 50\%$.

(3) As a laboratory is expected to perform the isolation and identification process to the same extent it performs it with patient specimens, the program must use the laboratory's type of service as determined in accordance with paragraph (a) of this section to determine the appropriateness of the laboratory's response.

(4) For antimicrobial susceptibility testing, laboratories must report those antimicrobial agents considered appropriate for the causative organism and infection site. Determination of which antimicrobial agents are appropriate must be based on a consensus document such as a National Committee on Clinical Laboratory Standards (NCCLS) publication. Grading is based on the number of correct responses divided by the number of appropriate drugs tested. A laboratory must indicate which antibiotics are included in its test panel; the program may evaluate the laboratory only for those antibiotics for which service is offered.

(5) A laboratory's shipment score for bacteriology is the score for organism identification or, if the laboratory also performs antimicrobial susceptibility testing, the score determined by dividing the total number of possible organisms a laboratory should have identified plus the number of correct antimicrobial agent responses by the number of correctly identified organisms plus the number of additional organisms reported plus the num-

ber of appropriate antimicrobial agents tested multiplied by 100 to establish a score for each sample in each shipment or testing event. For example, if a shipment or testing event contained three reportable organisms and a laboratory reported all three correctly, but reported one additional organism and made one error in antimicrobial susceptibility testing of three appropriate agents tested, its score for bacteriology would be: $(3 + 2)/(3 + 1 + 3) \times 100 = 71\%$.

§ 493.101 Mycobacteriology

(a) *Types of laboratories.* In mycobacteriology, there are three types of laboratories for proficiency testing purposes;

(1) Those that perform acid-fast stains and refer cultures to another laboratory for identification;

(2) Those that isolate and perform identification of Mycobacterium tuberculosis, but refer cultures other than M. tuberculosis to another laboratory for identification, perform antimycobacterial susceptibility tests on the organisms isolated, or both; and

(3) Those that isolate and identify all mycobacteria to the extent required for correct clinical diagnosis, perform antimycobacterial susceptibility tests on the organisms isolated, or both.

(b) *Program content and frequency of challenge.* To be approved for proficiency testing for mycobacteriology, the annual program must provide a minimum of six samples per quarter. The samples may be provided through mailed shipments or, at HHS' option, provided to HHS for on-site testing. An approved program must furnish HHS a description of samples that it plans to include in its program during each calendar year.

(c) *Evaluation of a laboratory's performance.* HHS approves only those programs that assess the accuracy of a laboratory's response in accordance with paragraphs (c) (1) through (3) of this section.

(1) The program determines the reportable organisms. To determine the accuracy of a laboratory's response, the program must compare the laboratory's response for each sample with the response which reflects agreement of at least 80% or more of ten or more referee laboratories or all participating laboratories agree. Sample scores must be averaged to determine the shipment or testing event score.

(2) Since laboratories may incorrectly report the presence of organisms in addition to the correctly identified principal organism(s), the grading system must provide a means of deducting credit for additional erroneous organisms reported. Therefore, the total number of correct responses divided by the number of organisms present plus the number of incorrect

organisms reported must be multiplied by 100 to establish a score for each sample in each shipment or testing event. For example, if a sample contained one principal organism and the laboratory reported it correctly but reported the presence of an additional organism, which was not present, the sample grade would be $1/(1 + 1) \times 100 = 50\%$.

(3) As a laboratory is expected to perform the isolation and identification process to the same extent it performs it with patient specimens, the program must use a laboratory's type of service as determined in accordance with paragraph (a) of this section to determine the appropriateness of the laboratory's response.

§ 493.103 Mycology

(a) *Types of laboratories.* In mycology, there are three types of laboratories for proficiency testing purposes that may perform different levels of service for yeasts, dimorphic fungi, dermatophytes, and aerobic actinomycetes:

(1) Those that perform direct examination and culture, recognizing the type of organism that is present and referring isolates to another laboratory for identification;

(2) Those that isolate and perform identification to the genus and, in some cases, the species level or to the extent required to differentiate recognized pathogens from others; and

(3) Those that isolate and identify organisms to the extent required for correct clinical diagnosis.

(b) *Program content and frequency of challenge.* To be approved for proficiency testing for mycology, the annual program must provide a minimum of six samples per quarter. The samples may be provided through mailed shipments or, or at HHS' option, may be provided to HHS for on-site testing. An annual program must include representatives of five major groups of organisms: yeast or yeast-like fungi; dimorphic fungi; dematiaceous fungi; dermatophytes; and saprophytes, including opportunistic fungi.

(1) An approved program must furnish HHS, during the calendar year, with a list of organisms that will be included in its annual program. Some of the organisms representative of the five major groups must vary from year to year. At least 50% of the samples must be mixtures of the principal organism and appropriate normal background flora. Other important emerging pathogens (as determined by HHS) and organisms commonly occurring in patient specimens must be included periodically in the program.

(2) An approved program may vary over time. As an example, the types

of organisms that might be included in an approved program over time are—

Candida albicans
Candida (other species)
Cryptococcus neoformans
Sporothrix schenchii
Exophiala jeanselmei
Fonsecaea pedrosoi
Acemonium sp.
Trichophyton sp.
Aspergillus fumigatus
Nocardia sp.
Blastomyces dermatitidis[2]
Zygomycetes sp.

(c) *Evaluation of a laboratory's performance.* HHS approves only those programs that assess the accuracy of a laboratory's response in accordance with paragraphs (c)(1) through (3) of this section.

(1) The program determines the reportable organisms. To determine the accuracy of a laboratory's response the program must compare the laboratory's response for each sample with the response that reflects agreement of at least 80% of ten or more referee laboratories or all participating laboratories agree. Sample scores must be averaged to determine the survey score.

(2) Since laboratories may incorrectly report the presence of organisms in addition to the correctly identified principal organism(s), the grading system must deduct credit for these additional erroneous organisms reported. Therefore, the total number of correct responses divided by the number of organisms present plus the number of incorrect organisms reported must be multiplied by 100 to establish a score for each sample in each shipment or testing event. For example, if a sample contained one principal organism and the laboratory reported it correctly but reported the presence of an additional organism, which was not present, the sample grade would be $1/(1 + 1) \times 100 = 50\%$.

(3) As a laboratory is expected to perform the isolation and identification process to the same extent it performs it with patient specimens, a laboratory's type of service as determined in accordance with paragraph (a) of this section must be used to determine the appropriateness of it response.

---

[2]Provided as a nonviable sample.

§ 493.105   Parasitology

(a) *Types of laboratories.* In parasitology there are three types of laboratories for proficiency testing purposes—

(1) Those that are able to recognize the presence of parasites but usually refer them to another laboratory for identification;

(2) Those that identify parasites to the extent required to establish a correct clinical diagnosis without performing permanent stains for identification; and

(3) Those that identify parasites to the extent required to establish a correct clinical diagnosis and perform permanent stains.

(b) *Program content and frequency of challenge.* To be approved for proficiency testing in parasitology, a program must provide a minimum of six samples per quarter. The samples may be provided through mailed shipments or, or at HHS option, may be provided to HHS for on-site testing. An annual program must include parasites that are commonly encountered in the United States as well as those recently introduced into the United States. Other important emerging pathogens (as determined by HHS) and organisms commonly occurring in patient specimens must be included periodically in the program.

(1) An approved program must furnish HHS, during the calendar year, with a list of organisms that will be included in its annual program. Samples must include both formalinized specimens and PVA (polyvinyl alcohol) fixed specimens and blood smears, as appropriate for a particular parasite and stage of the parasite. Samples must contain protozoa or helminths or a combination of parasites. Some samples must be devoid of parasites.

(2) An approved program may vary over time. As an example, the types of organisms that might be included in an approved program over time are—

Entamoeba histolytica
Entamoeba coli
Giardia lamblia
Endolimax nana
Dientamoeba fragilis
Iodamoeba butschlii
Chilomastix meanili
Hookworm
Ascaris lumbricoides
Strongyloides stercoralis
Trichuris trichiura
Enterobius vermicularis

Diphyllobothrium latum
Cryptosporidium sp.
Plasmodium falciparum

(c) *Evaluation of a laboratory's performance.* HHS approves only those programs that assess the accuracy of a laboratory's responses in accordance with paragraphs (c)(1) through (3) of this section.

(1) The program must determine the reportable organisms. To determine the accuracy of a laboratory's response, the program must compare the laboratory's response with the response that reflects agreement of at least 80% of ten or more referee laboratories or all participating laboratories. Sample scores must be averaged to determine the score for the shipment or testing event.

(2) Since laboratories may incorrectly report the presence of organisms in addition to the correctly identified principal organism(s), the grading system must deduct credit for these additional erroneous organisms reported. Therefore, the total number of correct responses divided by the number of organisms present plus the number of incorrect organisms reported must be multiplied by 100 to establish a score for each sample in each shipment or testing event. For example, if a sample contained one principal organism and the laboratory reported it correctly but reported the presence of an additional organism, which was not present, the sample grade would be $1/(1 + 1) \times 100 = 50\%$.

(3) As a laboratory is expected to perform the isolation and identification process to the same extent as it performs it with patient specimens, a laboratory's type of service as determined in accordance with paragraph (a) of this section must be used to determine the appropriateness of its response.

§ 493.107 Diagnostic Immunology

The subspecialties under the specialty of immunology for which a program may offer proficiency testing are syphilis serology and general immunology. Specific criteria for these subspecialties are found at §§ 493.109 and 493.111.

§ 493.109 Syphilis Serology

(1) *Program content and frequency of challenge.*—The annual program must provide a minimum of five samples per quarter and must provide samples that cover the full range of reactivity from highly reactive to non-

reactive. The samples may be provided through mailed shipments or, at HHS' option, may be provided to HHS for on-site testing.

(b) *Challenges per quarter.* The minimum challenges per quarter a program must offer for syphilis serology are five.

(c) *Evaluation of analyte or test performance.* HHS approves only those programs that assess the accuracy of a laboratory's responses in accordance with paragraphs (c)(1) through (3) of this section.

(1) To determine the accuracy of a laboratory's response for a qualitative syphilis test, the program must compare the laboratory's response for each challenge with the response that reflects agreement of at least 80% of ten or more referee laboratories or all participating laboratories agree. The proficiency testing program must indicate the minimum concentration that will be considered as indicating a positive response. For quantitative syphilis tests, the program must determine the correct response for each challenge by the distance of the response from the target value.

(2) After the target value has been established for each response, the program must determine the appropriateness of the response by using either fixed criteria or the number of standard deviations the response differs from the target values.

(3) The criterion for acceptable performance for syphilis serology is the target value ± 1 dilution or (positive or negative).

§ 493.111   General Immunology

(a) *Program content and frequency of challenge.*

To be approved for proficiency testing for immunology, the annual program must provide a minimum of two samples per quarter and must provide samples that cover the full range of reactivity from highly reactive to nonreactive. The samples may be provided through mailed shipments or, at HHS' option, may be provided to HHS for on-site testing.

(b) *Challenges per quarter.* The minimum number of challenges per quarter the program must provide for each analyte or test procedure is two.

Analyte or Test Procedure

Alpha-1 antitrypsin
Alpha-fetoprotein
Antinuclear antibody
Antistreptolysin O
Anti-human immunodeficiency virus (HIV)

Complement C3
Complement C4
Hepatitis markers (HBsAg, anti-HBc, HBeAg)
IgA
IgG
IgE
IgM
Infectious mononucleosis
Rheumatoid facotr
Rubella

(c) *Evaluation of a laboratory's analyte or test performance.* HHS approves only those programs that assess the accuracy of a laboratory's responses in accordance with paragraphs (c) (1) through (4) of this section.

(1) To determine the accuracy of a laboratory's response for qualitative immunology tests or analytes, the program must compare the laboratory's response for each challenge with the response that reflects agreement of at least 80% of ten or more referee laboratories or all participating laboratories agree. The proficiency testing program must indicate the minimum concentration that will be considered as indicating a positive response. For quantitative immunology analytes or tests, the program must determine the correct response for each challenge by the distance of the response from the target value.

(2) After the target value has been established for each response, the appropriateness of the response must be determined by using either fixed criteria or the number of standard deviations (SDs) the response differs from the target value.

(3) *Criteria for acceptable performance.* The criteria for acceptable performance are:

Analyte or Test	Criteria for acceptable performance
Alpha-1 antitrypsin......................	Target value ± 3 SD.
Alpha-fetoprotein........................	Target value ± 3 SD.
Antinuclear antibody	Target value ± 1 dilution or (pos. or neg.).
Antistreptolysin O......................	Target value ± 1 dilution or (pos. or neg.)
Anti-human Immunodeficiency Virus.	Reactive or nonreactive.

Coplement C3	Target value ± 3 SD.
Complement C4	Target value ± 3 SD.
Hepatitis (HBsAg, anti-HBc, HBeAg)	Reactive (positive) or nonreactive (negative).
IgA	Target value ± 3 SD.
IgE	Target value ± 3 SD.
IgG	Target value ± 3 SD.
IgM	Target value ± 3 SD.
Infectious mononucleosis	Target value ± 1 dilution or (pos. or neg.).
Rheumatoid factor	Target value ± 1 dilution or (pos. or neg.).
Rubella	Target value ± 1 dilution or (pos. or neg.).

§ 493.115　Chemistry

The subspecialties under the specialty of chemistry for which a proficiency testing program may offer proficiency testing are routine chemistry, endocrinology and toxicology. Specific criteria for these subspecialties are listed in 493.117 through 493.119.

§ 493.117　Routine Chemistry

Program content and frequency of challenge. To be approved for proficiency testing for chemistry, a program must provide a minimum of two samples per quarter. The annual program must provide samples that cover the clinically relevant range of values that would be expected in patient specimens. The specimens may be provided through mailed shipments or, at HHS' option, may be provided to HHS for on-site testing.

(b) *Challenges per quarter.* The minimum number of challenges per quarter a program must provide for each analyte or test procedure is two.

Analyte or Test Procedure

Alanine aminotransferase (ALT/SGPT)
Albumin
Alkaline phosphatase
Amylase
Aspartate aminotransferase (AST/SGOT)
Bilirubin, total

Blood gas pH
    pO_2
    pCO_2
Calcium, total
Chloride
Cholesterol, total
Cholesterol, high density lipoprotein
Creatine kinase
Creatine kinase, isoenzymes
Creatinine
Glucose
Iron, total
Lactate dehydrogenase (LDG)
LDH isoenzymes
Magnesium
Potassium
Sodium
Triglycerides
Urea Nitrogen
Uric Acid

(c) *Evaluation of a laboratory analyte or test performance.* HHS approves only those programs that assess the accuracy of a laboratory's responses in accordance with paragraphs (c) (1) through (3) of this section.

(1) To determine the accuracy of a laboratory's response for qualitative chemistry tests or analytes, the program must compare the laboratory's response for each challenge with the response that reflects agreement of at least 80% of ten or more reference laboratories or all participating laboratories. For quantitative chemistry tests or analytes, the program must determine the correct response for each challenge by the distance of the response from the target value.

(2) After the target value has been established for each response, the appropriateness of the response must be determined by using either fixed criteria (percentage difference from the target value) or the number of standard deviations (SDs) the response differs from the target value.

(3) *Criteria for acceptable performance*

The criteria for acceptable performance are—

Analyte or test	Criteria for acceptable performance
Alanine aminotransferase (ALT/SGPT).	Target value ± 20%.
Albumin.............,.....................	Target value ± 10%.
Alkaline phosphatase..................	Target value ± 3 SD.
Amylase...................................	Target value ± 3 SD.
Aspartate aminotransferase (AST/SGOT).	Target value ± 10%.
Bilirubin, total...........................	Target value ± .3 mg/dL or 20% (greater term).
Blood gas pCO2.........................	Target value ± 3 SD.
pCO2.........................	Target value ± 3 mm Hg or 8% (greater).
pH...........................	Target value ± .04.
Calcium, total...........................	Target value ± .8 mg/dL
Chloride..................................	Target value ± 5%.
Cholestrol, total........................	Target value ± 15%.
Cholestrol, high density lipoprotein.	Target value ± 3 SD.
Creatine kinase.........................	Target value ± 3 SD.
Creatine kinase isoenzymes.	MB elevated (+ or −) or Target value ± 3 SD.
Creatinine................................	Target value ± .2 mg/dL or 7% (greater term).
Glucose...................................	Target value ± 10%.
Iron, total...............................	Target value ± 20%.
Lactate dehydrogenase (LDH).	Target value ± 20%.
LDH isoenzyme.........................	LDH1/LDH2 (+ or −) or Target value ± 3 SD.
Magnesium	Target value ± 25%.
Potassium	Target value ± .3 mmol/L
Sodium...................................	Target value ± 3 mmol/L or ± 4 mmol/1 if target value is above 150 mmol/L
Triglycerides............................	Target value ± 3 SD.
Urea nitrogen...........................	Target value ± 2 mg/dL or 9% (greater).
Uric acid	Target value ± 17%.

§ 493.119   Endocrinology

(a) *Program content and frequency of challenge.* To be approved for proficiency testing for endocrinology, a program must provide a minimum of two samples per quarter. The annual program must provide samples that cover the clinically relevant range of values that would be expected in patient specimens. The samples may be provided through mailed shipments or, at HHS' option, may be provided to HHS for on-site testing.

(b) *Challenges per quarter.* The minimum number of challenges per quarter a program must provide for each analyte or test procedure is two.

Analyte or Test

Cortisol
Thyroid-stimulating hormone
Thyroxine

(c) *Evaluation of a laboratory's analyte or test performance.* HHS approves only those programs that assess the accuracy of a laboratory's responses in accordance with paragraphs (c)(1) through (3) of this section.

(1) To determine the accuracy of a laboratory's responses on qualitative tests or analytes, a program must compare the laboratory's response for each challenge with the response that reflects agreement of at least 80% of ten or more referee laboratories or all participating laboratories agree. For quantitative chemistry tests or analytes, the program must determine the correct response for each challenge by the distance of the response from this target value.

(2) After the target value has been established for each response, the appropriateness of the response must be determined by using either fixed criteria (percentage difference from the target value) or the number of standard deviations (SDs) the response differs from the target value.

(3) *Criteria for acceptable performance.* The criteria for acceptable performance are—

Analyte or test	Criteria for acceptable performance
Cortisol....................................	Target value ± 25%.
Thyroxine................................	Target value ± 2SD.
Thyroid-stimulating hormone.	Target value ± 3SD.

§ 493.121 Toxicology

(a) *Program content and frequency of challenge.* To be approved for proficiency testing for toxicology, the annual program must provide a minimum of two samples per quarter. The annual program must provide samples that cover the clinically relevant range of values that would be expected in specimens of patients on drug therapy and that cover the level of clinical significance for the particular drug. The samples may be provided through mailed shipments or, at HHS' option, may be provided to HHS for on-site testing.

(b) *Challenges per quarter.* The minimum number of challenges per quarter a program must provide or for each analyte or test procedure is two.

Analyte or Test Procedure

Alcohol (blood)
Blood lead
Carbamazepine
Digoxin
Ethosuximide
Gentamicin
Lithium
Phenobarbital
Phenytoin
Primidone
Procainamide (and metabolite)
Quinidine
Theophylline
Valproic Acid

(c) *Evaluation of a laboratory's analyte or test performance.* HHS approves only those programs that assess the accuracy of a laboratory's responses in accordance with paragraphs (c)(1) through (3) of this section.

(1) To determine the accuracy of a laboratory's responses for qualitative tests or analytes, the program must compare the laboratory's response for each challenge with the response that reflects agreement of at least 80% of ten or more referee laboratories or all participating laboratories. For quantitative chemistry tests or analytes, the program must determine the correct response for each challenge by the distance of the response from the target value.

(2) After the target value has been established for each response, the appropriateness of the response must be determined by using either fixed

criteria (percentage difference from the target value) or the number of standard deviations (SDs) the response differs from the target value.

(3) *Criteria for acceptable performance.* The criteria for acceptable performance are:

Analyte or test	Criteria for acceptable performance
Alcohol, blood...........................	Target value ± 25%.
Blood lead	Target value ± 15% or 6 mcg/dL (greater).
Carbamazephine.........................	Target value ± 25%.
Digoxin	Target value ± 20% or ± 2 ng/mL (greater term).
Ethosuximide.............................	Target value ± 20%.
Gentamicin................................	Target value ± 25%.
Lithium	Target value ± .2 mmol/L
Phenobarbital	Target value ± 20%.
Phenytoin...................................	Target value ± 25%.
Primidone	Target value ± 25%.
Procainamide (and metabolite).	Target value ± 25%.
Quinidine...................................	Target value ± 25%.
Theophylline..............................	Target value ± 25%.
Valproic Acid	Target value ± 25%.

§ 493.125 Hematology (Including Routine Hematology and Coagulation)

(a) *Program content and frequency of challenge.* To be approved for proficiency testing for hematology, a program must provide a minimum of two samples per quarter. The annual program must provide samples that cover the full range of values that would be expected in patient specimens. The samples may be provided through mailed shipments or, at HHS' option, may be provided to HHS for on-site testing.

(b) *Challenges per quarter.* The minimum number of challenges per quarter a program must provide for each analyte or test procedure is two.

Analyte or Test Procedure

Cell identification
White cell differential
Erthyrocyte count
Hematocrit

Hemoglobin
Leukocyte count
Platelet count
Fibrinogen
Partial thromboplastin time
Prothrombin time

(c) *Evaluation of a laboratory's analyte or test performance.* HHS approves only those programs that assess the accuracy of a laboratory's responses in accordance with paragraphs (c) (1) through (3) of this section.

(1) To determine the accuracy of a laboratory's responses for qualitative tests or analytes, the program must compare the laboratory's response for each challenge with the response that reflects agreement of at least 80% of ten or more referee laboratories or all participating laboratories. For quantitative hematology tests or analytes, the program must determine the correct response for each challenge by the distance of the response from the target value.

(2) After the target value has been established for each response, the appropriateness of the response is determined using either fixed criteria (percentage difference from the target value) or the number of standard deviations (SDs) the response differs from the target value.

(3) *Criteria for acceptable performance.* The criteria for acceptable performance are:

Analyte or test	Criteria for acceptable performance
Cell identification......................	80% consensus on identification.
White cell differentiation..............	Target ± 3 SD.
Erythrocyte count	Target ± 3 SD or 6% (lesser).
Hematocrit..............................	Target ± 3 SD or 6% (lesser).
Hemoglobin.............................	Target ± 3 SD or 5% (lesser).
Leukocyte count........................	Target ± 3 SD or 10% (lesser).
Platelet count..........................	Target ± 3 SD or 25% (lesser).
Fibrinogen.............................	Target ± 3 SD.
Partial thromboplastin time.	Target ± 3 SD or ± 15% (greater).
Prothrombin time......................	Target ± 3 SD or ± 15% (greater).

§ 493.129 Cytology: Gynecologic Examinations

(Option 1—Onsite PT)

(a) *Program content and frequency of challenge.* To be approved for proficiency testing for gynecologic examinations (Pap smears) in cytology, a program must provide onsite testing. The testing may be at the laboratory or at another site. The program must provide (5 to 12)[3] challenges per testing event (1 to 4)[3] times per year. Each testing event must include normal challenges, infectious agents, benign reactive processes, premalignant processes, and malignant processes.

(b) *Evaluation of an individual's performance.* HHS approves only those programs that assess the responses of each individual engaged in the examination of cytologic preparations. The program must assess the accuracy of each individual's response by using slides that have been referenced.

(1) A slide with atypical results is referenced if there has been a confirmation by tissue biopsy.

(2) A slide with negative results is referenced if there is 95 per cent consensus agreement. The reference review must be conducted before testing the laboratory.

(Option 2—Mailed PT)

(a) *Program content and frequency of challenge.* To be approved for proficiency testing for gynecologic examinations (Pap smears) in cytology, a program must provide (5–12) challenges per shipment (1 to 4) times per year. Each shipment must include normal challenges, infectious agents, benign reactive processes, premalignant processes, and malignant processes.

(b) *Evaluation of an individual's performance.* HHS approves only those programs that assess the responses of each individual engaged in the examination of cytologic preparations. The program must assess the accuracy of each individual's response by using slides that have been referenced.

(1) A slide with atypical results is referenced if there has been a confirmation by tissue biopsy.

(2) A slide with negative results is referenced if there is a 95 per cent consensus agreement. The reference review must be conducted before testing the laboratory.

§ 493.153 Immunohematology

(a) *Types of laboratories.* In immunohematology, there are three types of laboratories for proficiency testing purposes—

---

[3]We are inviting comments on the number of challenges and their frequency and are proposing what we consider to be realistic ranges.

(1) Those that perform ABO and/or Rh typing but do not perform irregular antibody detection or identification;

(2) Those that perform ABO and/or Rh typing, compatibility testing, irregular antibody detection and crossmatching but do not perform antibody identification; and

(3) Those that perform ABO and/or Rh typing, compatibility testing, irregular antibody detection and crossmatching and also perform antibody identification.

(b) *Program content and frequency of challenge.* To be approved for proficiency testing for immunohematology, a program must provide a minimum of two samples per quarter. The annual program must provide samples that cover the full range of interpretation that would be expected in patient specimens. The samples may be provided through mailed shipments or, at HHS' option, may be provided to HHS for on-site testing.

(c) *Challenges per quarter.* The minimum number of challenges per quarter a program must provide for each analyte or test procedure is two.

Analyte or Test Procedure

ABO grouping
Rh typing
Antibody detection
Antibody identification

Crossmatch compatibility

(d) *Evaluation of a laboratory's analyte or test performance.* HHS approves only those programs that assess the accuracy of a laboratory's response in accordance with paragraphs (d)(1) through (3) of this section.

(1) To determine the accuracy of a laboratory's response, a program must compare the laboratory's response for each challenge with the response that reflects agreement of at least 90% of ten or more referee laboratories or all participating laboratories.

(2) After the target value has been established for each response, the appropriateness of the response is determined using either fixed criteria (percentage difference from the target value) or the number of standard deviations (SDs) the response differs from the target value.

(3) *Criteria for acceptable performance.* The criteria for acceptable performance are—

Analyte or test	Criteria for acceptable performance
ABO grouping	100% accuracy on 95% consensus.

Rh typing	100% accuracy on 95% consensus.
Antibody detection	100% accuracy on 90% consensus.
Antibody identification	80% accuracy on 90% consensus
Crossmatch compatibility.	100% accuracy on 95% consensus for incompatible crossmatches and no more than one false incompatible result for a compatible sample.

SUBPART E—PATIENT TEST MANAGEMENT

§ 493.201   Condition: Patient Test Management

The laboratory must maintain and employ a system that provides for proper receipt and processing of patient specimens and accurate reporting of patient test results.

(a) *Standard; Procedures for specimen submission.* The laboratory must have written policies and procedures regarding collection, labeling, preservation or fixation, and transportation of specimens that, when followed, assure accurate and reliable test results. The laboratory must follow these practices and make available to clients instructions for specimen collection, handling, preservation and transportation as a means of ensuring that specimens submitted are satisfactory for testing.

(b) *Standard; Specimen requisition.* The laboratory must perform tests only at the written or electronic request of an authorized person and maintain records of test requisitions for at least two years. The laboratory must assure that the requisition includes—

(1) The patient's name or other method of specimen identification to assure accurate reporting of results;

(2) The name or other suitable identifier of the person who ordered the test or the name of the clinical laboratory submitting the specimen;

(3) The date of specimen collection and the time of specimen collection, if pertinent;

(4) The source of specimen and type of test ordered; and

(5) For cytology specimens—

(i) Patient age;

(ii) Pertinent clinical information; and

(iii) For Pap smears, the last menstrual period and indication whether the patient is at risk for developing cervical cancer or its precursors.

(c) *Standard; Specimen records.* The laboratory must maintain a system of reliable specimen identification, document each step in processing patient specimens and testing to assure accurate test reporting, and retain records of patient testing for at least two years. This system must provide—

(1) The laboratory number or other identification of the specimen;

(2) The patient's name or other method of specimen identification to assure accurate reporting of results;

(3) The name or other suitable identifier of the person who ordered the test or the name of the clinical laboratory submitting the specimen;

(4) The date of specimen collection; the time of specimen collection, if pertinent; and the date of specimen receipt in the laboratory;

(5) The source of specimen and type of test ordered;

(6) The condition or disposition of specimens that do not meet the laboratory's criteria for specimen acceptability; and

(7) The records and dates of performance of each step in patient testing leading to the final report to assure proper identification and reliable reporting of test results.

(d) *Standard; test report.* The laboratory report must be sent promptly to the authorized person or laboratory that initially requested the test and a legally reproduced record of each test result must be preserved by the testing laboratory for a period of at least two years after the date of reporting. For cytology, test reports must be maintained at least ten years after the date of reporting.

(1) The legally reproduced copies of test reports must be filed in the laboratory in a manner that permits ready identification and accessibility.

(2) The results of transcripts of laboratory tests or procedures must be released only to authorized persons.

(3) Pertinent "normal" ranges, as determined by the laboratory performing the tests, must be available to the authorized person who ordered or who utilizes the test results.

(4) The laboratory must establish special reporting procedures for potential life-threatening laboratory results or panic values. In addition, the laboratory must immediately alert the individual requesting the test when any result indicating a life-threatening condition is obtained.

(5) The laboratory must indicate on the test report any information regarding the condition or disposition of specimens that do not meet the laboratory's criteria for acceptability.

(6) The laboratory must upon request make available to clients a list of test methods employed by the laboratory and a basis for the listed "normal" ranges. In addition, information that may affect the interpretation of test results, such as test interferences, if known, and performance claims including, where applicable, detection limits, sensitivity, specificity, accuracy, precision and validity of test measurement and other pertinent test characteristics must be provided to the individual requesting the test. Updates on testing information must be provided to clients whenever changes occur that affect the test results or interpretation of test results.

(7) The test report must include the name and address of each laboratory performing each test.

(e) *Standard; referral of specimens.* The laboratory may refer specimens for testing only to a laboratory that is Medicare-approved or CLIA-licensed (or exempted from CLIA-licensure) for the appropriate specialty or subspecialty.

(1) The authorized person who orders a test or procedure must be notified by the referring laboratory of the name and address of each laboratory that performs a laboratory test.

(2) If the referring laboratory interprets or revises in any way the test results provided by the testing laboratory, the referring laboratory must notify the authorized person who requested the test or procedure and the testing laboratory. The referring laboratory must maintain a legally reproduced copy of such interpretations, alterations or revisions and of the notice to the client and testing laboratory.

(3) The referring laboratory may permit each testing laboratory to send the test directly to the authorized person who initially requested the test. In such case, the referring laboratory must maintain a legally reproduced copy of each testing laboratory's report.

(4) The test report must include the name and address of each testing laboratory.

SUBPART F—QUALITY CONTROL

§ 493.221   Condition: General Quality Control

The laboratory must impose and practice quality control procedures that provide and assure accurate, reliable and valid test results and reports and that meet the standards in §§ 493.223 through 493.239 of this subpart.

§ 493.223   Standard; Facilities

The laboratory must be constructed, arranged and maintained to ensure adequate space, facilities and essential utilities for the performance and reporting of tests.

§ 493.225   Standard; Adequacy of Methods and Equipment

The laboratory must employ methodologies and equipment that provide accurate and reliable test results and reports.

(a) The laboratory must have sufficient equipment and instruments to perform the type and volume of testing offered by the laboratory.

(b) The instrumentation used must be capable of providing test results within the laboratory's stated performance characteristics. These performance characteristics include detection limits, precision, accuracy, specificity, sensitivity as well as freedom from interferences and related test variables.

(c) Test methods are performed in a manner that permits the laboratory to provide test results within the laboratory's stated performance characteristics, including precision, accuracy, sensitivity, specificity, detection limits, as well as freedom from interference and related test variables. In determining test methodology, the laboratory must consider factors such as utilizing the appropriate test system to give the performance characteristics specified by the laboratory, assuring a statistically valid number of counts to give accurate and reliable test results for systems such as cell counters, radioactive counters, spectrophotometers and other equipment for which this is a critical variable.

(d) The laboratory must have adequate systems in place to report results in an accurate and reliable manner and within the time frames established by the laboratory.

§ 493.227   Standard; Temperature and Humidity Monitoring

Temperature and humidity must be maintained and monitored within an acceptable range of values to assure—
   (a) Proper storage of specimens, reagents and supplies; and
   (b) Accurate and reliable test performance and reporting.

§ 493.229   Standard; Labeling of Testing Supplies

(a) Reagents, solutions, culture media, controls and calibrators and other materials must be labeled to indicate—
   (1) Identity, and, when significant, titer, strength or concentration;
   (2) Recommended storage requirements; and
   (3) Preparation or expiration date and other pertinent information.

(b) The laboratory may not use materials that have exceeded their expiration date, are of substandard reactivity, or have deteriorated.

§ 493.231 Standard; Procedure Manual

(a) Personnel examining specimens and performing related procedures within a specialty or subspecialty have available in the testing area complete written instructions and descriptions related to—

(1) The current analytical methods used by personnel;

(2) Specimen processing procedures;

(3) Preparation of solutions, reagents, and stains;

(4) Calibration procedures;

(5) Microscopic examination;

(6) Quality control procedures;

(7) Quality assurance policies;

(8) Limitations in methodologies;

(9) Actions to be followed when quality control results deviate from expected values or patterns;

(10) Procedures for reporting patient results;

(11) Pertinent literature references; and

(12) Alternative methods for performing tests or storing the test specimens in the event that a test system becomes inoperable.

(b) Procedures and changes in procedures must be approved and signed and dated by the current director of the laboratory.

(c) The laboratory must maintain, for a period of up to two years, records of each procedure it uses and the span of time the procedure was in use.

(d) Textbooks may be used as supplements to these written descriptions but may not be used in their place.

§ 493.223 Standard; Equipment Maintenance and Function Checks

The laboratory establishes and employs policies and procedures for—

(a) The proper maintenance of equipment, instruments and test systems by—

(1) Defining its preventive maintenance program for each instrument and piece of equipment based on the manufacturer's instructions. If the laboratory choose to perform preventive maintenance less frequently than the manufacturer recommends or the manufacturer does not specify a frequency, the laboratory must document the validity of its preventive maintenance program; and

(2) Documenting the performance of its preventive maintenance program.

(b) Performing function checks on equipment, including spectrophotometers, radioactive counters, cell counters, automated analyzers, centrifuges, densitometer and data processors to assure proper performance and the production of accurate and reliable test results by—

(1) Calibrating, recalibrating, or rechecking each instrument or device or test system at least once each day of use;

(2) Performing the function checks with at least the frequency specified by the manufacturer unless the laboratory has data that document that a lesser frequency will not alter the performance characteristics of the system. The laboratory must establish performance criteria for each test or procedure if the manufacturer of the test system or equipment has not specified the type of maintenance and function checks to perform; and

(3) Performing all necessary baseline or background checks on radioactive counters, cell counters, refractometers, spectrophotometers and other equipment requiring such measurements. The laboratory may not report test results unless the background or baseline checks are within acceptable limits.

§ 493.235   Standard; Validation of Methods

The laboratory must have a written protocol and documentation for the validation of each method that verifies that the method produces test results within the laboratory's stated performance characteristics.

(a) The linearity of each quantitative method, if applicable, must be established.

(b) In the case of qualitative and screening tests, the laboratory must establish and define the basis for specifying reportable results as positive, negative, degree of reactivity, or in accordance with another reporting system. The laboratory must follow these established limits in reporting test results.

(c) A method used by the laboratory must be validated before it is used and documentation of the validation must be available for the period during which the procedure is used by the laboratory or for two years, whichever is longer.

(d) The laboratory must have documentation of the level of precision, accuracy, sensitivity, and specificity that the laboratory claims for each method in use and for which it reports results.

(e) The laboratory must maintain documentation verifying that test systems perform according to the laboratory's specifications. This documentation must be available to the authorized persons ordering or receiving test results.

(f) The laboratory must define the basis for reporting patient test results.

(g) The laboratory may not report patient test results if it does not have data to verify the specified test performance characteristics and reporting limits.

§ 493.237   Standard; Frequency of Quality Control

The laboratory must perform quality control at the frequencies specified in this section unless another frequency is specified in §§ 493.241 through 493.315 or HHS approves a lesser frequency in Appendix C of the State Operational Manual (HCFA Pub. 7).

(a) The laboratory must establish and document a schedule for calibration, recalibration or calibration verification of each automated and manual method.

(1) The laboratory must perform procedural calibration or recalibration at least once every six months using a complete range of calibrators and, in addition, when any of the following occur:

(i) A complete change of reagents for a procedure is introduced. If all of the reagents for a test are packaged together, the laboratory is not required to recalibrate for each package of reagents, provided the reagents are received in the same shipment and contain the same lot number;

(ii) There is major preventive maintenance or replacement of critical parts, such as an excitor lamp;

(iii) Controls begin to reflect an unusual trend or are outside of acceptable limits;

(iv) The manufacturer's recommendation specify more frequent recalibration; or

(v) The laboratory's established schedule requires more frequent recalibration.

(2) The number of calibrators the laboratory uses to verify calibration (recalibration) varies by method—

(i) For methods in which a linear relationship exists between concentration and direct instrument reading, at least three points and a zero or minimum value are required; and

(ii) For methods in which a nonlinear relationship exists between concentration and direct instrument readings, at least five points and a zero or minimum value are required.

(3) The calibrators must cover the entire range of expected patient values to be reported for the test procedures.

(4) For patient values above the maximum calibration point or below the minimum calibration point—

(i) The laboratory must report the patient results as greater than the upper limit or less than the lower limit or an equivalent designation; or

(ii) The laboratory must dilute the sample and the diluted sample must fall within the linear range of the method if results are to be reported. If a dilution method is employed, the laboratory must be able to provide evidence that the dilution process can yield accurate, reliable and valid test results.

(b) For quantitative tests, the laboratory must include two calibrator samples, one calibrator sample and one control sample, or two control samples in each run of unknown samples when these reference samples are available.

(1) A run is an interval within which the accuracy and precision of a measuring system is expected to be stable but must not exceed a period of 24 hours and must not exceed the manufacturer's specification for including controls.

(2) The laboratory must use the calibrator samples, the control samples, or combination thereof, and monitor both the abnormal and normal range of reportable patient values.

(i) If calibrators are not used, two controls of different concentrations must be used;

(ii) If controls are not used, two calibrators of different concentrations must be used. Two separate dilutions from a stock calibrator must be prepared or a calibrator and a sample spiked with a calibrator must be used:

(iii) Other exceptions apply as HHS approves in Appendix C of the State Operations Manual (HCFA Pub. 7); and

(iv) If calibrators and controls are not available, the laboratory must have a mechanism to assure the quality, accuracy and precision of the test results.

(3) A laboratory does not have to meet the requirements of this section when calibrators and controls are not available because of the nature of the instruments or the tests performed.

(4) For qualitative tests, the laboratory must include a positive and negative control with each run of specimens unless otherwise specified by HHS under the provision of these regulations.

(c) The laboratory must determine its statistical limits for each lot number of controls through repetitive testing. The laboratory may use the manufacturer's limits, provided they are verified by the laboratory, represent the actual within-laboratory analytical range expected and the stated limits correspond to the methods employed by the laboratory. Acceptable limits for unassayed materials must be established by the laboratory through concurrent testing with an assayed or laboratory-defined control material.

(d) Initially, the laboratory must check each batch or shipment of reagents, discs, stains, antisera and identification system (systems using two or more substrates and antigen detection systems) for positive and negative reactivity.

(e) Each day of use, unless otherwise specified in this subpart of Appendix C of the State Operations Manual (HCFA Pub. 7), the laboratory must test staining materials for intended reactivity by concurrent application to smears of microorganisms with predictable staining characteristics.

(f) Each day of use, the laboratory must test direct antigen detection systems using positive and negative control organisms that evaluate both the extraction and reaction phases.

(g) The laboratory must check each batch or shipment of media for sterility, ability to support growth and, if appropriate, selectivity/inhibition and/or biochemical response. The laboratory may use a commercial manufacturer's quality control checks of media if the laboratory has documentation to verify that the manufacturer has used the quality assurance practices that have been approved by HHS in Appendix C of the State Operations Manual (HCFA Pub. 7). The laboratory must document that the physical characteristics of the media are not compromised and report any deterioration in the media to the manufacturer. The laboratory must follow the manufacturer's specifications for using the media and be responsible for the test results.

(h) A batch of media (solid or liquid) must—

(1) Consist of all tubes, plates, or containers of medium prepared at the same time and in the same laboratory; or

(2) If received from an outside source or commercial supplier, all of the plates, tubes or containers must have the same lot numbers and be received in a single shipment.

(i) Results must not be reported unless all controls are within acceptable limits.

§ 493.239 Standard; Remedial Actions

The laboratory must establish and employ policies and procedures for actions to be taken when—

(a) Test systems do not meet established criteria including—

(1) Quality control results are outside of acceptable limits;

(2) Equipment or methodologies perform outside of established operating limits;

(3) Test results are outside of acceptable linear limits; and

(4) Proficiency test results are not within acceptable limits;

(b) It cannot test samples within specified times that it has established. The laboratory must establish and follow criteria for referring or for storing specimens. The laboratory must notify the authorized person who ordered the test if the laboratory cannot run a specimen because test results may be affected;

(c) It detects errors in the reported patient results. The laboratory must establish and follow polices for correcting reporting errors and notify the authorized person ordering and utilizing the test results promptly; or

(d) It does not report test results within its established timeframes.

§ 493.240   Standard; Quality Control—Records

The laboratory must document all quality control activities and retain records for at least two years.

(a) The laboratory must maintain records of the following—

(1) Preventive maintenance checks;

(2) Equipment function checks;

(3) Procedural calibrations; and

(4) Validation procedures.

(b) The laboratory must maintain records of each step in the processing and testing of quality control samples to assure that the quality control samples are tested in the same manner as patient samples; the laboratory must document quality control results.

(c) The laboratory must maintain records of remedial activities taken when test systems do not meet established criteria and samples cannot be tested within timeframes established by the laboratory.

§ 493.241   Condition: Quality Control—Specialties and Subspecialties

The laboratory must establish and follow policies and procedures for an acceptable quality control program that include verification and assessment of accuracy, measurement of precision and detection of error for all analyses and procedures performed by the laboratory. In addition to the general requirements specified in §§ 493.221 through 493.240, the laboratory must meet the applicable requirements of §§ 493.243 through 493.315 for each specialty and subspecialty for which the laboratory is licensed (CLIA) or approved (Medicare and Medicaid). Failure to meet any of the applicable conditions in §§ 493.243 through 493.315 will result in the loss of approval, licensure, or letter of exemption for the entire speciality to which the condition applies; failure to meet any of the standards in §§ 493.243 through 493.315 will result in the loss of approval, licensure or exemption from licensure for the subspecialty to which the standard applies.

§ 493.243 Condition: Microbiology

The laboratory must meet the applicable quality control requirements in §§ 493.221 through 493.240 and in §§ 493.245 through 493.253 for the subspecialties for which it is Medicare or Medicaid approved or CLIA-licensed under the specialty or microbiology.

§ 493.245 Standard; Bacteriology

To meet the quality control requirements for bacteriology, the laboratory must comply with the applicable requirements in §§ 493.221 through 493.240 and with paragraphs (a) through (d) of this section.

(a) The laboratory must check positive and negative reactivity with control organisms—

(1) Each time of use for fluorescent stains and DNA probes based on radioisotopes methods;

(2) Each day of use for catalase, coagulase, and oxidase reagents;

(3) Each week of use for Gram and acid-fast stains, bacitracin, optochin, ONPG, XV, strips, discs and X, V discs or strips; and

(4) Each month of use for antisera.

(b) For antimicrobial susceptibility tests, the laboratory must check each new batch of media and each lot of antimicrobial discs before, or concurrent with, initial use, using improved reference organisms.

(c) The laboratory's zone sizes or minimum inhibitory concentration for reference organisms must be within established limits.

(d) Each day tests are performed, the laboratory must use the appropriate control organism(s) to check the procedure, unless the laboratory can establish precision and accuracy to be within the limits established by HHS in Appendix C of the State Operations Manual (HCFA Pub. 7).

§ 493.247 Standard; Mycobacteriology

To meet the quality control requirements for mycobacteriology, the laboratory must comply with the applicable requirements in §§ 493.221 through 493.240 and with paragraphs (a) through (d) of this section.

(a) Each day of use, the laboratory must check the iron uptake test with at least one acid-fast organism that produces a negative reaction and check all other reagents used for Mycobacteria identification with at least one acid-fast organism that produces a positive reaction.

(b) The laboratory must check fluorescent stains for positive and negative reactivity each time of use.

(c) The laboratory must check each week of use acid-fast stains with an acid-fast organism that produces a positive reaction.

(d) For susceptibility tests performed on *Mycobacterium tuberculosis* isolates, the laboratory must check the procedure each week of use with a control strain of *Mycobacterium tuberculosis*.

§ 493.249   Standard; Mycology

To meet the quality control requirements for mycology, the laboratory must comply with the applicable requirements in §§ 493.221 through 493.240 and with paragraphs (a) through (c) of this section.

(a) Each day of use, the laboratory must check the nitrate reagent with a peptone control.

(b) Each week of use, the laboratory must check acid-fast stains for positive and negative reactivity.

(c) For susceptibility tests, the laboratory must test each drug each day of use with at least one control strain that is susceptible to the drug. The laboratory must establish control limits.

§ 493.251   Standard; Parasitology

To meet the quality control requirements for parasitology, the laboratory must comply with the applicable requirements of §§ 493.221 through 493.240 and with paragraphs (a) through (c) of this section.

(a) The laboratory must have available a reference collection of slides, photographs or gross specimens for identification of parasites available and use it in the laboratory for appropriate comparison with diagnostic specimens.

(b) The laboratory must use a calibrated ocular micrometer for determining the size of ova and parasites, if size is a critical parameter.

(c) Each month of use, the laboratory must check permanent stains using fecal samples.

§ 493.253   Standard; Virology.

To meet the quality control requirements for virology, the laboratory must comply with the applicable requirements in §§ 493.221 through 493.240 and with paragraphs (a) through (c) of this section.

(a) The laboratory must have available host systems for the isolation of viruses and test methods for the identification of viruses that cover the entire range of viruses that are etiologically related to clinical diseases for which services are offered.

(b) The laboratory must maintain records that reflect the systems used and the reactions observed.

(c) In tests for the identification of viruses, the laboratory must employ uninoculated cell or cell substrate controls to detect erroneous identification results.

§ 493.255   Condition: Diagnostic Immunology

The laboratory must meet the applicable quality control requirements in §§ 493.221 through 493.240 and §§ 493.257 through 493.259 for the subspecialities for which it is Medicare or Medicaid approved for CLIA-licensed under the speciality of diagnostic immunology.

§ 493.257   Standard; Syphilis Serology

To meet the quality control requirements for syphilis serology, the laboratory must comply with the applicable requirements in §§ 493.221 through 493.240 and with paragraphs (a) through (e) of this section.

(a) For laboratories performing syphilis testing, the equipment, glassware, reagents, controls, and techniques for tests for syphilis must conform to manufacturers' specifications.

(b) The laboratory must run serologic tests on unknown specimens concurently with a positive serum control, known titer or controls of graded reactivity plus a negative control unless otherwise specified by HHS in Appendix C of the State Operations Manual (HCFA Pub. 7).

(c) The laboratory must employ controls for all test components to ensure reactivity and uniform dosages.

(d) The laboratory may not report test results unless the predetermined reactivity pattern is observed.

(e) All facilities transfusing blood and blood products or serving as referral laboratories for these facilities must meet the syphilis serology testing requirements of 21 CFR 640.5(a).

§ 493.259   Standard; General Immunology

To meet the quality control requirements for general immunology, the laboratory must comply with the applicable requirements in §§ 493.221 through 493.240 and with paragraphs (a) through (d) of this section.

(a) The laboratory must run serologic tests on unknown specimens concurrently with a positive serum control, known titer or controls of graded reactivity plus a negative control unless otherwise specified by HHS in Appendix C of the State Operations Manual (HCFA Pub. 7).

(b) The laboratory must employ controls for all test components (antigens, complement, erythrocyte indicator systems, etc.) to ensure reactivity and uniform dosages.

(c) The laboratory may not report test results unless the predetermined reactivity pattern is observed.

(d) All facilities transfusing blood and blood products or serving as referral laboratories for these facilities must meet:

(1) The HIV testing requirements of 21 CFR 610.45; and

(2) Hepatitis testing requirements of 21 CFR 610.40.

§ 493.261   Condition: Chemistry

The laboratory must meet the applicable quality control requirements in §§ 493.221 through 493.240 and §§ 493.263 through 493.267 for the subspecialties for which it is Medicare or Medicaid approved or CLIA-licensed under the specialty of chemistry.

§ 493.263   Standard; Routine chemistry

To meet the quality control requirements for routine chemistry, the laboratory must comply with the applicable requirements in §§ 493.221 through 493.240 and with paragraphs (a) and (b) of this section.

(a) For blood gas analyses, the laboratory must include two calibrators and a control material each eight hours of testing in addition to including a calibrator or control each time patients are tested.

(b) For urinalysis qualitative or screening tests, the laboratory must include a positive control each day of testing to check the reactivity of each constituent.

§ 493.265   Standard; Endocrinology

To meet the quality control requirements for endocrinology, the laboratory must comply with the applicable requirements contained in §§ 493.21 through 493.240.

§ 493.267   Standard; Toxicology

To meet the quality control requirements for toxicology, the laboratory must comply with the applicable requirements in §§ 493.221 through 493.240.

§ 493.269   Condition: Hematology

To meet the quality control requirements for hematology, the laboratory must comply with the applicable requirements in §§ 493.221 through 493.240 and with paragraphs (a) through (b) of this section.

(a) For each eight hours of operation, the laboratory must have two levels of control except for manual cell counts, in which one level of control is required for each eight hours of operation.

(b) The laboratory must run tests for coagulation in duplicate.

§ 493.270   Condition: Pathology

The laboratory must meet the applicable quality control requirements in §§ 493.221 through 493.240 and §§ 493.271 through 493.274 for the subspecialties for which it is Medicare or Medicaid approved or CLIA-licensed under the specialty of pathology.

§ 493.271   Standard; Cytology

To meet the quality control requirements for cytology, the laboratory must comply with the applicable requirements in §§ 493.221 through 493.240 and paragraphs (a) and (h) of this section.

(a) The laboratory must assure that—

(1) All gynecologic smears are stained using a Papanicolaou staining method;

(2) Gynecologic and non-gynecologic specimens are stained separately; and

(3) Staining solutions are discarded or filtered after staining body fluids.

(b) The individual providing technical supervision of cytology must assure that—

(1) All genecological smears interpreted to be in the "suspicious" or positive category are confirmed by the technical supervisor in cytology. The report must be signed to reflect the review or, if a computer report is generated, it must reflect an electronic signature authorized by the technical supervisor.

(2) All nongynecological cytological preparations, positive and negative, are reviewed by the technical supervisor in cytology. The report must be signed to reflect supervisory review or, if a computer report is generated, it must reflect an electronic signature authorized by the technical supervisor.

(3) Provision is made for documenting and evaluating each cytotechnologist's side examination performance, including feedback on the suspicious or abnormal cases referred to the technical supervisor in cytology, and on each cytotechnologist's performance evaluations through the rescreening of negative cases.

(c) The laboratory must establish and follow a program designed to detect errors in the performance of cytological examinations and the reporting of results.

(1) The laboratory must establish a program that includes on a regular basis a review of slides screened by each individual responsible for the examination of slides; records of initial examinations and rescreening results must be available. The review must meet the requirements of paragraph (c)(1)(i) or (c)(1)(ii) of this section.

(i) At least ten percent of the gynecologic cases reviewed by each cytotechnologist interpreted to be negative must be rescreened by a second cytotechnologist or the technical supervisor in cytology before reporting patient results; or

(ii) All gynecologic cases that are interpreted to be negative and that are from patients who are identified as having a high probability of developing cervical cancer, as defined in § 493.201(b)(5)(iii), must be rescreened by a second cytotechnologist or the technical supervisor before reporting patient results.

(2) The laboratory must compare clinical information with cytology reports and, for all abnormal cytology reports, must compare the cytology report with the histopathology report, if available, and determine the causes of any discrepancies.

(3) The laboratory must review all prior cytologic specimens, if available, for each abnormal cytology result.

(4) The laboratory must establish and document an annual statistical evaluation of the number of cytology cases examined, number of specimens

processed by specimen type, volume of patients reported by diagnosis, false-negative and false-positive rates, number of unsatisfactory specimens submitted by each physician or laboratory and number of complaints received from individuals ordering or receiving test reports.

(5) The laboratory must evaluate each cytotechnologist's individual case reviews against its overall statistical rates, document any discrepancies, including reasons for the deviation, and document corrective action, if appropriate.

(d) The laboratory report must—

(1) Distinguish between inadequate smear and negative results;

(2) Contain narrative discriptions for any abnormal or suspicious results;

(3) Include the presence of endometrial cells if endometrial cells are present out of cycle;

(4) Indicate evidence of viral infection if present; and

(5) Contain appropriate provisions for follow-up recommendations.

(e) Corrected reports issued by the laboratory must indicate basis for correction.

(f) The laboratory must retain all negative slides for five years from the date of examination.

(g) The laboratory must retain all abnormal slides and slides from patients with a history of suspicious or abnormal cytology results for ten years from the date of examination.

§ 493.273 Standard; Histopathology

To meet the quality control requirements for histopathology, a laboratory must comply with the applicable requirements in §§ 493.221 through 493.240 and paragraphs (a) through (c) of this section.

(a) The laboratory must control all special stains for intended level of reactivity by use of positive slides and records must document the quality control checks of the stain materials.

(b) The laboratory must retain stained slides at least ten years from the date of examination and retain specimen blocks at least two years from the date of examination.

(c) The laboratory must retain remnants of tissue specimens in a fixative solution until the portions submitted for microscopic examination have been examined and a diagnosis made by an individual qualified under § 493.403(b)(2) of this part. In addition, an individual who meets the requirements of § 493.403(b)(3) may examine and provide reports for specimens for skin pathology; an individual meeting the requirements of

§ 493.407(b)(4) may examine and provide reports for specimens for oral pathology.

(d) The laboratory must utilize acceptable terminology of a recognized system of disease nomenclature in reporting results.

(e) The laboratory must report results of all biopsy-confirmed cases of cervical cancer to an established cancer registry.

§ 493.274   Standard; Oral Pathology

To meet the quality control requirements for oral pathology, the laboratory must comply with the applicable requirements in §§ 493.221 through 493.240 and 493.273.

§ 493.275   Condition: Radiobioassay

To meet quality control requirements for radiobioassay, the laboratory must meet the specific requirements of §§ 493.221 through 493.240 of this subpart.

§ 493.277   Condition: Histocompatibility

In addition to meeting the requirements for general quality control in §§ 493.221 through 493.240, for quality control for general immunology in § 493.259 of this subpart and for immunohematology in § 493.281 of this subpart, the laboratory must comply with the applicable requirements in paragraphs (a) through (d) of this section.

(a) For renal allotransplantation the laboratory must meet requirement of paragraphs (a)(1) through (23) of this section.

(1) The laboratory must have available and follow criteria for selecting appropriate patient serum samples for crossmatching;

(2) The laboratory must have available results of final crossmatches before an organ or tissue is transplanted.

(3) The laboratory must have available and follow criteria for the technique used in crossmatching;

(4) The laboratory must have available and follow criteria for preparation of donor lymphocytes for crossmatching.

(5) The laboratory must have available and follow criteria for reporting crossmatch results.

(6) The laboratory must have available serum specimens for all potential transplant recipients at initial typing, for periodic screening, for pretrans-

plantation crossmatch and following sensitizing events, such as transfusion and transplant loss.

(7) The laboratory's storage and maintenance of both recipient sera and reagents must—

(i) Be at an acceptable temperature range for sera and components;

(ii) Use a temperature alarm system and have an emergency plan for alternate storage; and

(iii) Be well-organized with all specimens properly identified and easily retrievable.

(8) The laboratory's reagent typing sera inventory (applicable only to locally constructed trays) must indicate source, bleeding date and identification number, and volume remaining.

(9) The laboratory must properly label and store cells, complement, buffer, dyes, etc.

(10) The laboratory must type all potential transplant recipient cells.

(11) The laboratory must type cells from organ donors referred to the laboratory.

(12) The laboratory must have available and follow criteria for the preparation of lymphocytes for HLA-A, B and DR typing.

(13) The laboratory must have available and follow criteria for selecting typing reagents, whether locally or commercially prepared.

(14) The laboratory must have available and follow criteria for the assignment of HLA antigens.

(15) The laboratory's reagent tray or trays for typing recipient and donor cells must be adequate to define all HLA-A, B and DR specifications as required to determine splits and cross-reactivity.

(16) The laboratory must have a written policy that it follows that establishes when antigen redefinition and retyping are required.

(17) The laboratory must screen recipient sera for preformed antibodies with a suitable lymphocyte panel that assures that—

(i) Potential transplant recipient serum are screened for HLA-A and B content; and

(ii) Screening is performed on initial typing of living related donors and cadaver organs. The laboratory must also screen at monthly intervals after initial screening and following sensitizing events.

(18) The laboratory must use a suitable cell panel for screening patient sera (antibody screen), a screen that contains all the major HLA specificities and common splits—

(i) If the laboratory does not use commercial panels, it must maintain a list of individuals for fresh panel bleeding; and

(ii) If the laboratory uses frozen panels, there must be a suitable storage system.

(19) The laboratory must use testing such as the mixed lymphocyte culture to determine cellularly defined antigens.

(20)(i) If the laboratory reports the patient's ABO blood grouping and Rh typing, it must perform the testing in accordance with § 493.281 of this subpart.

(ii) If the laboratory performs an ABO blood grouping to purify cell population, a control for the ABO must be included.

(21) The laboratory must include positive and negative controls on each tray. The laboratory must determine the reactivity of cell panels used for antibody detection by testing the various components of the antigen panel with positive and negative controls when the panel is used. The entire panel does not have to be tested at one time.

(22) The laboratory must, at least once each month, give each individual performing tests a previously tested specimen as an unknown to verify his or her ability to reproduce test results. The laboratory must maintain records of the results for each individual.

(23) The laboratory must participate in at least one national or regional cell exchange program, if available, or develop an exchange system with another laboratory in order to validate interlaboratory reproducibility.

(b) For laboratories performing only transfusions, bone marrow transplants, and other nonrenal transplantation, the laboratory must meet all the requirements specified in this section except for the performance of mixed lymphocyte cultures.

(c) For laboratories performing only disease-associated studies, the laboratory must meet all the requirements specified in this section except for the requirements for those concerning the performance of mixed lymphocyte cultures.

(d) For laboratories performing tests for organ transplantation, the laboratory must test the donor for HIV reactivity using the same protocols as required under § 493.301 of this part for the transfusion of blood and blood products, unless the organ recipient (or an individual authorized to act on his or her behalf) waives the tests because of medical circumstances.

§ 493.279 Condition: Medical Cytogenetics

To meet the quality control requirements for cytogenetics, the laboratory must comply with the applicable requirements of §§ 493.221 through 493.240 and with paragraphs (a) through (d) of this section.

(a) The laboratory examination of X and Y chromatin counts must be based on an examination of an adequate number of cells.

(b) The laboratory must have records that document the number of cells counted, the number of cells karyotyped, the number of chromosomes counted for each metaphase spread, and the quality of the banding; that the resolution is sufficient to support the reported results; and that an adequate number of karyotypes are prepared for each patient.

(c) The laboratory also must have policies and procedures for assuring an adequate patient sample identification during the process of accessioning, photographing or other image reproduction technique, cell preparation and photographic printing, and storage and reporting of results or photographs.

(d) The laboratory report must include the summary and interpretation of the observations and number of cells counted and the use of appropriate nomenclature.

§ 493.281 Condition: Immunohematology

To meet the quality control requirements for immunohematology, the laboratory must comply with the applicable requirements in §§ 493.221 through 493.240 and with paragraphs (a) through (d) of this section.

(a) The laboratory must perform ABO and $Rh_0(D)$ grouping, antibody detection and identification and compatibility testing in accordance with 21 CFR Part 606 (with the exception of 21 CFR 606.20a, Personnel) and 21 CFR 640 et seq.

(b) The laboratory must perform ABO grouping by testing unknown red cells with anti-A and anti-B grouping serums. For confirmation of ABO grouping, the unknown serum must be tested with known A_1 and B red cells.

(c) The laboratory must determine the $Rh_0(D)$ group by testing unknown red cells with anti-D (anti-Rh_0) blood grouping reagent.

(d) The laboratory must employ a control system capable of detecting false positive Rh test results.

§ 493.301 Condition: Transfusion Services and Bloodbanking

If a facility provides services for the transfusion of blood and blood products, the facility must ensure that there are facilities for procurement, safekeeping and transfusion of blood and blood products and that blood products provided are readily available. This condition is met by complying with the standards in §§ 493.303 through 493.315 of this subpart.

§ 493.303 Standard; Immunohematological Testing, Processing, Storage and Transmission of Blood and Blood Products

In addition to the requirements in this section, the facility must also meet the applicable quality control requirements in §§ 493.221 through 493.240 of this part.

(a) Blood and blood product collection, processing and distribution must comply with 21 CFR Part 640 and 21 CFR Part 606, and the testing laboratory must be Medicare-approved.

(b) Dating periods for blood and blood products must conform to 21 CFR 610.53.

§ 493.305 Standard; Facilities

The facility must maintain, at a minimum, proper blood storage facilities under the adequate control and supervision of the pathologist or other doctor of medicine or osteopathy meeting the qualifications in § 493.403(b)(7).

§ 493.307 Standard; Arrangement for Services

In the case of services provided outside the blood bank, the facility must have an agreement reviewed and approved by the director that governs the procurement, transfer and availability of blood.

§ 493.309 Standard; Provision of Testing

There must be provision for prompt blood grouping, antibody detection, compatibility testing and for laboratory investigation of transfusion reactions, either through the facility or under arrangement with an approved facility on a continuous basis, under the supervision of a pathologist or other doctor of medicine or osteopathy.

§ 493.311 Standard; Storage Facilities

The blood storage facilities must have an adequate temperature alarm system that is regularly inspected.

§ 493.313   Standard; Retention of Transfused Blood

According to the facility's established procedures, samples of each unit of transfused blood must be retained for further testing in the event of reactions. The facility must promptly dispose of blood not retained for further testing that has exceeded its expiration date.

§ 493.315   Standard; Investigation of Transfusion Reactions

The facility, according to its established procedures, must promptly investigate all transfusion reactions occurring in its own facility for which it has investigational responsibility and make recommendations to the medical staff regarding improvements in transfusion procedures. The facility must document that all necessary remedial actions are taken to prevent future recurrences and that all policies and procedures are reviewed to assure that they are adequate to ensure the safety of individuals being transfused within the facility.

SUBPART G—PERSONNEL

§ 493.401   Condition: Laboratory Director

The laboratory must be directed by an individual who is qualified to provide overall management and supervision of laboratory services.

§ 493.403   Standard; Laboratory Director Qualifications

The laboratory director must be qualified to provide day-to-day management of the laboratory personnel and test performance.
  (a) The director must be—
  (1) A pathologist or other doctor of medicine or osteopathy;
  (2) A laboratory specialist with a doctoral degree from an accredited university in physical, chemical or biological sciences;
  (3) An individual who before [the effective date of these regulations] was properly qualified and served as a laboratory director in accordance with the requirements specified in Appendix A to this part;
  (4) An individual who qualifies under State law to direct the laboratory; or

(5) An individual who has been approved by a certification board acceptable to HHS.

(b) If the laboratory performs services in any of following specialties or subspecialties, specific qualifications are required for the individual providing technical supervision.

(1) Cytology—In the case of tests limited to cytology, the individual is a physician who—

(i) Is certified in anatomic pathology by the American Board of Pathology or the American Osteopathic Board of Pathology;

(ii) Is certified by the American Society of Cytology; or

(iii) Possesses qualifications that are equivalent to those required for certification by the Boards specified in paragraph (b)(1)(i) of this section; and

(iv) Has had three months' experience in cytology.

(2) Histopathology—In the case of tests limited to histopathology, the individual is a physician who meets the requirements of paragraphs (b)(1)(i) or (b)(1)(iii) of this section.

(3) Dermatopathology—In the case of tests limited to dermatopathology, the individual—

(i) Is a physician who meets the requirements of paragraph (b)(1)(i) or (b)(1)(iii) of this section;

(ii) Is certified in dermatopathology by the American Board of Dermatology, the American Osteopathic Board of Dermatology, the American Board of Pathology, or the American Osteopathic Board of Pathology; or

(iii) Possesses qualifications that are equivalent to those required for certification as specified by the Boards specified in paragraph (b)(2)(i) of this section.

(4) Oral pathology—In the case of tests limited to oral pathology, the individual—

(i) Is a physician who meets the requirements of paragraph (b)(1)(i) or (b)(1)(iii) of this section;

(ii) Is certified in oral pathology by the American Board of Oral Pathology; or

(iii) Possesses qualifications that are equivalent to those required for certification as specified by the Boards specified in paragraph (b)(3)(i) of this section.

(5) Histocompatibility—In the case of tests limited to histocompatibility, the individual—

(i) Holds an earned doctoral degree in a biological science or is a physician; and

(ii) Has had four years of experience in immunology, two of which have been in histocompatibility testing.

(6) Cytogenetics—The individual—

(i) Holds an earned doctoral degree in a biological science or is a physician; and

(ii) Has had four years of experience in immunology or genetics, two of which have been in cytogenetics.

(7) Transfusion services and blood banking—The individual is a pathologist or other doctor of medicine or osteopathy with training and experience in transfusion services.

§ 493.405   Standard; Laboratory Director Responsibilities

The laboratory director must be responsible for the overall technical supervision of the laboratory personnel, for the performance and reporting of testing procedures and for assuring compliance with the applicable regulations.

(a) The laboratory director must provide technical supervision of the laboratory services and assure that technical supervision is provided by individuals as required under § 493.403(b).

(b) The laboratory director must—

(1) Assure that tests, examinations and procedures are properly performed, recorded and reported;

(2) Assure that the laboratory maintains an ongoing quality assurance program;

(3) Assure that when tests are being performed there is a supervisor on the premises who meets the qualifications of § 493.409; and

(4) Assure compliance with the applicable regulations.

(c) The laboratory director must ensure that the staff—

(1) Has the appropriate education, experience and training to perform and report laboratory tests promptly and proficiently;

(2) Is sufficient in number for the scope and complexity of the services provided;

(3) Receives regular in-service training appropriate for the type and complexity of the laboratory services offered; and

(4) Maintains competency to perform test procedures and report the results promptly and proficiently.

§ 493.407   Condition: Laboratory Supervision

(a) Laboratory services must be supervised by an individual who meets the requirements of this subpart and is on the laboratory premises at all times when tests are being performed.

(b) Exception: When there is an emergency outside regularly scheduled hours of duty, an individual who qualifies as a supervisor under this section is not required to be on the premises, provided that the supervisor reviews the emergency test results during the next duty period and that a record is maintained to reflect the actual review.

§ 493.409   Standard; Laboratory Supervisor Qualifications

The laboratory supervisor is qualified to provide onsite day-to-day supervision of laboratory test performance and test reporting. The supervisor—

(a) Qualifies as a laboratory director under § 493.401 of this part;

(b) Has earned a bachelor's degree in medical technology from an accredited college or university;

(c) Has successfully completed three years of academic study (a minimum of 90 semester hours or equivalent) in an accredited college or university that met the specific requirements for entrance into a school of medical technology accredited by an accrediting agency approved by HHS and has successfully completed a course of training of at least 12 months in such a school of medical technology;

(d) Has earned a bachelor's degree in one of the chemical, physical, or biological sciences and, in addition, has at least one year of pertinent full-time laboratory experience, training, or both, in the specialty or subspecialty in which the individual performs tests;

(e) Has successfully completed three years (90 semester hours or equivalent) in an accredited college or university with the following distribution of courses;

(1) *For those whose training was completed before September 15, 1963.* At least 24 semester hours in chemistry and biology courses of which:

(i) At least six semester hours were in inorganic chemistry *and* at least three semester hours were in other chemistry courses, and

(ii) At least 12 semester hours were in biology courses pertinent to the medical sciences; *or*

(2) *For those whose training was completed after September 14, 1963.*

(i) Sixteen semester hours in chemistry courses that are acceptable toward a major in chemistry, including at least six semester hours in inorganic chemistry;

(ii) Sixteen semester hours in biology courses that are pertinent to the medical sciences and are acceptable toward a major in the biological sciences; and

(iii) Three semester hours of mathematics; and

(3) Has experience, training, or both, covering several fields of medical laboratory work of at least one year and of such quality as to provide him or her with education and training in medical technology equivalent to that described in paragraphs (a) and (b) of this section;

(f) Met the requirements of Appendix B to this part before [effective date of these regulations]; or

(g) Has qualified as a technologist in a Medicare-approved or CLIA-licensed laboratory before July 1, 1971 in accordance with the requirements of Appendix B to this part.

§ 493.411 Standard; Laboratory Supervisor Duties

The laboratory supervisor, under the general direction of the laboratory director, supervises laboratory personnel, test performance, and test reporting.

§ 493.413 Condition: Personnel Performing Cytology Services

All cytological preparations must be examined by an individual meeting the qualifications of § 493.403(b)(1) or the requirements of § 493.415.

§ 493.415 Standard; Cytotechnologist Qualifications

The cytotechnologist—

(a) Has successfully completed two years in an accredited college or university with at least 12 semester hours in science, eight hours of which are in biology, and

(1) Has had 12 months of training in a school of cytotechnology accredited by an accrediting agency approved by HHS; or

(2) Has received six months of formal training in a school of cytotechnology accredited by an accrediting agency approved by HHS and six months of full-time experience in cytotechnology in a laboratory acceptable to the pathologist who directed the formal six months of training; or

(b) Met the requirements of Appendix C to this part before [the effective date of these regulations].

§ 493.417   Standard; Cytotechnologist Duties

The cytotechnologist must—

(a) Document the gynecologic and non-gynecologic cases examined; and

(b) Record slide interpretation results of each gynecologic and non-gynecologic case reviewed.

§ 493.419   Condition: Technical Personnel

The laboratory staff must be technically competent to perform test procedures and report test results promptly and proficiently.

SUBPART H—QUALITY ASSURANCE

§ 493.451   Condition—Quality Assurance

The laboratory must establish and follow policies and procedures for an ongoing quality assurance program designed to monitor and evaluate quality; identify and correct problems; assure the accurate, reliable and prompt reporting of test results; and assure the adequacy and competency of the staff.

(a) *Standard.* The laboratory must have an ongoing system under which it monitors and evaluates quality control and proficiency testing data for the purpose of substantiating that all tests performed and reported by the laboratory conform to the laboratory's specified performance criteria. These criteria include: precision, accuracy, detection limits, interferences, linearity, sensitivity, specificity, validity and adequacy.

(b) *Standard.* The laboratory must have a mechanism for assuring the accurate and timely reporting of test results. Reporting times must be within the acceptable time periods established by the laboratory.

(c) *Standard.* The laboratory must have a mechanism for assuring that—

(1) All quality control data are reviewed;

(2) Patient test results are not reported when control values are outside the acceptable range established by the laboratory; and

(3) Actions are taken to correct the problems which led to the unsatisfactory quality control results and the corrective actions are documented.

(d) *Standard.* The laboratory must have a mechanism for assuring that corrective action is taken and is documented on all unacceptable proficiency testing results.

(e) *Standard.* The laboratory must have a mechanism to assure that specimens are not tested unless the laboratory's established criteria for acceptability are met and that the authorized person ordering the test is notified of the condition of specimens not meeting the laboratory's criteria for a satisfactory specimen suitable for testing or any limitations on the reliability of the test results.

(f) *Standard*. The laboratory must have policies and procedures for an ongoing program to assure that employees are competent, and maintain their competency, to perform their duties as specified by the laboratory. Policies and procedures may include direct observation of routine patient test performance as well as analysis of unknowns, monitoring the reporting of test results, or other activities identified by the laboratory. The laboratory must evaluate employee performance by—

(1) Retesting specimens of previously analyzed specimens, internal blind proficiency test samples, or external proficiency test samples (that have already been reported to approved proficiency testing programs) to assess the performance levels of each staff member responsible for performing and supervising testing; or

(2) Enrolling in external proficiency testing programs to the extent that there are programs available to cover all analyses performed, to assess an individual's laboratory performance. (The proficiency test samples are in addition to those required in Subpart C of this part.) For cytology, the laboratory may insert into the workload slides from previously reported cases as blind samples or may arrange to exchange cases with another laboratory for the purpose of rescreening slides and comparing results.

(g) *Standard*. The laboratory must have a mechanism for assessing problems identified during quality assurance reviews and discussing them with the staff. The laboratory must take necessary corrective actions to prevent recurrences and make available to HHS documentation of corrective action.

(h) *Standard*. The laboratory must evaluate all data analysis and test reporting systems to assure that the systems perform according to specifications and provide accurate and reliable reporting, transmittal, storage and retrieval of data.

(i) *Standard*. The laboratory must establish and follow policies and procedures to assure that all complaints and problems reported by individuals or facilities who order, receive and use its test results are investigated and that all necessary corrective actions are instituted and documented.

(j) *Standard*. The laboratory must maintain records of its quality assurance program and document all corrective actions taken to remedy problems it has identified.

SUBPART I—INSPECTION

§ 493.501 Condition: Inspection

HHS or its designees may conduct an unannounced inspection of any laboratory at any time during its hours of operation. HHS may deny approval to a laboratory for a period of at least one year for violation of any of the requirements of this part or of the Social Security Act, subject to the appeal rights specified in Part 498 of this chapter.

(a) *Standard.* The laboratory may be required, as part of this inspection, to—

(1) Test samples (including proficiency testing samples) or perform procedures as HHS requires;

(2) Allow an interview of all employees of the laboratory;

(3) Allow employees to be observed performing tests (including proficiency testing specimens provided by the inspection team), data analysis and reporting; and

(4) Provide copies to HHS of all records and data it requires.

(b) *Standard.* All records and data must be readily accessible and retrievable within a reasonable time frame during the course of the inspection. All records must be available for at least two years unless other time frames are specified in this part or HHS specifies a different interval in Appendix C of the State Operations Manual (HCFA Pub. 7).

(c) *Standard.* The laboratory must provide all information and data needed to make a determination of the laboratory's status; this information may be requested before, during or after the inspection.

(d) *Standard.* The laboratory must notify HHS within 30 days of the effective date of all changes in directors, ownership and control, location, test specialties and subspecialties offered, and hours of operation.

(e) *Standard.* The laboratory must successfully participate in an approved proficiency testing program for up to one year before inspection, issuance of a CLIA license (or letter of exemption) or Medicare or Medicaid approval can take place, if HHS so requires. The laboratory must submit the results of the testing to HHS before inspection, approval or licensure of the facility can take place.

(f) *Standard.* HHS may reinspect a laboratory at anytime necessary to evaluate the ability of the laboratory to provide accurate and reliable test results.

SUBPART J—CLIA REQUIREMENTS

§ 493.701   Basis and Scope.

(a) This subpart applies to laboratories engaged in the laboratory examination of, or other laboratory procedures relating to, human specimens solicited or accepted in interstate commerce, directly or indirectly, for the purpose of providing information for the diagnosis, prevention, or treatment of any disease or impairment, or the assessment of the health, of human beings. All screening procedures are included as well as quantitative testing of specimens for the presence or absence of any substance, pathogen or other analytes.

(b) This subpart does not apply to—

(1) Any laboratory that performs 100 or fewer tests during any calendar year; however, the laboratory must—

(i) Hold an unrevoked or unsuspended letter of exemption for low volume from HHS;

(ii) Provide information to HHS upon request, permit inspections, and make records available as required by § 493.501 of this part for licensed laboratories; and

(iii) Perform testing that HHS has determined poses no significant threat to public health;

(2) Any laboratory operated by a licensed physician, osteopath, dentist, or podiatrist, or group of these individuals in any combination who performs laboratory tests or procedures solely as an adjunct to the treatment of the practitioner's or practitioners' own patients;

(3) Any laboratory performing tests or other procedures solely for the purpose of determining whether to write an insurance contract or determine eligibility or continued eligibility for insurance payments; and

(4) Any laboratory exempted under section 353(1) of the Public Health Service Act.

§ 493.702   Definitions

As used in this subpart—

"Act" means the Public Health Service Act, as amended, 42 U.S.C. 201, et seq., also known as the Clinical Laboratories Improvement Act of 1967 (CLIA).

§ 493.704 Licensure Application and Issuance

(a) *Licensure application.* (1) An application for the issuance or renewal of a license must be made for each laboratory location by the owner, director or authorized representative of the laboratory on the form or forms prescribed by HHS.

(2) The application for renewal of a license may not be submitted less than 30 days nor more than 60 days before the expiration date of the license.

(b) *Licensure issuance or renewal.* (1) As a part of the review of the application for issuance or renewal of licensure, HHS may require the laboratory to furnish additional information needed to consider the application. HHS also reviews the results of an onsite inspection of the laboratory's premises, performance in proficiency testing and compliance with this part. If HHS determines that the laboratory complies with the standards and other requirements of CLIA and provides consistent performance of accurate and reliable test procedures and services, HHS issues an initial or a renewal license with respect to one or more specialties or subspecialties as specified in §§ 493.243 through 493.281 of this part.

(2) HHS issues initial or renewal licenses for a period of at least one year. If no changes occur that affect the licensure status of the laboratory, HHS notifies the laboratory of its continued approval for at least another year. If changes occur that affect the licensure status of the laboratory, HHS issues a revised license reflecting those changes.

(3) If HHS does not issue or renew a license (in whole or in part), HHS gives the laboratory reasonable notice and issues a statement of grounds on which it proposes to not issue or renew the license or any part of it. The laboratory is also given an opportunity to request a hearing in accordance with the provisions of Part 498 of this chapter.

(4) If a laboratory applies for licensure in any specialty or subspecialty for which a license has been revoked or application for a license has been denied by HHS, licensure for that specialty or subspecialty will not be approved until at least one year elapses from the effective date of the adverse action. HHS may waive this one year period if the laboratory submits good cause for the waiver. A laboratory that requests reinstatement after this one year period must provide assurance that it complies with this subpart.

(c) *Exception.* These standards for issuance and renewal of licenses do not apply to accredited laboratories if—

(1) HHS determines that the standards applied by an accrediting organization are equal to or more stringent than the requirements of CLIA and of these regulations;

(2) The accrediting organization assures that its standards are met by the laboratory;

(3) The accrediting organization and accredited laboratories make available to HHS all records and information required by these regulations and permit inspections as required by HHS.

(4) The laboratory holds an unrevoked and unsuspended letter of exemption issued in accordance with § 493.710 of this subpart.

§ 493.706 Revocation and Suspension of Licenses and Letters of Exemption; Notice

(a) A laboratory license or letter of exemption may be revoked or suspended whenever HHS, after reasonable notice and opportunity for a hearing to the owner or director of the laboratory as provided in Part 498 of this chapter, finds—

(1) In the case of a license, that the owner, director or any employee of the laboratory has committed any of the actions specified in section 353(e) of CLIA or has not met the requirements of this part; or

(2) In the case of a letter of exemption, that the laboratory is no longer eligible for its letter of exemption.

(b) Any notice issued under paragraph (a) of this section will contain a statement of the proposed action and of the grounds upon which HHS proposes to act.

(c) If HHS proposes to suspend a license or letter of exemption the notice will state—

(1) The period of such proposed suspension or the action required to end the suspension; and

(2) That the license or letter will be revoked if the appropriate remedial action is not taken within the suspension period.

(d) If HHS proposes to revoke a license or letter of exemption, the notice will state the specialty or subspecialty with respect to which the license or letter of exemption will no longer apply.

§ 493.708 Approval of Accreditation and State Licensure Programs; Notice

(a) Approval of accreditation and State licensure programs is based on HHS' determination that these programs have requirements at least as stringent as those contained in CLIA. HHS, in making this evaluation, considers each program's standards, standards enforcement and survey procedures related to: quality control; maintenance of records, equipment and facilities; qualifications of personnel; proficiency testing; program administration related to renewal of accreditation; frequency and comprehensiveness of onsite inspections; and maintenance and availability of data and records related to accredited laboratories.

(b) In filing an application for approval, the accrediting organization or State must initially provide all information and data HHS determines necessary to determine if a program can be approved and thereafter must provide all information needed by HHS to determine a program's continued approval. The accrediting organization or State licensure program must—

(1) Provide information regarding the accreditation or licensure, specialties or subspecialties for which accreditation or licensure is applicable which accreditation or licensure is applicable at a frequency required by HHS; and

(2) Notify HHS within five days of any changes in accreditation or licensure.

(c) HHS may require an accrediting organization or State licensure program to sign a written agreement specifying the terms of approval.

(d) If HHS determines at any time that the accrediting organization's or State licensure program's requirements are no longer at least as stringent as CLIA requirements, HHS will notify the accrediting organization or State licensure program and provide a reasonable period of time for revision. If the organization or State licensure program does not provide satisfactory evidence on a timely basis of its continued acceptability, HHS will notify the accrediting organization or State licensure program of the bases for revoking approval. The notice will state that the provisions of section 353 of CLIA and 42 CFR Part 493 requiring licensure will apply to all its accredited or licensed laboratories effective 30 days after the date the notice is received. HHS will also notify each laboratory affected by this determination that its exemption from CLIA licensure is not in effect 30 days after the date the notice is received by the accreditation organization or State licensure program.

§ 493.710   Letter of Exemption

(a) HHS may issue a letter of exemption to a laboratory provided that—

(1) The laboratory owner or authorized representative of the laboratory signs an agreement to permit inspections as required by HHS and makes available records and other information HHS requires; and

(2) The laboratory submits an application form provided by HHS that

certifies that the laboratory is accredited or licensed by an approved organization or State licensure program and specifies the specialties and subspecialties for which the laboratory is accredited or licensed and the date or dates of accreditation or licensure.

(b) If a laboratory fails to comply with the requirements of this part, the laboratory will no longer be eligible for a letter of exemption and is subject to the revocation and suspension procedures described in § 493.706 of this subpart.

APPENDIX A—LABORATORY DIRECTOR QUALIFICATION REQUIREMENTS BEFORE [EFFECTIVE DATE OF REGULATIONS]*

With respect to individuals first qualifying as laboratory directors before July 1, 1971, an individual who was responsible for the direction of a clinical laboratory for 12 months between July 1, 1961, and January 1, 1968, and, in addition, met one of the following requirements in paragraphs (a) through (d) by July 1, 1971, is qualified to be a laboratory director under § 493.401.

(a) The individual was a physician and after graduation had at least 4 years of pertinent full-time clinical laboratory experience;

(b) The individual held a master's degree from an accredited institution with a chemical, physical, or biological science as a major subject and after graduation had at least four years of pertinent full-time clinical laboratory experience;

(c) The individual held a bachelor's degree from an accredited institution with a chemical, physical, or biological science as a major subject and after graduation had at least six years of pertinent full-time clinical laboratory experience; or

(d) The individual achieved a satisfactory grade through an examination conducted by or under sponsorship of the U.S. Public Health Service on or before July 1, 1970.

Note: The January 1, 1968 date for meeting the 12 months laboratory direction requirement in this appendix may be extended one year for each year of full-time clinical laboratory experience obtained before January 1, 1968 required by State law for a clinical laboratory director license. An exception to the July 1, 1971 qualifying date was made provided the individual

---

* Appendixes A-C on pp. 399–400 apply to this document (*HCFA Proposed Revision*) only.

requested approval of his or her qualifications by October 21, 1975 and had been employed in an approved clinical laboratory for at least three of the five years preceding the date of submission of his or her qualifications.

APPENDIX B—TECHNOLOGIST REQUIREMENTS BEFORE JULY 1, 1971

With respect to individuals first qualifying as technologist before July 1, 1971, an individual who met the following requirements in paragraphs (a) and (b) is now qualified to be a supervisor under § 493.409.

(a) The individual was performing the duties of a clinical laboratory technologist at any time between July 1, 1961, and January 1, 1968, and

(b) The individual has had at least ten years of pertinent clinical laboratory experience before January 1, 1968. This required experience may be substituted by education as follows:

A minimum of 30 semester hours of credit from an approved school of medical technology or toward a bachelor's degree from an accredited institution with a chemical, physical, or biological science as his or her major subject is equivalent to two years of experience. Additional education is equated at the rate of 15 semester hours of credit for one year of experience.

(c) Before [effective date of these regulations], the individual achieved a satisfactory grade in a proficiency examination approved by HHS.

APPENDIX C—CYTOTECHNOLOGIST REQUIREMENTS BEFORE [EFFECTIVE DATE OF REGULATIONS]

With respect to individuals first qualifying as a cytotechnologist before July 1, 1969, an individual who met one of the following requirements in paragraph (a) and (b) is qualified to be cytotechnologist under § 493.415.

(a) Before January 1, 1969, the individual has:

(1) Been graduated from high school;

(2) Completed six months of training in cytotechnology in a laboratory directed by a pathologist or other physician recognized as a specialist in cytology; and

(3) Completed two years of full-time supervised experience in cytotechnology.

(b) Before [effective date of these regulations], the individual achieved a satisfactory grade in a proficiency examination approved by HHS.
(Catalog of Federal Domestic Programs No. 13.714—Medical Assistance Program; No. 13.773, Medicare—Hospital Insurance Program; No. 13.774, Medicare—Supplementary Medical Insurance Program)

Dated: June 22, 1988.

William L. Roper,
Administrator, Health Care Financing Administration.

Approved: June 24, 1988.

Otis R. Bowen,
Secretary.

BIBLIOGRAPHY

Alvarez, R.J., et al. (1982). QC in microbiology labs: Essential for productivity. ASQC Congress Transactions—Detroit, pp. 737–742.

American Public Health Association. (1985). Standard methods for the examination of water and wastewater. American Public Health Association, Washington, D.C.

American Society for Quality Control, Statistics Division. (1983). Glossary and tables for statistical quality control. American Society for Quality Control, Milwaukee, WI.

Association of Official Analytical Chemists. (1990). Official methods of analysis, volumes 1 and 2. Association of Official Analytical Chemists, Arlington, VA.

Belk, W.P. and Sunderman, F.W. (1947). Survey of the accuracy of clinical analysis in clinical laboratories. *Am. J. Clin. Pathol.* 17:858–861.

Bell, M.R. (1989). Laboratory accreditation and quality system accreditation—a merging of the ways. STP 1057, pp. 120–143. American Society for Testing and Materials, Philadelphia, PA.

Belsky, J.S. (1991). Design of HVAC systems for laboratories. *Pharm. Eng. 11*:4, pp. 31–35.

Boothe, R. (1990). Who defines quality in service industries? *Qual. Prog. 23*:2, pp. 65–67.

Borghese, R.N. (1990). Quality auditing as a tool. *Biopharm. 10*:14–16.

Bossert, J. (ed.). (1988). Procurement Quality Control. Quality Press, Milwaukee, WI.

Bossert, J. (1990). Quality Function Deployment. Quality Press, Milwaukee, WI.

Butler, E.H. (1986). Speech at Annual Meeting of Society for Quality Assurance (1985). Reprinted in Soc. Qual. Assur. Newsletter.

Cardner, R.Y. (1984). GMP compliance auditing. *Pharm. Technol. 13*:6, pp. 36–38.

Chapman, K.G. 1990. Validating to protect the competitive edge. *Sci. Comput. Autom., 3*:5.

Daniel, A. (1990). Proficiency testing. *Med. Dev. Diag. Indus. 12*:5, pp. 34–37.

Dept. of Health and Human Services, and Natl. Inst. Health. (1984). Biosafety in Microbiological and Biomedical Laboratories. U.S. Government Printing Office, Washington, D.C.

Environmental Protection Agency. (1978). Manual for the interim certification of laboratories involved in analyzing public drinking water supplies—criteria and procedures, U.S. Dept. of Commerce PB-287 118, Washington, D.C.

Environmental Protection Agency (1988). Drinking Water Regulations Under the Safe Drinking Water Act, Fact Sheet. Criteria and Standards Division, Office of Drinking Water, Environmental Protection Agency, Washington, D.C.

Environmental Protection Agency. (1990). Good automated laboratory practices implementation manual—draft. Environmental Protection Agency, IRM, Research Triangle Park, NC.

Etnyre-Zacher, P. and Miller, S.M. (1990). An educational program for physicians' office laboratory personnel. *Am. Clin. Lab. 9*:1, pp. 10–17.

Federal Register. (1978). Human and veterinary drugs: good manufacturing practices and proposed exemptions for certain OTC products. *43*:190, pp. 45014–45089.

Federal Register. (1978). Nonclinical laboratory studies: good laboratory practices regulations: *43*:247, pp. 59986–60025.

Federal Register. (1979). Current good manufacturing practice in manufacturing, processing, packing, or holding human food. *44*:112, pp. 33238–33248.

Federal Register. (1988). Proposed rules: Medicare, Medicaid and CLIA programs; revision of the clinical laboratory regulations for the Medicare, Medicaid and the clinical laboratories improvement act of 1967 Programs. *53*:151.

Food and Drug Administration. (1983). FDA Inspection Operations Manual. U.S. Govt. Printing Office, Washington, D.C.

Food and Drug Administration Compliance Program Guidance Manual. (1991). Good Laboratory Practice (Nonclinical Laboratories). FDA No. 7348.808.

Garfield, F.M. (1984). Quality assurance principles for analytical laboratories. Association of Official Analytical Chemists, Inc., Arlington, VA.

Gladhill, R.L. (1989). Advantages of laboratory accreditation. ASTM STP 1057, pp. 19–23. American Society for Testing and Materials, Philadelphia, PA.

Guerra, J. (1986). Validation of analytical methods by FDA laboratories. *Pharm. Technol. 10*:3, pp. 74–84.

Health, Education and Welfare. (1980). Requirements of laws and regulations enforced by the U.S. Food and Drug Administration. HEW Publication No. (FDA) 79-1042, Washington, D.C.

Henry, J.B. (1984). Clinical diagnosis and management. Chapters 6, 53 and 55. W.B. Saunders Co., Philadelphia, PA.

Illinois State EPA. (1983). Certification and Operation of Environmental Laboratories. Title 35, Subtitle A, Chapt. II, Part 183.

Johnson, R.W. (1990). Quality assurance of tissue culture media used in the biotechnology industry. *Biopharm. 2*:40–44.

Lord, T. (1989). Microbes and cGMPs. *Pharm. Technol. 13*:6, pp. 36–38.

Marquardt, D., et al. (1991). Vision 2000: The strategy for ISO 9000 series standards in the 90s. *Quality Progress, 24*:5, pp. 25–31.

Miller, J.M. and Wentworth, B.B. (ed.). (1985). Methods for quality control in diagnostic microbiology. American Public Health Association, Washington, D.C.

Mills, C.A. (1989). The quality audit. Quality Press and McGraw Hill, Inc., Milwaukee, WI.

Motiska, P.J. and Shilliff, K.A. (1990). Ten precepts of quality. *Qual. Prog. 23*:2, pp. 27–28.

National Committee for Clinical Laboratory Standards. (1988). Labeling of laboratory prepared materials. Document GP4-P. NCCLS, Villanova, PA.

National Committee for Clinical Laboratory Standards. (1985). Selecting and evaluating a referral laboratory. NCCLS Document GP9-P, Villanova, PA.

National Committee for Clinical Laboratory Standards. (1988). Preparation and testing of reagent water in the clinical laboratory. NCCLS Document C3-T2, Villanova, PA.

National Committee for Clinical Laboratory Standards. (1990). Quality assurance for commercially prepared microbiological culture media. NCCLS Document MJ22-A, Villanova, PA.

Organisation for Economic Cooperation and Development. (1990). Good laboratory practice in the testing of chemicals. Organisation for Economic Cooperation and Development.

Computer System Validation Committee. (1986). Validation concepts for computer systems used in the manufacture of drug products. *Pharm. Technol. 10*:5, pp. 24–34.

Pharmaceutical Manufacturer's Association's Computer System Validation Committee. (1990). Computer system validation—staying current: change control. *Pharm. Technol. 14*:1, pp. 20–75.

Ratliff, T. (1990). The laboratory quality assurance system. Van Nostrand Reinhold, New York, NY.

Robinson, C.B. (1990). Auditing a quality system for the defense industry. Quality Press, Milwaukee, WI.

Robinson, C.B. (1990). Auditing a quality system. *Qual. Prog. 4*:49–52.

Rosander, A.C. (1989). The quest for quality in services. Quality Press, Milwaukee, WI.

Schock, H.E. (ed). (1989). Accreditation practices for inspections, tests, and laboratories. ASTM STP 1057. American Society for Testing and Materials, Philadelphia, PA.

Singer, D.C., et al. (1981). Standard operating procedures for the microbiology laboratory. *Cosm. Technol. 3*:40–49.

Snyder, J.W. (1981). Quality control in clinical microbiology. API Species 5:2, pp. 13–23, Analytab Products Inc., Plainview, NY.

Speck, M.L. (ed) (1984). Compendium for the microbiological examination of foods. American Public Health Association, Washington, D.C.

Taylor, J.K. (1989). Quality assurance of chemical measurements. Lewis Publishers, Chelsea, MI.

Taylor, J.K. (1985). Handbook for SRM users. NBS Publication 260-100, National Institute of Technology and Standards, Washington, D.C.

United States Pharmocopeia XXII. (1990). National Formulary XVII. Unites States Pharmocopeial Convention, Rockville, MD.

Wadsworth Center for Laboratories and Research. (1989). Proficiency testing program. New York State Department of Health, Albany, NY.

WHO Technical Report Series No. 323. (1966). General requirements for manufacturing establishments and control laboratories. World Health Organization, Geneva, Switzerland.

Willborn, W. (1987). Audit standards—a comparative analysis. Quality Press, Milwaukee, WI.

Willborn, W. (1983). Compendium of audit standards. Quality Press, Milwaukee, WI.

INDEX

Note: Reprinted materials from Appendixes A to F are not indexed here.